SAP PRESS e-books

Print or e-book, Kindle or iPad, workplace or airplane: Choose where and how to read your SAP PRESS books! You can now get all our titles as e-books, too:

- By download and online access
- For all popular devices
- And, of course, DRM-free

Convinced? Then go to www.sap-press.com and get your e-book today.

ABAP® Cookbook

SAP PRESS is a joint initiative of SAP and Rheinwerk Publishing. The know-how offered by SAP specialists combined with the expertise of Rheinwerk Publishing offers the reader expert books in the field. SAP PRESS features first-hand information and expert advice, and provides useful skills for professional decision-making.

SAP PRESS offers a variety of books on technical and business-related topics for the SAP user. For further information, please visit our website: *www.sap-press.com*.

Kiran Bandari
Complete ABAP (3rd Edition)
2022, 912 pages, hardcover and e-book
www.sap-press.com/5567

Baumbusch, Jäger, Lensch
ABAP RESTful Application Programming Model (2nd Edition)
2025, 576 pages, hardcover and e-book
www.sap-press.com/6161

Colle, Dentzer, Hrastnik
Core Data Services for ABAP (3rd Edition)
2024, 754 pages, hardcover and e-book
www.sap-press.com/5642

Brian O'Neill, Jelena Perfiljeva
ABAP: An Introduction (2nd Edition)
2020, 684 pages, hardcover and e-book
www.sap-press.com/4955

Paul Hardy
ABAP to the Future (4th Edition)
2022, 877 pages, hardcover and e-book
www.sap-press.com/5360

Fabian Lupa, Sven Treutler

ABAP® Cookbook

Practical Recipes for Modern Programming

Editor Megan Fuerst
Acquisitions Editor Hareem Shafi
German Edition Editor Janina Schweitzer
Translation Winema Language Services, Inc.
Copyeditor Julie McNamee
Cover Design Graham Geary
Photo Credit Shutterstock: 1321600256/©Alexander Raths
Layout Design Vera Brauner
Production Hannah Lane
Typesetting SatzPro, Germany
Printed and bound in Canada, on paper from sustainable sources

ISBN 978-1-4932-2777-8
1st edition 2026
1st German edition published 2025 by Rheinwerk Verlag, Bonn, Germany

© 2026 by:
Rheinwerk Publishing, Inc.
2 Heritage Drive, Suite 305
Quincy, MA 02171
USA
info@rheinwerk-publishing.com
+1.781.228.5070

Represented in the E. U. by:
Rheinwerk Verlag GmbH
Rheinwerkallee 4
53227 Bonn
Germany
service@rheinwerk-verlag.de
+49 (0) 228 42150-0

Library of Congress Cataloging-in-Publication Control Number: 2025040738

All rights reserved. Neither this publication nor any part of it may be copied or reproduced in any form or by any means or translated into another language, without the prior consent of Rheinwerk Publishing.

Rheinwerk Publishing makes no warranties or representations with respect to the content hereof and specifically disclaims any implied warranties of merchantability or fitness for any particular purpose. Rheinwerk Publishing assumes no responsibility for any errors that may appear in this publication.

"Rheinwerk Publishing" and the Rheinwerk Publishing logo are registered trademarks of Rheinwerk Verlag GmbH, Bonn, Germany. SAP PRESS is an imprint of Rheinwerk Verlag GmbH and Rheinwerk Publishing, Inc.

All screenshots and graphics reproduced in this book are subject to copyright © SAP SE, Dietmar-Hopp-Allee 16, 69190 Walldorf, Germany.

SAP, ABAP, ASAP, Concur Hipmunk, Duet, Duet Enterprise, ExpenseIt, SAP ActiveAttention, SAP Adaptive Server Enterprise, SAP Advantage Database Server, SAP ArchiveLink, SAP Ariba, SAP Business ByDesign, SAP Business Explorer (SAP BEx), SAP BusinessObjects, SAP BusinessObjects Explorer, SAP BusinessObjects Web Intelligence, SAP Business One, SAP Business Workflow, SAP BW/4HANA, SAP C/4HANA, SAP Concur, SAP Crystal Reports, SAP EarlyWatch, SAP Fieldglass, SAP Fiori, SAP Global Trade Services (SAP GTS), SAP GoingLive, SAP HANA, SAP Jam, SAP Leonardo, SAP Lumira, SAP MaxDB, SAP NetWeaver, SAP PartnerEdge, SAPPHIRE NOW, SAP PowerBuilder, SAP PowerDesigner, SAP R/2, SAP R/3, SAP Replication Server, SAP Roambi, SAP S/4HANA, SAP S/4HANA Cloud, SAP SQL Anywhere, SAP Strategic Enterprise Management (SAP SEM), SAP SuccessFactors, SAP Vora, TripIt, and Qualtrics are registered or unregistered trademarks of SAP SE, Walldorf, Germany.

All other products mentioned in this book are registered or unregistered trademarks of their respective companies.

No part of this book may be used or reproduced in any manner for the purpose of training artificial intelligence technologies or systems. In accordance with Article 4(3) of the Digital Single Market Directive 2019/790, Rheinwerk Publishing, Inc. expressly reserves this work from text and data mining.

Contents at a Glance

Contents

Preface

Continuous development of ABAP

ABAP continues to evolve, and it has been doing so at a rapid pace for several years now. With the *ABAP RESTful application programming model*, SAP has been offering an increasingly mature framework for development since 2018. More and more applications in the standard SAP system are being transferred to this programming model. In combination with *ABAP core data services* (ABAP CDS) and *SAP Fiori elements*, application behavior is modeled in a separate syntax and not implemented directly in ABAP, as in previous programming models. At the same time, *ABAP Cloud* is playing an increasingly important role as a development model for modern applications that is available in both on-premise and cloud systems. With the *clean core approach*, ABAP Cloud is also being promoted in on-premise systems, which means that the role of classic ABAP is increasingly changing.

Target group

Perhaps you've already been able to familiarize yourself with the techniques mentioned in your work as a developer and are also familiar with ABAP in cloud systems. Or maybe you haven't yet migrated to SAP S/4HANA, so you've been limited technologically to ABAP 7.50 with SAP ERP 6.0 EhP8, and you're wondering what lies ahead and what has happened in the past 10 years.

The level of knowledge of ABAP developers couldn't be further apart at the moment. In addition, the upgrade problem still exists for many companies. The issues of training and continuous education, as well as enabling developers for project deployment are correspondingly difficult. In this book, we want to meet you where you are and introduce you to the latest technical solutions for recurring problems, regardless of where you're in the spectrum described. Our book is therefore intended for ABAP developers who are interested in modern solutions and want to develop their software to a high standard, prepare for its possible operation in the cloud, and adhere to the clean core approach.

Goals of this book

You won't find any marketing buzzwords in this book—it was written by developers for developers. We introduce you to technologies that enable you to solve recurring problems using state-of-the-art methods and compare them with previous solution approaches. It's particularly important to us that you can use the solutions in both on-premise and cloud systems. We'll therefore limit ourselves to the range of functions available in SAP S/4HANA 2023 using ABAP Cloud, which means that you can use the solutions presented in all current ABAP runtime environments: SAP S/4HANA 2023 (on-premise version), SAP S/4HANA Cloud Private Edition 2023, SAP

S/4HANA Cloud Public Edition, and the SAP BTP ABAP environment. However, many newer functions are heavily release-dependent, so there may be restrictions on ABAP releases below 7.58.

Prerequisites

To be able to follow the contents of this book optimally, you should be familiar with the basics of the ABAP programming language, including ABAP objects, and have gained some hands-on experience. In addition, you should already be familiar with the *ABAP development tools for Eclipse*, that is, the Eclipse-based ABAP development environment (successor to the *ABAP Workbench*) as this is the only development environment available for all runtime environments and fully supports the object types used there. Nevertheless, you can still find tips and tricks for using the ABAP development tools as well as step-by-step instructions for the individual techniques. This is because switching from Transaction SE80 to the ABAP development tools often involves a significant learning curve, especially for experienced ABAP developers. Instructions for installing the ABAP development tools can be found in Appendix C.

Structure of the book

The chapters in this book each address a specific problem and present an up-to-date solution. As a rule, the chapters can be read independently of each other. That means you can either read the book from start to finish to get a comprehensive overview, or jump straight to the chapter you need if you're looking for specific information on a certain topic. Throughout the chapters, we use an overarching sample application that establishes the reference to real-life business. We took particular care to choose an example that was realistic, not too simple, but also not too complex. The individual chapters expand this sample application step-by-step so that you can follow our development steps.

Chapter 1 (Introduction to Modern ABAP Development) serves as an introduction to the technical concepts that are relevant in modern ABAP-based applications and positions ABAP development in the current SAP product portfolio.

In **Chapter 2** (The Application Scenario), we describe the business example we'll use as a guide in the subsequent chapters and explain how the sample application is structured from a technical point of view.

Chapter 3 (Handling System Fields and Runtime Information) is the first chapter in which specific techniques are presented—in this case, accessing information in the ABAP runtime environment using modern classes and statements.

In **Chapter 4** (Table Analysis), you'll learn how to give users, key users, or application support staff access to table content in an SAP Fiori app. You'll also learn how you as a developer can analyze database table contents using the ABAP development tools.

Chapter 5 (Table Maintenance Using Business Configuration Maintenance Objects) describes how you can develop a maintenance interface for customizing settings in the form of an SAP Fiori app. Such an app can be regarded as the successor to the maintenance dialogs created via the table maintenance generator.

Chapter 6 (Application Logs) deals with the logging of user actions and process steps with the application log. You'll get to know the new object type and the new classes that enable you to replace the old function modules for logging.

In **Chapter 7** (Change Documents), we describe class-based change logging, which you can use to track changes to master and transaction data, as well as their integration into ABAP RESTful application programming model business objects.

In **Chapter 8** (Lock Objects), you'll learn about the new class-based API for locking objects as well as the optimistic and pessimistic locking concept. You'll also learn how to integrate lock objects into applications based on the ABAP RESTful application programming model.

Chapter 9 (Number Range Objects) shows how you can use the new, class-based access option to draw numbers for primary key fields or check them. You'll also gain an insight into the program-based maintenance of number range objects and intervals.

In **Chapter 10** (Background Processing), we present the application jobs, the cloud-enabled successor concept to the classic, report-based batch jobs. Application jobs are used for the regular execution of applications.

In **Chapter 11** (Email Dispatch), you'll learn how to send emails from the SAP system—including attachments and using new classes.

Chapter 12 (Parallelizing Application Logic) shows you how you can gain performance in your programs by executing subtasks in parallel with multiple processes—without having to create your own function modules directly in ABAP objects.

The uploading of files is described in **Chapter 13** (File Upload). You'll learn how you can use the ABAP RESTful application programming model and SAP Fiori elements to enable users to upload files and consume the file content in your applications.

In **Chapter 14** (Using Excel Files), you'll learn how to read and create XLSX files using the XCO library. The file upload and download described in the previous chapter is used for this purpose.

Chapter 15 (Documenting Development Objects) presents the current options for documenting development objects using ABAP Doc and knowledge transfer documents.

Chapter 16 (Authorizations) deals with the assignment of authorizations. You'll learn how to authorize access to your applications in ABAP Cloud.

In **Chapter 17** (Using APIs), we move away from our sample application and refer to the standard SAP system. You'll learn how to find shared objects and APIs, as well as how to call them in the correct way. In the context of ABAP Cloud, the search for reusable and released objects is very relevant because unreleased objects can no longer be used.

The details of current extension techniques are described in **Chapter 18** (Extensions in ABAP Cloud). You'll learn how to extend applications that are based on the ABAP RESTful application programming model with your own functions and fields using the developer and key user extensibility.

Finally, we provide an outlook in **Chapter 19** (Outlook). We take a look at where we currently are in terms of technological development, where there are still gaps in the coverage of new technologies, and what new functions can be expected in future releases.

Information boxes

The highlighted boxes in this book contain information that's good to know and useful but outside the context. To help you immediately identify the type of information contained in the boxes, we've assigned icons to each box:

[»]
- Boxes marked with this icon contain information on *related topics*.

[!]
- This icon refers to *specifics* that you should consider. It also *warns* about frequent errors or problems that can occur.

[+]
- The *tips* marked with this icon provide specific recommendations that can make your work easier.

Use of the Sample Application

In the materials accompanying the book at *www.sap-press.com/6198*, you'll find the sample application, which is built up successively in the individual chapters. You can also find it on GitHub at *http://s-prs.co/v619800*.

Details on the use of the individual development objects are explained in Appendix A. The introduction to each of the following chapters mentions which objects are used. As far as technically possible, there is a separate subpackage for each chapter. The assignment of chapters and packages can also be found in Appendix A.

The sample application enables us to explain the individual techniques described in this book in a hands-on manner. However, you shouldn't use it directly for real-life production scenarios. For example, for presentation reasons, we've omitted translatable texts in some places. We've also shortened some of the error handling to focus on the actual topic of the respective chapter.

Acknowledgments

Writing a book for the first time is a project in itself. We would therefore like to express our special thanks to Rheinwerk Verlag and specifically to Janina Schweitzer for her support during the writing process and for accompanying us "from kick-off to go-live." We also want to thank Björn Schulz and Sebastian Freilinger-Huber for their feedback on the individual chapters and technologies. Thanks also to the abapGit community, whose open source tool gave us the opportunity to collaborate on our sample application and make it easily available to you.

Acknowledgments from Fabian Lupa

I would particularly like to thank Patrick Holdschlag for the constant professional and technical exchange on SAP topics and the review work in this book project. A big thank you also goes to Leon Requardt for the in-depth discussions on the ABAP RESTful application programming model and thoughts on best practices. Thank you, Elena, for your understanding, your patience, and your support.

Acknowledgments from Sven Treutler

I wrote this book so that my cat Marie can continue to enjoy her life of luxury.

Thank you, Grandma, Grandpa, Marion, Werner, Toni, and Patrick for always believing in me.

I would like to thank you, mom and dad, for teaching me the value of education and curiosity from an early age.

But my special thanks go to you, Antonia. Your patience, love, and ass-kicking have carried me through the many hours of writing.

We both hope that this book provides you with a practical introduction to current software development with ABAP and that it has aroused your interest in discovering old technologies in a new guise or even completely new technical possibilities. Let's go!

Fabian Lupa and **Sven Treutler**

Chapter 1
Introduction to Modern ABAP Development

In modern ABAP development in SAP S/4HANA and in the SAP BTP ABAP environment, there are some new fundamental technical concepts. In this chapter, you'll learn how to write ABAP code that runs in all runtime environments—whether in a cloud or an on-premise solution. You'll learn which technical underpinnings are relevant for that. In particular, we look at the language version and the application scenarios.

Modern ABAP development is based on many technical concepts of the ABAP platform that are still comparatively new. These include the ABAP language version, the ABAP RESTful application programming model, and release contracts. In addition, there are now multiple different runtime environments for ABAP, each of which has different requirements. The question arises as to how you can implement applications that are stable, low-maintenance during upgrades, and cloud-capable. This chapter provides you with an introductory overview before the following chapters go into more detail.

Structure of this chapter

Section 1.1 takes a look at the role of ABAP within the overall software development in the SAP context. In Section 1.2, we then present the new ABAP Cloud development model and describe the aspects it's made up of. We'll explain these in more detail in the subsequent sections.

Section 1.3 deals with the ABAP language version as a technical option for differentiating between classic ABAP and ABAP Cloud. Section 1.4 deals with calling and using objects available in the standard SAP system. In Section 1.5, we briefly discuss the ABAP development tools as a development environment, before introducing the programming model for transactional applications, the ABAP RESTful application programming model, in Section 1.6.

Section 1.7 deals with current SAP products that can be used to develop with ABAP. In particular, we look at the application scenarios for ABAP Cloud. In Section 1.8, you can find out what changes have been made to the ABAP release count and how you can establish the connection between cloud and on-premise releases. Section 1.9 deals with restrictions you may have to deal with depending on your release or the runtime environment to be supported.

1.1 The Role of ABAP in SAP Development

Continuous development of ABAP

Modern ABAP applications differ so much technologically from applications based on earlier programming models that these applications can hardly be compared with each other. With *ABAP core data services* (ABAP CDS) and the *ABAP RESTful application programming model*, applications are modeled declaratively with a separate syntax and are no longer implemented with classic ABAP statements. The implementation components also make generous use of functions from *modern ABAP*. This concept refers to the functional expressions and built-in functions that were added to the ABAP language scope from release 7.40 onward. User interfaces (UIs) are no longer implemented with ABAP at all, but are only described in the ABAP stack using *annotations*. With all of these changes, some people may wonder whether this is still ABAP today.

Non-ABAP technologies in the SAP environment

At the same time, other technologies outside the ABAP stack have become established in SAP development in recent years:

- **SAPUI5**
  This is a JavaScript-based frontend framework for developing modern, web-based UIs.
- **SAP Cloud Application Programming Model**
  As an ABAP alternative, this programming model is available in JavaScript and Java and is lightweight and scalable in cloud environments.
- **SAP Build**
  This product portfolio can be used to automate processes or develop entire applications using low-code/no-code approaches.

So what role do ABAP developers play in companies and projects today? Do they deal with all of these technologies and become SAP or full-stack developers, or do they focus on one area and specialize?

Keeping pace with technological change

How do companies keep up with these developments? Many development teams are still in the process of migrating their systems to SAP S/4HANA and have therefore not yet actively experienced the technological change of the past 10 years. Many people are also used to the fact that not much happens in ABAP development. Whereas an ABAP developer with in-depth module knowledge used to be able to implement requirements more or less on their own, frontend and backend developers are now needed to cover the various technology stacks. Even a cloud architect may be required as well.

However, the requirements today are often the same as in the past. The solutions aren't necessarily better, more stylish, or faster at first glance, but are primarily more modern and more broadly based in terms of technology. And maintenance can be much more complex than it used to be due to the integration of web technologies. It can be correspondingly difficult to argue to managers why more staff or more training is required to achieve the same results.

Maturity of technologies

However, we've now reached a point where the new and broader-based technologies are increasingly mature and have also been rolled out across the board in the core product, SAP S/4HANA. Transaction VA01 for sales order creation, for example, one of the most complex transactions in the system, has finally been given a successor based on the ABAP RESTful application programming model and SAP Fiori elements, instead of continuing to be embarrassingly embedded in the SAP Fiori launchpad via SAP GUI for HTML and merely visually prettified with an SAP Fiori theme. With *SAP Fiori elements*, you can also develop UIs (again) without having to retrain as a web developer. And, using *ABAP Cloud*, applications can be developed that run in all SAP solutions based on the current ABAP release—whether in the cloud or as an on-premise version.

To this end, various ABAP platform technologies and tools, some of which are more than 30 years old, have been replaced by successors, modernized, or at least provided with object-oriented facades. It's important to get to know these familiar concepts in a new guise to be able to develop in cloud systems or to adhere to the clean core concept for on-premise solutions.

Developer enablement

For newcomers to ABAP development, on the other hand, there are many current concepts that manage without the burden of legacy applications and outdated programming paradigms. From the perspective of *developer enablement*, it's currently particularly interesting to see who has internalized the new technologies faster or who is more open to them—the experienced developer or the newcomer. In upskilling measures, this range of prior knowledge, but also expectations, must be given special consideration.

1.2 The New Development Model for ABAP

ABAP Cloud

The new development model recommended by SAP today is called ABAP Cloud. Before you shy away from the term *cloud* and question the relevance of the book for your development, note that ABAP Cloud is also available as an on-premise version. The misleading name merely indicates that the applications developed using the model are generally cloud-capable, but not that they necessarily have to be executed in the cloud. Before we come to how this independence from the specific runtime environment can be technically implemented, we first want to say something about the development model concept.

Development model

A *development model* is understood here as a holistic development methodology. This includes the following elements:

- The programming language itself (e.g., the available syntax)
- The software architecture, which is implemented using programming models or specific frameworks
- The development environment, that is, the surrounding tooling for delivery and versioning
- The management and use of dependencies using stable interfaces

In this book, we'll frequently use the phrase "in ABAP Cloud" to mean that the specific technology or procedure is available and recommended in the development model.

ABAP Cloud components

ABAP Cloud uses the familiar ABAP technology of the ABAP platform, but deliberately restricts it to the following aspects:

- **ABAP language version**
  In ABAP Cloud, the use of the ABAP language version called *ABAP for cloud development* is mandatory.
- **Use of objects**
  Only repository objects released via application programming interfaces (APIs) can be used in ABAP Cloud. This applies to both procedure calls and type definitions.
- **Development environment**
  The ABAP development tools are used as the development environment in ABAP Cloud.
- **Programming model**
  The ABAP RESTful application programming model is used as a programming model for transactional applications.

In some cases, these rules are ensured by the syntax check or the runtime environment. For example, the classic ABAP Workbench can't be used in the SAP BTP ABAP environment and in SAP S/4HANA Cloud Public Edition due to a lack of SAP GUI support. The use of unreleased objects results in syntax errors in the ABAP language version.

Classic ABAP

As a contrast to ABAP Cloud, there's *classic ABAP*. In classic ABAP, there are no restrictions with regard to the aspects mentioned, and the entire spectrum of ABAP technology is available, including the latest techniques. For this reason, you can also use all the solutions shown in the book in classic ABAP or combine them with older technologies.

Requirements in the cloud

The definition of a new development model was simply necessary for the cloud runtime environments, that is, for the *SAP BTP ABAP environment* and *SAP S/4HANA Cloud Public Edition*. A way had to be described to develop applications that meet the requirements of cloud environments. For example, a `CALL SCREEN` would address an unavailable SAP GUI and an `OPEN DATASET` would address an unavailable file system. A `CLIENT SPECIFIED/USING CLIENT` would override the tenant concept. A direct `UPDATE` statement for an SAP standard table would jeopardize data consistency.

In addition, the aforementioned cloud systems are regularly and automatically updated to the next release. A customer's own coding must therefore be upgrade-stable, as it's not intended that upgrades will be blocked if necessary code adjustments haven't yet been made. It's not possible to react to changed or removed objects in the standard SAP system. In addition to promoting current technologies and application architectures, ABAP Cloud solves all of these problems by usually flagging them as syntax errors and thus preventing them during implementation.

Why an on-premise deployment?

While there's no technical need for the aforementioned restrictions in the SAP S/4HANA on-premise solution, they may still be desirable here. If you voluntarily adhere to the rules specified by ABAP Cloud, this will make it easier for you to perform the next upgrade, for example, because in cloud environments this upgrade would have taken place automatically. This on-premise perspective is also important for the clean core concept, which we'll discuss in more detail in Section 1.7. You also have the option of outsourcing your ABAP-based applications to the cloud later.

1.3 The ABAP Language Version

Specification of the language version

A core component of ABAP Cloud is the restriction of the language version to *ABAP for cloud development*. The ABAP language version specifies rules for the syntax and object types used. It's specified for each repository object. You may already be familiar with the corresponding field from the Unicode conversion. In the program attributes, it was initially used to differentiate between Unicode-capable and non-Unicode-capable programs. This was marked with a checkbox.

The language version known today as *standard ABAP* originated there and appropriately has the technical ID X, derived from the ticked checkbox for activating the Unicode checks. The **Boolean** field was later reinterpreted and can now contain the values listed in Table 1.1.

ABAP Language Version	ID
Standard ABAP	X
ABAP for cloud development	5
ABAP for key users	2

Table 1.1 ABAP Language Versions

The ABAP language version ABAP for cloud development has the technical ID 5. This can be understood as an allusion to the fact that this version follows classic ABAP, which was also referred to as *ABAP/4*. Other language versions exist, but are obsolete so not mentioned here.

ABAP language version settings

You can set the ABAP language versions at the following levels:

- The supported repository object
- The package as default value for new objects
- The software component

Maintenance at object level

To maintain the repository object, open this object in the ABAP development tools, and select the **General** section in the **Properties** view. You'll find the **ABAP Language Version** field for the ABAP language version at the bottom (see Figure 1.1).

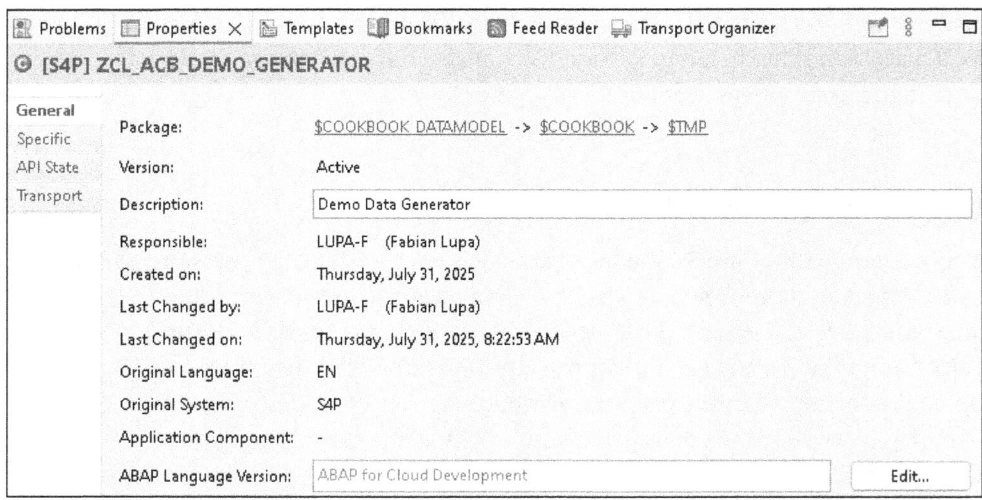

Figure 1.1 Maintaining the ABAP Language Version for a Global Class

If you click on the **Edit** button, a dialog box opens with a selection of the ABAP language versions available for the current system and the selected object type (see Figure 1.2).

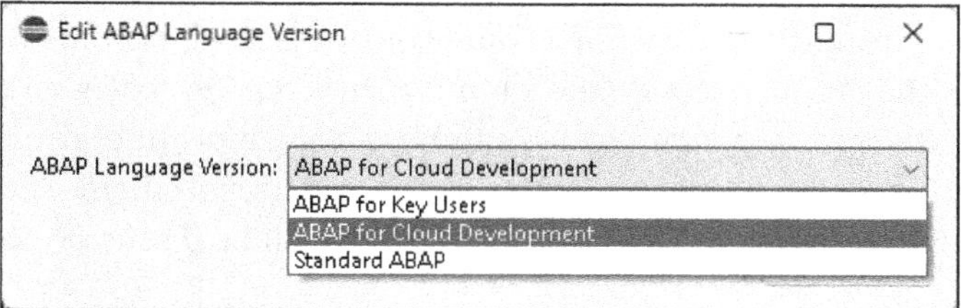

Figure 1.2 Selecting the ABAP Language Version

Maintenance at package level

When a new development object is created in a package, the ABAP language version is preset with the version that is stored in the package as the default ABAP language version. You can change this in the package maintenance in the **General Data** section (see Figure 1.3).

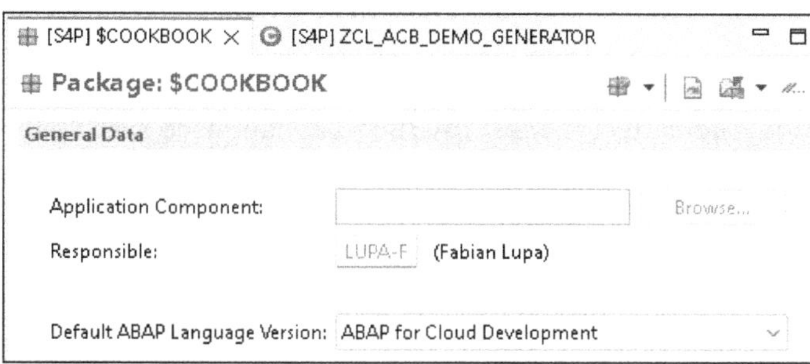

Figure 1.3 Maintaining the Default ABAP Language Version in the Package

[!]

Subsequent Change of the ABAP Language Version

Note that a subsequent change of the ABAP language version can make the syntax check stricter and the object may then have to be adapted.

[!]

No Restriction for Child Objects

Note that a subsequent change of the default ABAP language version in the package doesn't mean that all development objects assigned to the package will also be changed automatically. Likewise, the setting at package level doesn't restrict which ABAP language versions are permitted within the package.

Maintenance in the software component

In ABAP Cloud, the *software component* has become more important and can now also be used more easily by customers. It's used to group development objects from a delivery perspective. For example, the release and patch levels are defined at the software component level. You're probably familiar with this from the SAP_BASIS software component, whose release corresponds to the ABAP release. One or more *structure packages* are assigned to a software component, which can have a small or more complex hierarchy of main and development packages underneath them. The behavior of software components in relation to APIs will be discussed in Chapter 17.

Creating a software component

You can create and maintain custom software components in the on-premise version of SAP S/4HANA and in SAP S/4HANA Cloud Private Edition. In these products, maintenance is carried out classically via a report. To do this, you need to run the RSMAINTAIN_SWCOMPONENTS report in SAP GUI, for example, via Transaction SA38. When creating or changing a software component, you can choose between the ABAP language versions; **Standard ABAP** and **ABAP for Cloud** (see Figure 1.4). In contrast to the setting at package level, the **ABAP for Cloud** choice is associated with restrictions. In this case, all packages assigned to the software component, including the subordinate development objects, must use the ABAP language version for cloud development when creating a new one.

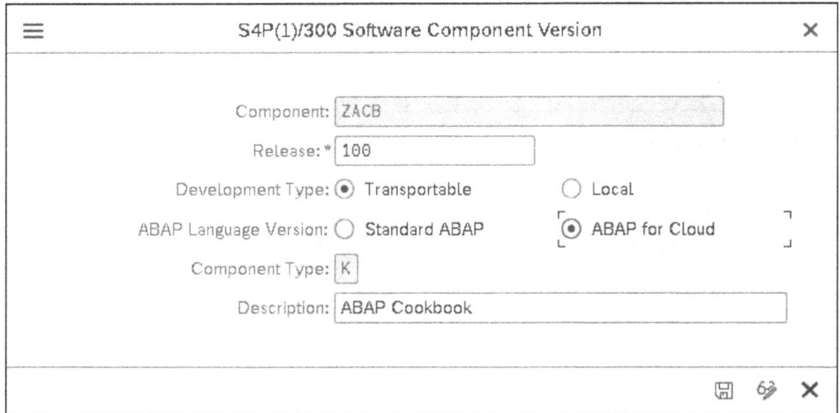

Figure 1.4 Maintaining the ABAP Language Version in the Software Component

[!]

Transport of Software Components

The software components maintained via the RSMAINTAIN_SWCOMPONENTS report aren't automatically recorded in transports. In subsequent systems, however, the entry of the software component is mandatory in some places: Some APIs, such as those for providing fixed domain values, filter on the basis of the ABAP language version in the associated software component. For this reason, you should make sure that you manually include your newly created or changed entry in a transport request. You can find the relevant function in the aforementioned report via the following menu path: **Software Component • Transport**. Alternatively, you can click on the **Transport software component entry** button. Only the administrative entry for the software component is transported via the VERS object type. This isn't a delivery transport of the associated development objects.

[«]

Designing Software Components and Package Hierarchies

The design of software components and package hierarchies isn't an easy task. You should therefore plan your structures carefully in advance. In SAP S/4HANA and SAP S/4HANA Cloud Private Edition, maintenance of these elements is optional. You can also continue to use the predefined software component for customer developments: HOME. However, this can lead to issues with some APIs available in ABAP Cloud, which ensure that no calls break through the ABAP Cloud context based on the software component assignment. You should maintain software components in SAP S/4HANA Cloud Public Edition and in the SAP BTP ABAP environment because they also have a controlling function in SAP Cloud Transport Management.

In SAP S/4HANA Cloud Public Edition and in the SAP BTP ABAP environment, custom software components are maintained via the SAP Fiori app called Manage Software Components (app ID F3562). Only the ABAP language version ABAP for cloud development is permitted in these two runtime environments.

Effects of the ABAP language version

The ABAP language version has a profound effect on the available repository objects. In the course of the new ABAP language version, SAP had the opportunity to switch off obsolete concepts and language components. As a result, many object types are no longer available in the ABAP for cloud development language version. You can't create a type group or a search help with this language version. There are more modern successor concepts for both property types. Even the ABAP report is no longer permitted in ABAP for cloud development. To run ABAP statements directly in the system, you can use ABAP console applications with the IF_OO_ADT_CLASSRUN interface for demonstration purposes.

Syntax errors and warnings

There are also syntactical changes. Not all ABAP statements are permitted in the ABAP for cloud development language version. For example, both the WRITE and OPEN DATASET statements cause syntax errors. The same applies to the USING CLIENT/CLIENT SPECIFIED addition to ABAP SQL statements. Other statements, such as ADD or the definition of a classic exception, generate syntax warnings (see Figure 1.5).

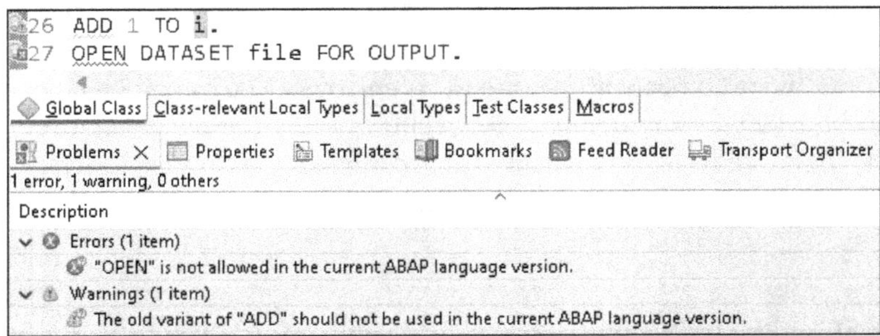
ADD 1 TO i.
OPEN DATASET file FOR OUTPUT.
Global Class | Class-relevant Local Types | Local Types | Test Classes | Macros
Problems | Properties | Templates | Bookmarks | Feed Reader | Transport Organizer
1 error, 1 warning, 0 others
Description
Errors (1 item)
"OPEN" is not allowed in the current ABAP language version.
Warnings (1 item)
The old variant of "ADD" should not be used in the current ABAP language version.

Figure 1.5 Examples of Syntax Warnings and Errors Caused by the ABAP Language Version

Syntax errors are also output when using nonreleased objects, which we'll describe in the following section.

1.4 APIs Released via Release Contracts

Accessing nonreleased APIs

You'll also receive syntax errors if you use objects that haven't been released. Figure 1.6 shows a read access via ABAP SQL to database table BUT000 in the standard SAP system. This results in a syntax error. In ABAP Cloud, only released APIs can be used. The ABAP for cloud development language version ensures this in the syntax check. In the specific example, the released object to be used as an alternative is named directly in the error message: the **CDS Entity I_BUSINESSPARTNER**.

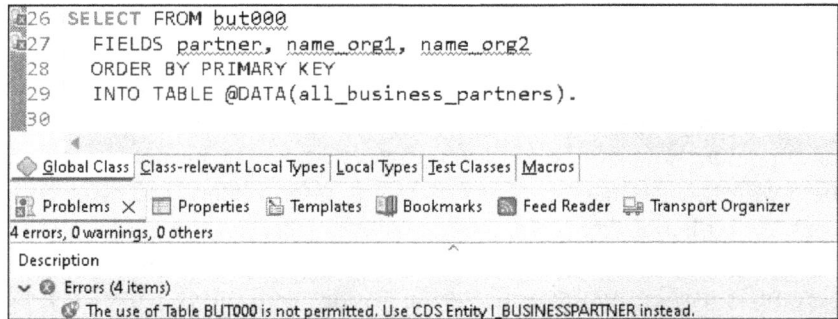

```
SELECT FROM but000
  FIELDS partner, name_org1, name_org2
  ORDER BY PRIMARY KEY
  INTO TABLE @DATA(all_business_partners).
```

Global Class | Class-relevant Local Types | Local Types | Test Classes | Macros

Problems | Properties | Templates | Bookmarks | Feed Reader | Transport Organizer

4 errors, 0 warnings, 0 others

Description

Errors (4 items)

The use of Table BUT000 is not permitted. Use CDS Entity I_BUSINESSPARTNER instead.

Figure 1.6 Syntax Error When Using a Nonreleased Object

This mechanism guarantees that you'll only receive third-party objects, that is, objects from software components other than the current one, for which the object provider promises that they won't become incompatible during upgrades. This is generally a very positive development because differentiating between externally usable and internal objects is made much easier. In terms of operation, however, this means that various objects known from the standard SAP system aren't released. These include popular data elements and function modules. Chapter 17 describes how to deal with this situation and how you can find and correctly call released APIs.

Release status

The release status of a development object is described via a separate repository object type named APIS. This type manages the lifecycle of the object provided as an API, including its deprecation and any replacement objects. However, you have nothing directly to do with the APIS object type. You can see the settings made in the **Properties** view of a repository object in the ABAP development tools in the **API State** section (see Figure 1.7).

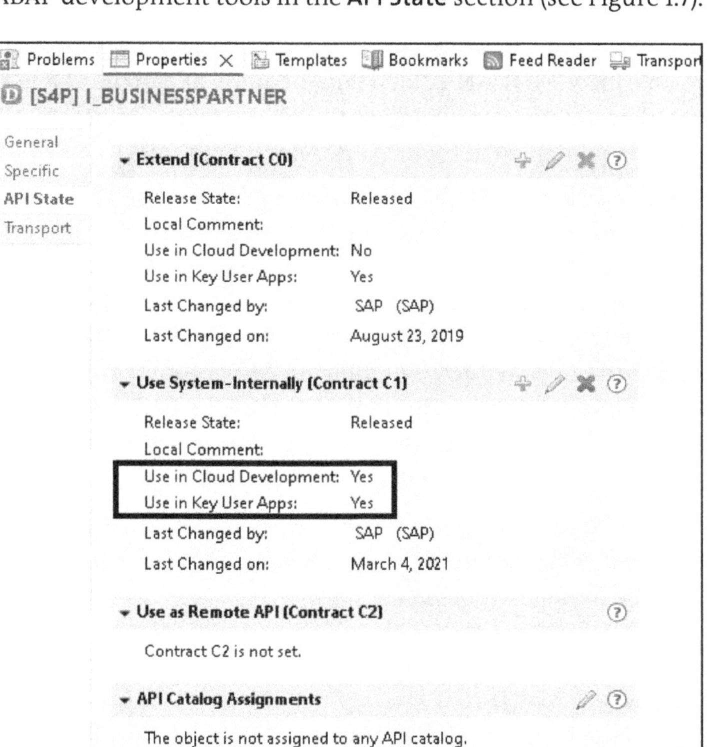

Figure 1.7 API State in the Properties View

Release contracts

For use in ABAP Cloud, the *C1 contract* is relevant, which guarantees a stable interface for use within the system. This *release contract* represents an agreement on stable use and further development between the object provider and the object user. The provider can only make compatible changes to the development object so that existing users aren't affected. On the other hand, the user guarantees that they communicate exclusively via the released objects with the offered component and don't rely solely on internal details. You can recognize the usage release for ABAP Cloud by the **Use in Cloud Development** indicator.

1.5 Development Environment

ABAP development tools in ABAP Cloud

ABAP Cloud requires the ABAP development tools as the development environment. This Eclipse-based environment has been available since 2012 and provides many advantages for classic ABAP development compared to the SAP GUI–based ABAP Workbench. For example, the functions include extensive refactoring actions and quick fixes, automatic syntax checking, and the display of development object documentation. ABAP development tools are required for ABAP Cloud for several reasons. SAP GUI isn't available in SAP S/4HANA Cloud Public Edition and in the SAP BTP ABAP environment. In addition, many of the object types used in ABAP Cloud can only be maintained in the ABAP development tools.

Joule in the ABAP development tools

While this book was being written, another ABAP development tools–exclusive feature was added with release 2502. The *Joule* AI wizard can simplify your day-to-day development work by explaining legacy code, suggesting code snippets, or generating unit tests directly in the ABAP development tools. These features are already available in the SAP BTP ABAP environment and in SAP S/4HANA Cloud Public Edition—and exclusively in the ABAP development tools. The long-term license policy for these functions isn't yet known at the time of writing. Current information on licensing and availability can be found in SAP Note 3571857.

If you don't yet use ABAP development tools, we recommend that you start changing over now. Information on their installation and connection to your system can be found in Appendix C.

1.6 Programming Model

ABAP RESTful application programming model

Part of the ABAP Cloud development model is also the specification of a programming model. The ABAP RESTful application programming model is promoted for transactional applications, which are primarily used to collect, change, and display master and transaction data. It's constantly being expanded and covers the requirements of both developers and users in various ways. We'll only provide a brief overview of this programming model here. The individual components are explained based on the sample application in Chapter 2.

Code pushdown with ABAP CDS

From a technical perspective, the ABAP RESTful application programming model initially implements the *code pushdown*. This involves making optimal use of the SAP HANA database by performing calculations on the database wherever possible and reducing the number of queries. This is achieved by using the *ABAP core data services* (ABAP CDS) on which the programming model is based. In the standard SAP system on SAP S/4HANA, extensive CDS views are already available via the *virtual data model* (VDM), which you can reuse.

Implementation scenarios

The ABAP RESTful application programming model framework takes a large part of the implementation work off your hands. In what is called the *managed scenario*, standard operations for creating, changing, deleting, or listing business object instances are implemented generically by the framework and don't have to be implemented manually by you for each application. Deviations from the standard implementation and extensions can be easily integrated into the behavior definition via *determinations*, *actions*, and *validations* as well as via the declarative configuration of individual components.

If that isn't sufficient, parts of the ABAP RESTful application programming model application can be implemented *unmanaged*. In this case, you need to replace the standard logic with your separately implemented logic, for example, for number assignment or lock logic. In extreme cases, you can also implement the entire business object yourself, including the standard operations (*unmanaged scenario*). This is particularly useful if you want to encapsulate existing applications using the ABAP RESTful application programming model.

You'll see what these different customization options look like in several chapters of this book. We'll show you how an application based on an ABAP RESTful application programming model is structured via a sample application in Chapter 2.

Consuming ABAP RESTful application programming model objects

The consumption of business objects developed using the ABAP RESTful application programming model is possible via multiple technologies. The actual implementation of the business logic is completely independent of the access protocol via the various abstraction layers. The business object can be used via HTTP using the Open Data Protocol (OData)-based interface in read and write scenarios, as an API, or to display a UI. At the same time, it can be addressed locally within the system via the *Entity Manipulation Language* (EML). ABAP RESTful application programming model business objects therefore represent a central access point, regardless of the usage context.

Nontransactional applications

Implementation approaches are also available in ABAP Cloud for nontransactional use cases. You can use HTTP services, for example, to implement your own web service directly without framework support. Application jobs enable you to define and implement actions to be executed periodically.

1.7 Usage Scenarios for ABAP Cloud

We've already mentioned the current ABAP-based products from SAP and their differences in several places. In this section, we want to revisit them centrally and explain in particular the terms we use in this book to differentiate between them.

ABAP-based products

SAP currently offers the following ABAP-based products and thus ABAP runtime environments:

- SAP S/4HANA
- SAP S/4HANA Cloud Private Edition
- SAP S/4HANA Cloud Public Edition
- SAP BTP ABAP environment

Other products, such as industry solutions or products in the analytics or customer experience environment, aren't explicitly addressed in this book. However, the contents of this book should be transferable to these products if they are release-equivalent.

SAP S/4HANA

SAP S/4HANA is the classic ABAP-based on-premise ERP solution from SAP. Its name doesn't formally contain any information about the provision, but you'll often find it as SAP S/4HANA (on-premise). This is the further development and direct successor to SAP ERP 6.0 based on SAP NetWeaver, with the ABAP platform as the technical basis.

With this product, you're flexible in terms of operation and can take responsibility for this yourself, even in your own data centers, or outsource it partially or completely. There are no restrictions on ABAP development in SAP S/4HANA. You can continue to use classic ABAP development as in SAP ERP 6.0 or use ABAP Cloud as an alternative or supplement from SAP S/4HANA 2022. SAP S/4HANA will follow a two-year release cycle from the 2023 release, meaning that the next release will be SAP S/4HANA 2025.

SAP S/4HANA Cloud Public Edition

SAP S/4HANA Cloud isn't a specific product name. However, this term is often used synonymously with the product name SAP S/4HANA Cloud Public Edition, formerly also known as SAP S/4HANA Public Cloud. SAP S/4HANA Cloud Public Edition is the counterpart to SAP S/4HANA, is offered exclusively as software as a service (SaaS), and has fundamental differences to SAP S/4HANA:

- SAP is fully responsible for the technical operation of the system, including updates.
- The SAP Fiori launchpad is the central point of entry; no SAP GUI is available.
- The configuration is carried out via *SAP Central Business Configuration*.
- Authorizations are assigned via *business users*, *business roles*, *business catalogs*, and *identity and access management (IAM) apps*.
- *Key user extensibility* and *developer extensibility* are available to extend the system. Both models can be used for developments in ABAP with the restricted ABAP language versions. You can't use classic ABAP on this system—ABAP Cloud is the available model.

  As you're in the cloud environment here anyway, other technology stacks are also directly available for extensions as part of *side-by-side extensibility* with the SAP Cloud Application Programming Model, SAP Build, or the SAP BTP ABAP environment.
- New releases for SAP S/4HANA Cloud Public Edition appear every six months and are installed automatically. For this reason, we won't provide any release numbers for this product in this book because no old releases of the cloud products can be in use.

SAP BTP ABAP environment

Like SAP S/4HANA Cloud Public Edition, the SAP BTP ABAP environment product that is only available in the cloud. It's a *platform as a service* (PaaS). From a developer perspective, the SAP BTP ABAP environment is comparable to SAP S/4HANA Cloud Public Edition. The only difference is that the business applications aren't available here and that only the technological basis is available. In this respect, you can also regard the SAP BTP ABAP environment as a Basis system or standalone ABAP platform.

Historically, the SAP BTP ABAP environment was the first ABAP runtime environment in the cloud. It's also known under the nickname of *Steampunk*. The ABAP runtime environment in SAP S/4HANA Cloud Public Edition, SAP S/4HANA, and SAP S/4HANA Cloud Private Edition is therefore also referred to as *Embedded Steampunk* because it's available directly in the system. Its formal name is now *SAP S/4HANA Cloud ABAP environment*. The SAP BTP ABAP environment is updated quarterly and is therefore the first environment to receive new functions in ABAP development.

SAP S/4HANA Cloud Private Edition

SAP S/4HANA Cloud Private Edition, formerly known as *SAP S/4HANA Private Cloud*, is largely an on-premise SAP S/4HANA system from a developer's perspective, but is operated by SAP or a certified hosting partner.

Access to client 000 or at the operating system level, for example, is restricted compared to the on-premise edition of SAP S/4HANA. Technically, however, this version is otherwise indistinguishable from the on-premise version. Classic ABAP and SAP GUI are available here alongside ABAP Cloud.

Products from a developer's perspective

From a developer's perspective, SAP S/4HANA and SAP S/4HANA Cloud Private Edition are largely similar and both development models can be used here, while in SAP S/4HANA Cloud Public Edition and SAP BTP ABAP environment, only ABAP Cloud is available as well as other approaches to authorizations, the transport system, and configuration apply. Just as the naming of ABAP Cloud could suggest that this development model is only available in cloud products, the term SAP S/4HANA Cloud could also suggest that SAP S/4HANA Cloud Public Edition and SAP S/4HANA Cloud Private Edition are similar. We'll therefore always state the product names explicitly in this book and not abbreviate them to avoid confusion.

Available ABAP language versions

The ABAP language versions available in the products are listed in Table 1.2.

ABAP Language Version/ Product	Standard ABAP	ABAP for Key Users	ABAP for Cloud Development
SAP S/4HANA	Available	Available as of 7.57/2022	Available as of 7.57/2022
SAP S/4HANA Cloud Private Edition	Available	Available as of 7.57/2022	Available as of 7.57/2022
SAP S/4HANA Cloud Public Edition	Not available	Available	Available
SAP BTP ABAP environment	Not available	Available	Available

Table 1.2 ABAP Language Versions and Their Availability

With the SAP S/4HANA and SAP S/4HANA Cloud Private Edition solutions, you have the choice. For new developments, however, we recommend weighing up whether ABAP Cloud is a useful alternative to classic ABAP for you. Using the new development model is an opportunity to carry out your developments regardless of the chosen runtime environment because ABAP Cloud is available in all products, while at the same time promoting best practices and modern technologies. This is particularly relevant for the clean core approach. However, you'll also incur expenses for the continued training of the development teams, and you'll have to take technological restrictions into account. A useful approach is to first familiarize yourself with the development model in the context of smaller projects with a more flexible time frame.

Clean core

The *clean core* concept simply describes the idea of not making any enhancements or even changes to the standard SAP system that could cause issues during the upgrade in case the standard SAP system changes. Without this kind of adjustment, upgrade projects are significantly simplified. In SAP S/4HANA Cloud Public Edition and the SAP BTP ABAP environment, upgrades are carried out automatically, so clean core is not only desirable but also necessary in the cloud context. ABAP developments carried out there are always clean core by definition because ABAP Cloud and the ABAP for cloud development language version don't allow any syntactical use or adaptation and no calling of nonreleased objects. Clean core is an extensive topic that is only mentioned in the context of ABAP Cloud in this book.

1.8 ABAP Releases On-Premise and in the Cloud

Availability of functions

Some of the functions or techniques presented in this book are highly dependent on the release. If you reprogram the examples to SAP S/4HANA 2022 or older, some functions may not yet be available. Especially in ABAP Cloud, where the use of all older APIs was prohibited by a syntax check, SAP has to deliver successor APIs step-by-step. Many such successor APIs are now available, but in real life, the focus is still often on the release notes and the road map.

Larger new functions are also constantly being added to the ABAP RESTful application programming model. To understand the release notes, it's advantageous if you understand the changed logic for the ABAP release number. This corresponds to the releases of the `SAP_BASIS` software component in the system.

[+]

Release Notes and Road Map

Release-specific information on ABAP can be found in the keyword documentation. The version of this documentation designed for ABAP Cloud is available at *http://s-prs.de/v1064800.*

Information on announced features can be found in the road map for the ABAP platform at *http://s-prs.de/v1064801.*

Development of ABAP releases

How the ABAP releases and the releases of the products or runtime environments using ABAP have developed in recent years is illustrated in Figure 1.8.

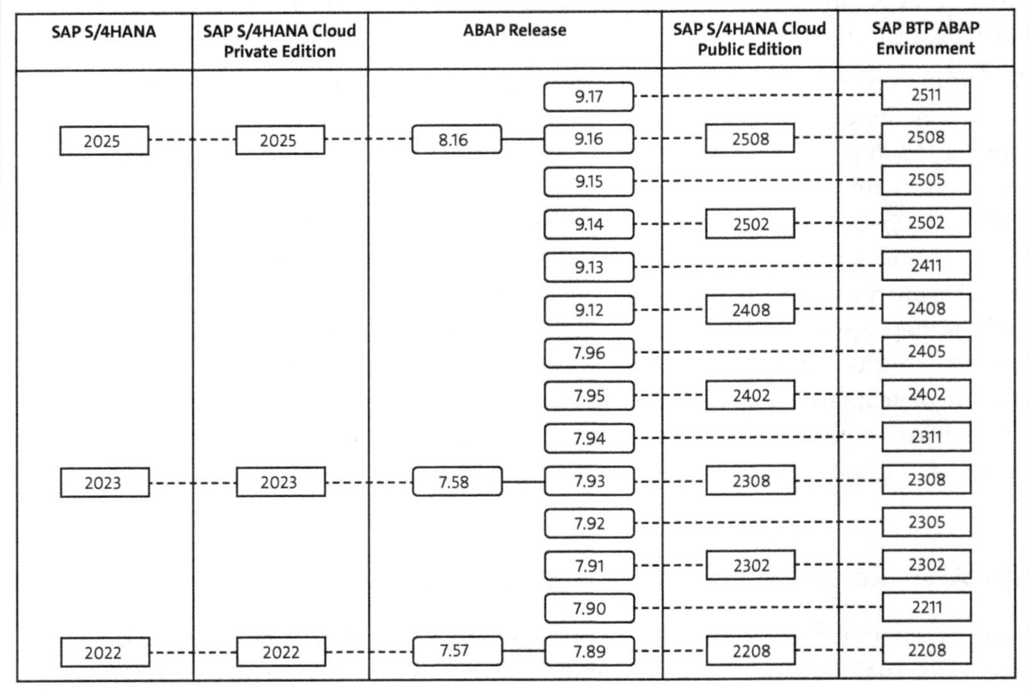

Figure 1.8 ABAP Releases and SAP Products

Figure 1.8 indicates the release cycles of the products. The SAP BTP ABAP environment with quarterly releases is updated most frequently, while SAP S/4HANA and SAP S/4HANA Cloud Private Edition are updated the least frequently with a release cycle of now two years. The year is stated in the release name, while the SAP BTP ABAP environment and SAP S/4HANA Cloud Public Edition use a combination of year and month. The on-premise and private cloud ABAP releases still followed the 7.5x numbering system until 2023; from 2025, their release numbers will start with 8.1x, beginning with 8.16. SAP S/4HANA Cloud Public Edition and the SAP BTP ABAP environment used the numbering 7.9x until release 2405 (originally started at 7.60); from 2408 these release numbers start with 9.1x.

1.9 Restrictions Depending on the Release and Runtime Environment

Although the use of ABAP Cloud is intended to ensure compatibility between the various runtime environments, problems can arise in some places if you want to support multiple runtime environments at the same time.

Release deviations

If you develop an application in the SAP BTP ABAP environment, you could be ahead of the functional scope in SAP S/4HANA Cloud Public Edition and would therefore have to limit yourself to the functional scope available there if you want to support both environments. In this case, you could be just one quarter ahead. In combination with SAP S/4HANA and SAP S/4HANA Cloud Private Edition, things look very different. Features provided over several years could be missing here, even if the current release of the product is used.

Release status of this book

Our book was written on the basis of SAP S/4HANA 2023 and ABAP 7.58. The cloud runtime environment is about 1.5 years ahead of this stage of development. This means, for example, that in Chapter 14, there was still no option for Excel file processing to edit existing Excel files, whereas this would already have been possible in the cloud. To use the `CL_ABAP_BEHAVIOR_SAVER_FAILED` class in Chapter 7, an SAP Note must first be imported so that the class is also released for use in ABAP Cloud.

Environment-specific behavior

Even with an equivalent ABAP release, there may be deviations or incompatibilities due to environment-specific behavior. For example, IAM apps are required for authorization maintenance in SAP S/4HANA Cloud Public Edition and in the SAP BTP ABAP environment. However, the associated `SIA` object types are only available in these environments and can't be maintained in SAP S/4HANA and SAP S/4HANA Cloud Private Edition. The situation is similar in the area of connectivity, where cloud-specific objects or configurations are sometimes required. Thus, if you want to support multiple environments for your product, it may be necessary to provide your environment-specific objects separately, even though you've adhered to all ABAP Cloud rules.

1.10 Summary

In this chapter we've dealt with modern ABAP development in general. You've become familiar with the role of ABAP in the current SAP product portfolio and now know the meaning of ABAP Cloud. You've seen in which products the development model is available or even mandatory.

A central component for the implementation of ABAP Cloud is the concept of ABAP language versions and release contracts, which ensure compliance with a restricted language scope and restricted object usage via the syntax check. The ABAP development tools and the ABAP RESTful application programming model are further key components of the ABAP Cloud development model.

Finally, this chapter has provided an overview of the ABAP releases and the associated products, and you now know what restrictions and hurdles you have to expect if you want to support multiple runtime environments at the same time.

Chapter 2
The Application Scenario

In this chapter, you'll get to know the sample application that will accompany you throughout this book. We'll add new functions to it in each chapter. This chapter first presents the basic structure of the application.

We've chosen a recipe portal as our application scenario and basis for the programming examples in this book. It provides various functions. In the following chapters, we'll use this application scenario to explain the new features in ABAP Cloud and demonstrate their use in real-life scenarios.

Structure of this chapter

Section 2.1 begins by describing the technical requirements that are placed on our recipe portal. Based on these requirements, we then present the technical framework and implementation. In Section 2.2, we describe how the domain, data element, and database table repository objects can be created, and we explain in Section 2.3 how an initial, simple application based on the ABAP RESTful application programming model can be generated from them. In Section 2.4, we expand this application to include associations that can be used to add ingredients to a recipe and write reviews. Finally, the recipe portal is presented from a user's perspective in Section 2.5.

Using the Associated Sample Application

All programming examples used in this book can be found in the download material for this book at *www.sap-press.com/6198* and in our Git repository at *http://s-prs.co/v619800*. The objects created in this chapter are located in the `DATAMODEL` subpackage.

2.1 Concept of the Sample Application

In this section, we present both the functional and the technical concept of our sample application—a recipe portal.

Use case diagram

Figure 2.1 presents the requirements of a recipe portal in the form of a Unified Modeling Language (UML) use case diagram. This diagram serves to provide a general overview of the interaction options between all actors and the system.

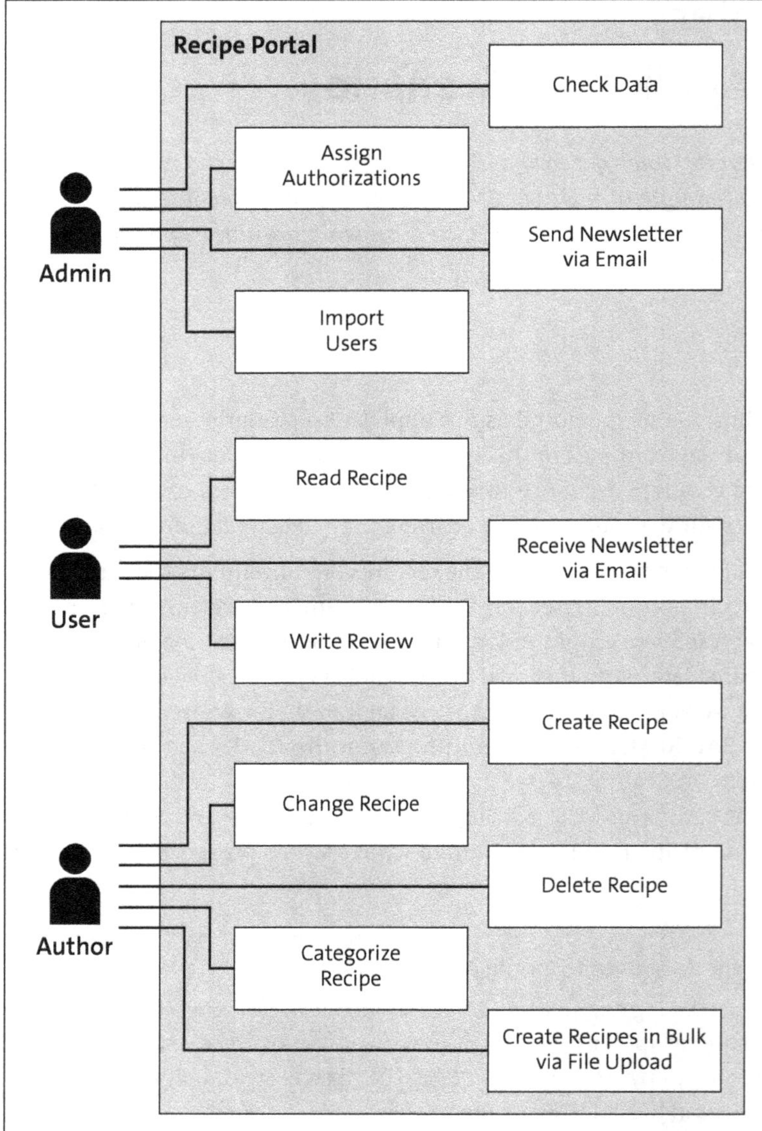

Figure 2.1 Use Case Diagram for the Recipe Portal

In our application scenario, there are three actors:

- **User**
  This is a person who uses the recipe portal for research purposes and for finding new inspirations for dishes.
- **Author**
  This is a special user who has extended interaction options, such as creating recipes with ingredient lists.

- **Admin**
  This actor takes on technical tasks so that the recipe portal can be operated.

User

The user uses the application for the following tasks:

- **Reading a recipe**
  The user can read the instructions and ingredients for a dish.
- **Receiving an email newsletter**
  The user receives regular notifications about which recipes have been added.
- **Writing a review**
  The user can enter a rating for each recipe in the recipe portal.

Author

The following use cases are assigned to the author:

- **Creating a recipe**
  The author can enter a recipe with all the ingredients and a text about the recipe in the recipe portal.
- **Changing a recipe**
  The author can change the ingredients for a recipe, such as varying the quantity of an ingredient or deleting an ingredient.
- **Deleting a recipe**
  The author can delete the recipe. The associated ingredients and reviews will then be deleted as well.
- **Categorizing a recipe**
  The author can assign a recipe to different categories.
- **Uploading large numbers of recipes via file upload**
  The author has the option of uploading multiple recipes to the recipe portal via file upload.

Admin

The admin is responsible for the following use cases:

- **Sending an email newsletter**
  The admin can determine the recipes that are to be sent in a newsletter.
- **Importing users**
  The admin can import large numbers of users.
- **Checking data**
  The admin must check the consistency of the data in the database tables.
- **Assigning authorizations**
  The admin can assign authorizations to specific users so that they can use more or different functions than other users.

Data model Based on the use case diagram in Figure 2.1, the technical concept for implementing the application scenario gets defined. In Figure 2.2, you can see the data model for the application scenario.

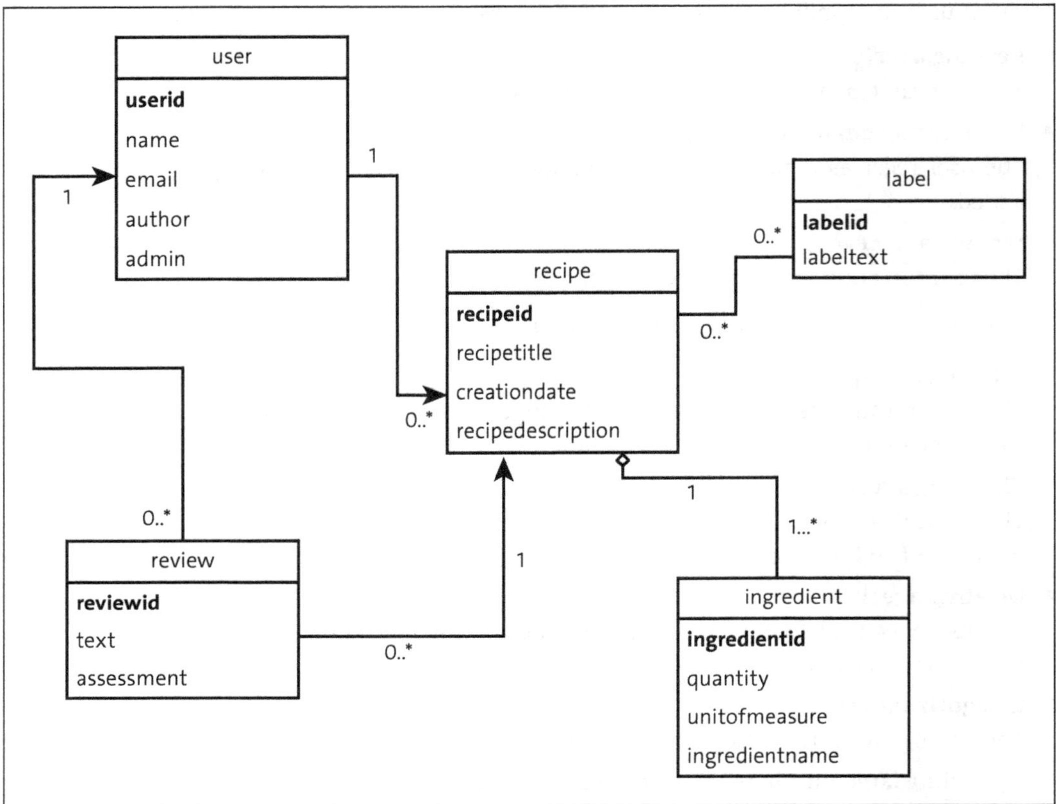

Figure 2.2 Data Model for the Recipe Portal

In this data model, we use the following business objects:

- **User**
  The user is an actor in the system. It can be a regular user, an author, or an admin.
- **Recipe**
  This central object represents step-by-step instructions on how to prepare a particular dish. In addition to the title of the dish, it also contains the individual instructions for preparing the dish.
- **Ingredient**
  This object represents the components of a recipe that are required to implement it. The object consists of the number, a unit of measure (e.g., oz), and the actual ingredient (e.g., flour).

- **Label**
  This object represents the categorization of recipes. A recipe can be assigned multiple labels, such as "dessert" and "vegan."
- **Review**
  This object represents the feedback for a dish. A user can enter rating texts for a recipe.

[«]

The Label Object

We'll look at the label in more detail in Chapter 5. In the following sections, we'll focus on the installation of the other objects.

2.2 Creating Dictionary Objects

This section describes the technical implementation of the individual objects of our sample application in the SAP system, based on the business concept from Section 2.1.

Framework parameters

The following framework parameters and naming conventions apply to the use of the recipe portal:

- Original language of the objects is EN.
- Technical designations are provided in English, but comments and language-dependent texts are in English.
- All object names in the application example contain the abbreviation ACB.

Creating dictionary objects

Once you've connected our development environment, ABAP development tools, to an SAP system, you can create the necessary objects in the *ABAP Dictionary* as described next. These objects are used to map the data model in the system. To do this, you must create domains, data elements, and database tables as global repository objects. The advantage of those global repository objects is that they can be used multiple times, while they need to be created only once. This means that changes to these objects only have to be made once, and all users can benefit from them.

Creating a package

Before the dictionary objects can be created, you must first create a package:

1. Select your project in the project explorer, right-click, and choose **New • ABAP Package** from the context menu (see Figure 2.3).

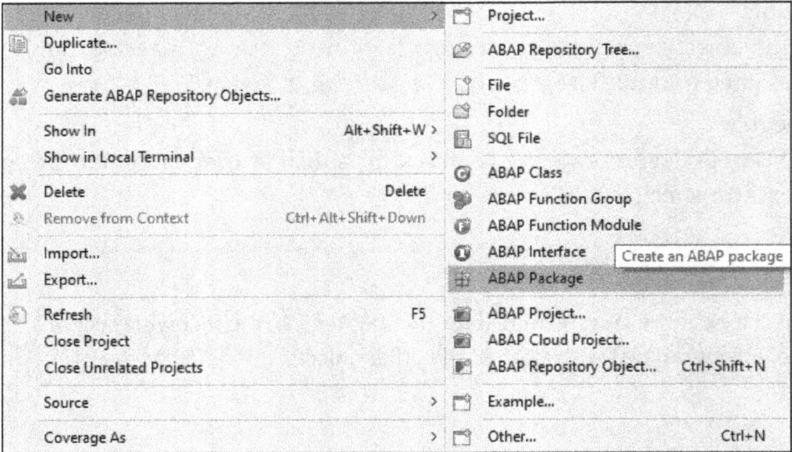

Figure 2.3 Creating a Package

2. In the dialog that opens (see Figure 2.4), you need to define the project properties with the values from Table 2.1.

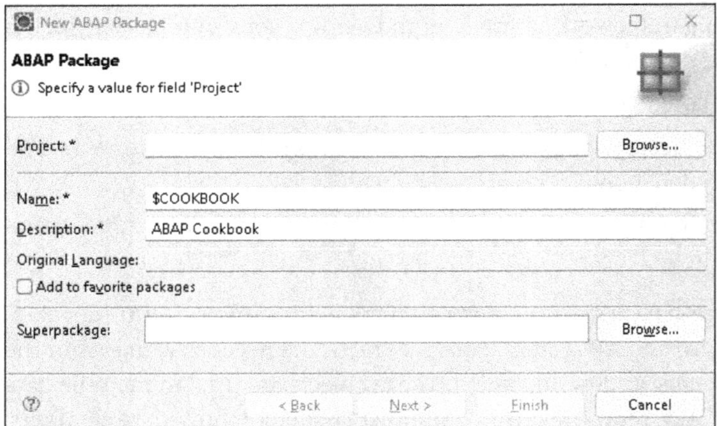

Figure 2.4 Properties of a Package

Field	Description	Value
Project	SAP system	This value will be entered automatically.
Name	Name of the package	$COOKBOOK
Description	Description of the package	Recipe Portal
Package Type	Type of package	Development

Table 2.1 Features of Our ABAP Package

3. Confirm the two dialog boxes that open next by clicking on the **Next** button. The $ character in the package name identifies our package as a local package that doesn't require any transport records.

ABAP language version

The $COOKBOOK package has now been created. It's now important to change the value in the **Default ABAP Language Version** field for this package to **ABAP for Cloud Development** (see Figure 2.5). This change ensures that all repository objects within the package must comply with the rules of the ABAP for cloud development language version setting.

The $COOKBOOK package must now be saved. It doesn't need to be activated. Now, we can create the repository objects.

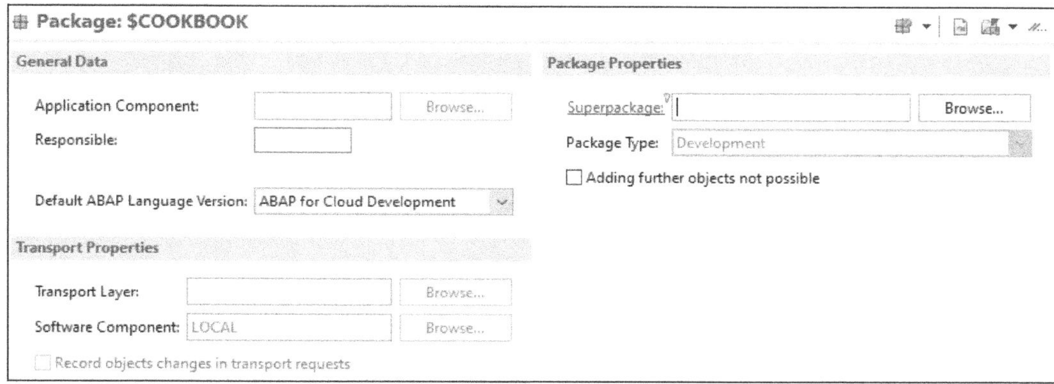

Figure 2.5 Created $COOKBOOK Package

Creating a domain

First, we want to create the domains for the sample application. The domain dictionary object defines the technical and semantic properties of data types. At this point, we create a domain for the Recipe business object as an example. You can proceed in the same way for the other objects presented in the previous section:

1. Right-click on the package for our application in the ABAP development tools. In the context menu that opens, select the **New • Other ABAP Repository Object** path (see Figure 2.6).

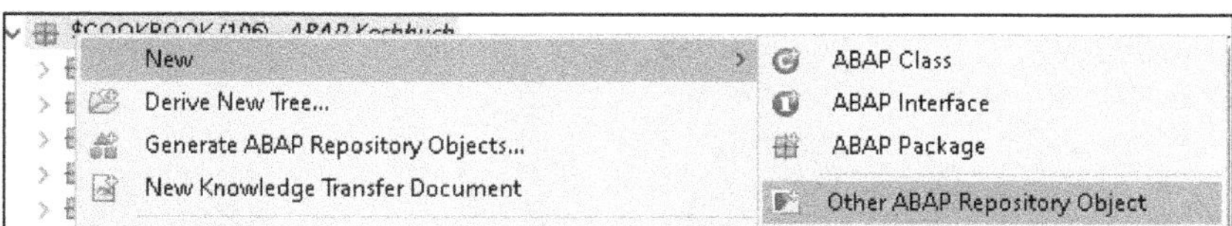

Figure 2.6 Creating a Domain via the Context Menu

2. In the search window that opens, search for the term "Domain", and then select it from the list (see Figure 2.7).

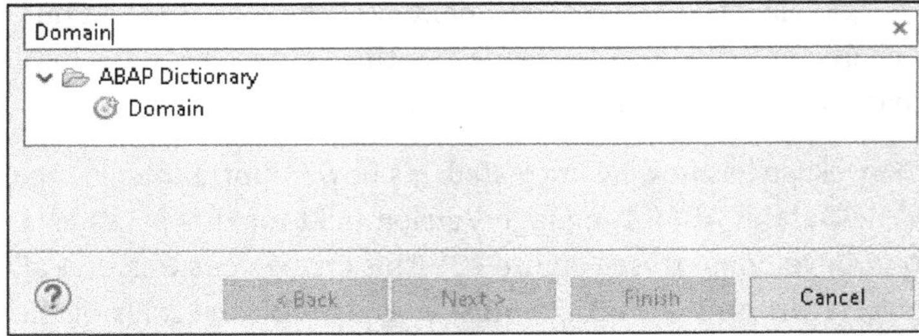

Figure 2.7 Selecting ABAP Dictionary Object "Domain"

3. The dialog box shown in Figure 2.8 opens. Enter an appropriate name and the corresponding description for the domain here. For our example, we entered `ZACB_RECIPE_ID`.

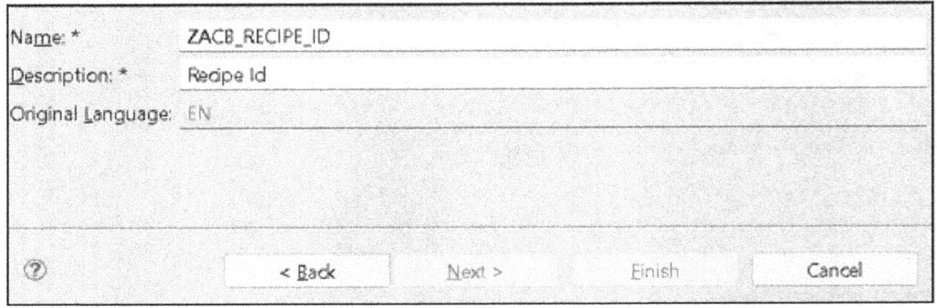

Figure 2.8 Popup Window for Creating a Domain

4. Click on the **Next** button.
5. The ABAP Dictionary editor opens where you can enter the technical properties of the domain (see Figure 2.9).

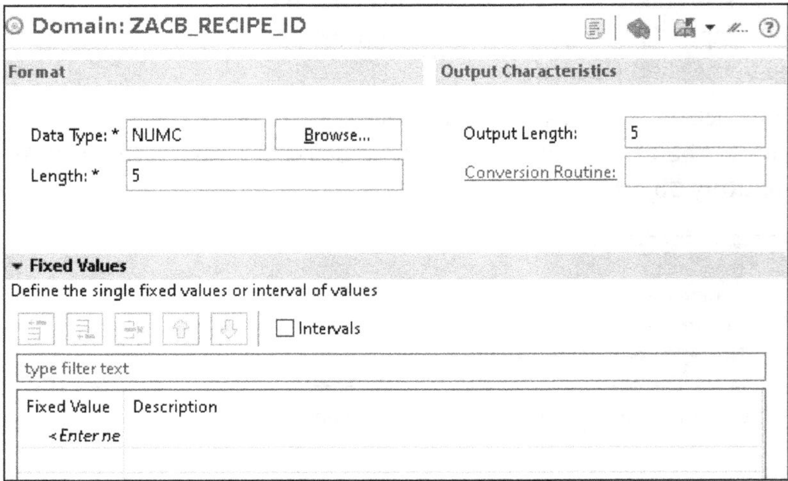

Figure 2.9 Properties of the Domain

Enter the following details:

- **Data Type**: Here, you can choose between 13 different basic data types. They each represent the way in which numbers (integers and floating point representations), text fields, or date and time information will be displayed.
- **Length**: The length of a field value can either be fixed via the data type or specified manually for text fields.

The semantic properties described are rudimentary properties. In addition to these, much more can be specified in the semantic properties of the domain, for example, that only lowercase letters or only specific values are allowed. These restrictions can be made using fixed values or by specifying a table.

6. Once you've made all the necessary entries, the domain still needs to be saved. To do this, press the Ctrl+S keyboard shortcut.
7. Finally, the domain must be activated by pressing Ctrl+F3.

Creating a data element

After creating the domains, you can create the *data elements* for the application. A data element is also a repository object and defines the semantic meaning of an object. It's used for database tables or for defining variables.

Here, we create a data element as an example to use it for fields in a database table:

1. Unlike when creating the domain, you must call the context menu for the **Dictionary** entry within the package structure (see Figure 2.10). Select the **New • Data Element** path.

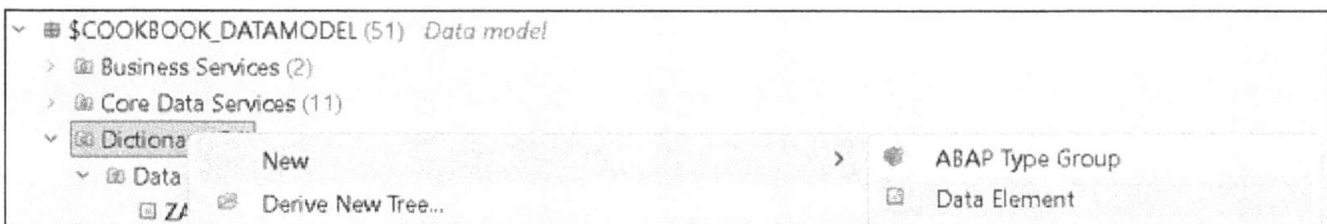

Figure 2.10 Creating a Data Element via the Context Menu

2. In the dialog that opens next (see Figure 2.11), you must enter a name and a description for the data element. The name can have a maximum of 30 characters and can consist of letters, numbers, and underscores. It must be introduced with a letter or a namespace prefix. Click on the **Next** button.

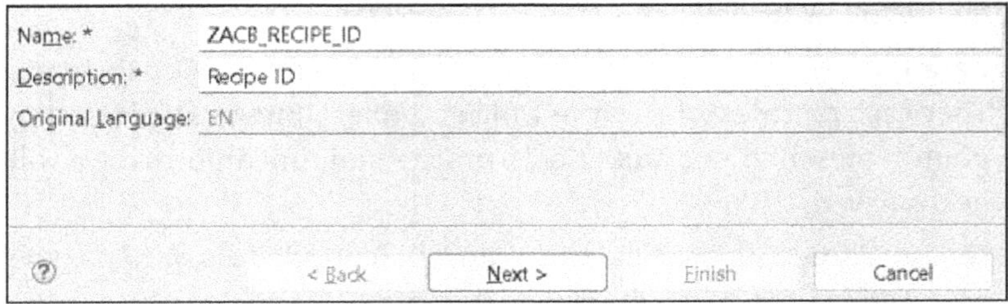

Figure 2.11 Creating a Data Element

3. Enter the following values in the ABAP Dictionary editor (see Figure 2.12):
   - **Category**: This field is used to specify the data type information.
   - **Type Name**: The name of the data element must be specified here. It can either be a domain provided by SAP or one that you've previously created yourself. In our example, we specify the ZACB_RECIPE_ID domain we created previously.
   - **Field Labels**: This area displays the texts that are displayed in the user interface (UI) when the data element is output. They are divided into four levels according to length. For our example, it's sufficient to enter the same value everywhere.

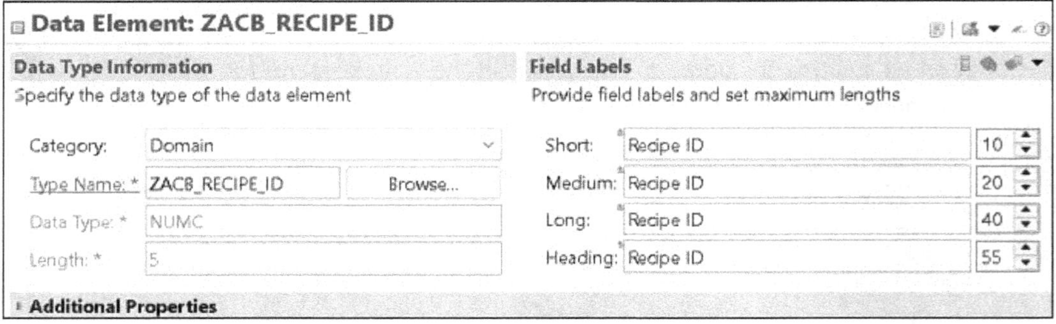

Figure 2.12 Technical Properties of a Data Element

4. Save the data element by pressing the Ctrl+S shortcut.
5. In the final step, the data element must be activated (Ctrl+F3).

Creating a database table

Once the domain and the data element have been created, you can create the database tables for the individual business objects in accordance with the data model described in the previous section. We'll show you this here using the database tables for the recipe as an example:

1. Open the context menu for the **Dictionary** entry within our project structure, and select the following entry: **New • Database Table** (see Figure 2.13).

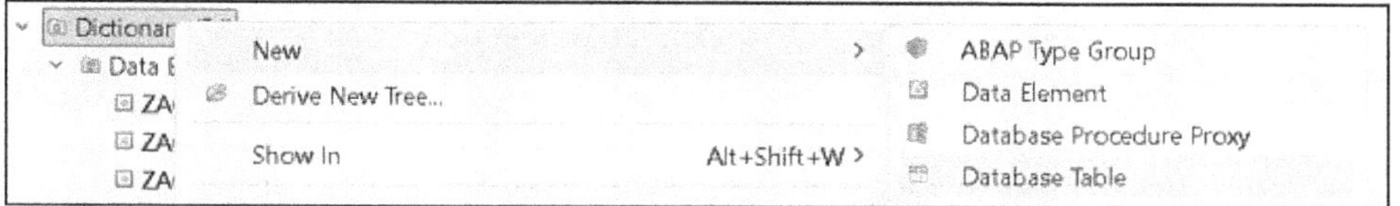

Figure 2.13 Creating a Database Table via the Context Menu

2. Enter a name and a description in the following dialog. Enter "ZACB_RECIPE" as the name for our database table for the recipes. Once all mandatory fields have been filled in, click on the **Next** button (see Figure 2.14).

Figure 2.14 Creating a Database Table

3. The ABAP Dictionary editor opens and generates an initial template for the database table in the form of annotations (see Listing 2.1):

```
@EndUserText.label : 'Recipes'
@AbapCatalog.enhancement.category : #NOT_EXTENSIBLE
@AbapCatalog.tableCategory : #TRANSPARENT
@AbapCatalog.deliveryClass : #A
@AbapCatalog.dataMaintenance : #RESTRICTED
define table zacb_recipe {

  key client : abap.clnt not null;
}
```

Listing 2.1 Generated Code for a Database Table

Properties of the database table

The name of the database table can be found in the following line:

```
define table <table name>
```

In the ABAP Dictionary editor, you can now add all other fields of the database table with the appropriate data element in the following format:

```
Field : data element
```

If a table field is a key, the Key attribute must also be added. Additional properties can be defined in the database table header via annotations (see Table 2.2).

Annotation	Description
@EndUserText.label	Description of the database table
@AbapCatalog.enhancement.category	Whether the table be extended
@AbapCatalog.tableCategory	The table category involved
@AbapCatalog.deliveryClass	Whether it's an application or Customizing table
@AbapCatalog.dataMaintenance	How the table can be changed

Table 2.2 Annotations for a Database Table

Once all entries for the table have been made, the ZACB_RECIPE database table looks like Listing 2.2.

```
@EndUserText.label : 'Recipes'
@AbapCatalog.enhancement.category : #NOT_EXTENSIBLE
@AbapCatalog.tableCategory : #TRANSPARENT
@AbapCatalog.deliveryClass : #A
@AbapCatalog.dataMaintenance : #RESTRICTED
define table zacb_recipe {
  key client            : abap.clnt not null;
  key recipe_id         : zacb_recipe_id not null;
  recipe_name           : zacb_recipe_name;
  recipe_text           : zacb_recipe_text;
}
```

Listing 2.2 Database Table ZACB_RECIPE

The database table only needs to be saved (Ctrl+S) and activated (Ctrl+F3).

2.3 Generating an ABAP RESTful Application Programming Model Application

Now that the dictionary objects have been created, you can generate an ABAP RESTful application programming model application. For this purpose, the data model of the application is exposed via an Open Data Protocol (OData) service, on the basis of which an initial simple user UI can already be output.

[«]

2

Using Generators

The ABAP development tools provide multiple generators that help you create boilerplate coding and objects by generating code blocks based on parameters. This allows you to reach your goal much faster for new developments compared to a manual object creation. However, the generators can create only entirely new objects. You can't use them to edit existing objects. You can't restart the generation process based on a previous result. For this reason, you should check the suggested parameter values carefully to avoid manual reworking. In CDS views and the ABAP RESTful application programming model in particular, many interdependent or interrelated objects are used. It's difficult to fundamentally change, rename, or delete these at a later date.

2.3.1 Generating an OData Service and a Virtual Data Model

Creating an OData UI service

You can proceed as follows to generate the OData service and the virtual data model of your application:

1. Right-click on the **ZACB_RECIPE** database table you've just created in the project structure, and select **Generate ABAP Repository Objects** from the context menu (see Figure 2.15).

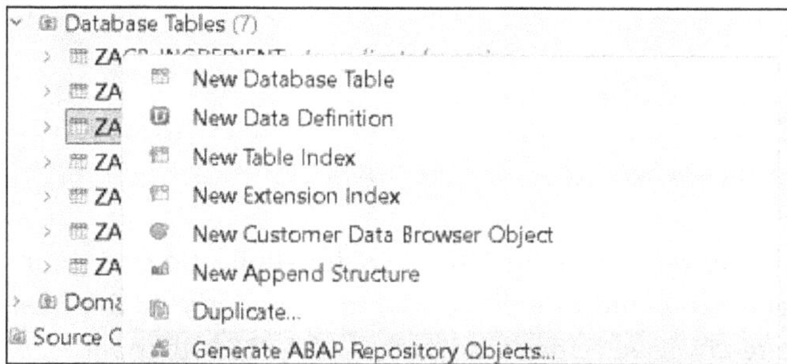

Figure 2.15 Selecting the "Generate ABAP Repository Objects" Item from the Context Menu

2. In the generator that opens next, select the **OData UI Service** object type, and click the **Next** button (see Figure 2.16).

   You'll then receive the error message shown in Figure 2.17, which informs you that the database table is missing two fields with an administrative function, which are used to record its last change.

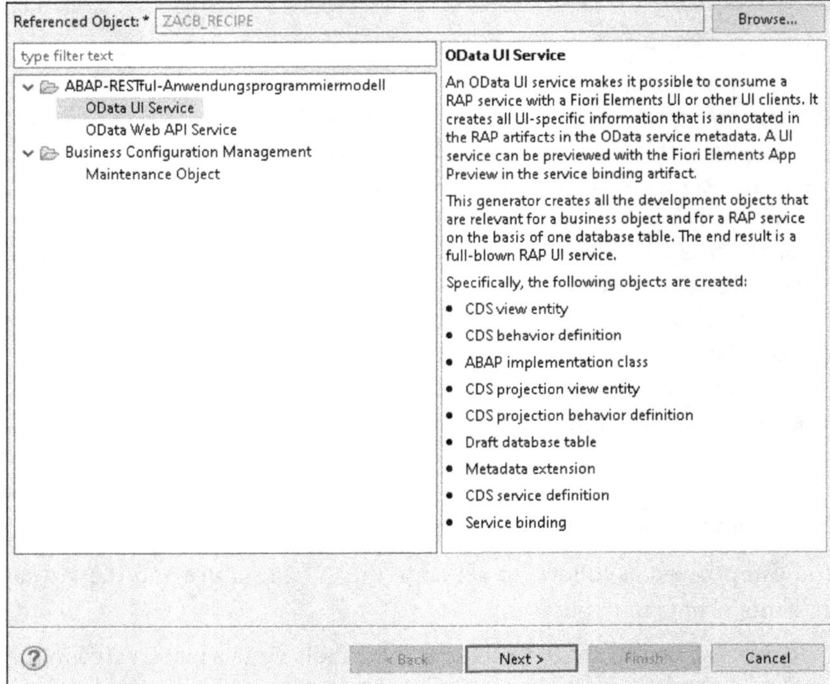

Figure 2.16 Popup Window of the Generator for an OData UI Service

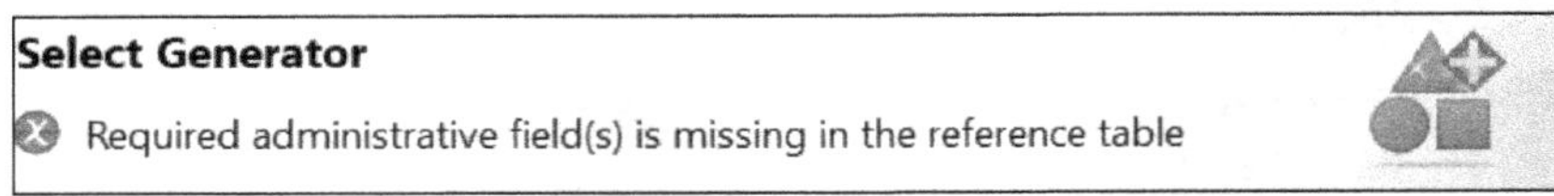

Figure 2.17 Error Message When Generating the OData UI Service

These fields are required within an ABAP RESTful application programming model application and must be added to all underlying database tables. If you already have experience with ABAP RESTful application programming model applications, you can also complete this step directly when creating the database tables.

In Listing 2.3, we extend the database table with the aforementioned administration fields.

```
define table zacb_recipe {

  key client         : abap.clnt not null;
  key recipe_id      : zacb_recipe_id not null;
  recipe_name        : zacb_recipe_name;
  recipe_text        : zacb_recipe_text;
  created_by         : abp_creation_user;
  created_at         : abp_creation_tstmpl;
```

```
  local_last_changed_by : abp_locinst_lastchange_user;
  local_last_changed_at : abp_locinst_lastchange_tstmpl;
  last_changed_at       : abp_lastchange_tstmpl;
  last_changed_by       : abp_lastchange_user;

}
```

Listing 2.3 Extension of Database Table ZACB_RECIPE

3. Activate the extended database table, and right-click to call the generator again.
4. In the dialog step that follows, you must specify a package. Select the package of our sample application, and confirm by clicking the **Next** button.
5. The subsequent dialog box (see Figure 2.18) displays the default names of the various objects in our ABAP RESTful application programming model application that are created when the OData UI service is generated. Check the suggested names on all layers listed in the **RAP Layers** section, and continue by clicking the **Next** button.

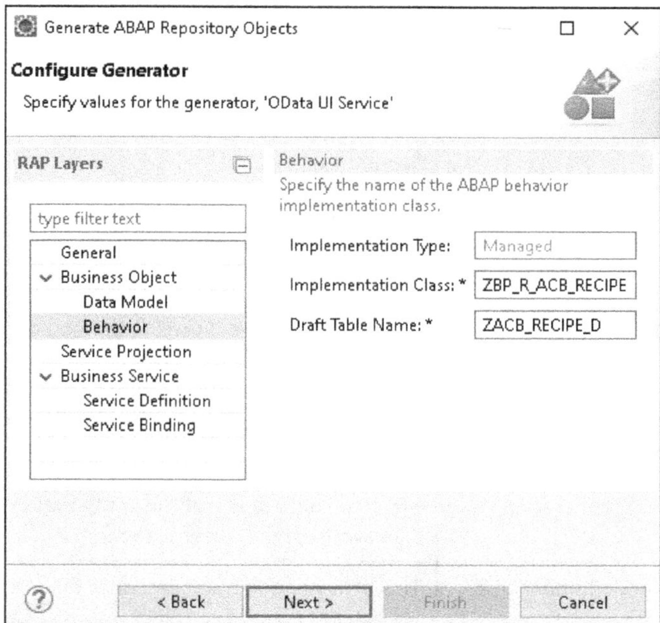

Figure 2.18 ABAP RESTful Application Programming Model Application Objects Created by the Generator

6. Enter a transport request in the step that follows. After that, you can start generating the objects by clicking the **Finish** button.

[!]

Naming the Objects

Remember to check the names of the generated objects. They should match the conventions you've chosen. If that's not the case, you need to correct the names directly at this point. A subsequent correction can still be made, but this results in more work.

2.3.2 Generated CDS Entities of the ABAP RESTful Application Programming Model Application

Generated CDS entities

Table 2.3 lists the individual objects of the ABAP RESTful application programming model application that are created via the generator.

Object Name	Explanation
Base entity ZACB_R_RECIPE	This data definition defines the data model of the root entity.
Basic behavior definition ZACB_R_RECIPE	This behavior definition describes the standard transactional behavior of the base entity. The behavior definition can be used directly in the managed scenario and also implements the draft concept of the ABAP RESTful application programming model (see Chapter 1, Section 1.6).
Behavior class ZBP_ACB_R_RECIPE	This ABAP class provides the implementation of the behavior defined in the behavior definition.
Draft table ZACB_RECIPE_D	This database table is used to temporarily store the draft data at runtime. It's managed by the ABAP RESTful application programming model framework.
Projection entity ZACB_C_RECIPE	This data definition is used to define the projected data model of the entity that is relevant for the current scenario.
Projection behavior definition ZACB_C_RECIPE	This behavior definition describes the behavior of the underlying base entity.
Metadata extension ZACB_C_RECIPE	The metadata extension is used to display the UI of the application via CDS annotations.

Table 2.3 Generated Objects Overview: ABAP RESTful Application Programming Model Application

Object Name	Explanation
Service definition ZACB_UI_RECIPE	A service definition is used to define the relevant entity sets for our service and also to provide aliases if required.
Service binding ZACB_UI_RECIPE_04	The service binding is used to make the generated service definition available as an OData UI service.

Table 2.3 Generated Objects Overview: ABAP RESTful Application Programming Model Application (Cont.)

Relationships between the objects

These objects were generated automatically. The relationships between the individual objects are illustrated in Figure 2.19. All objects are available to us as CDS entities.

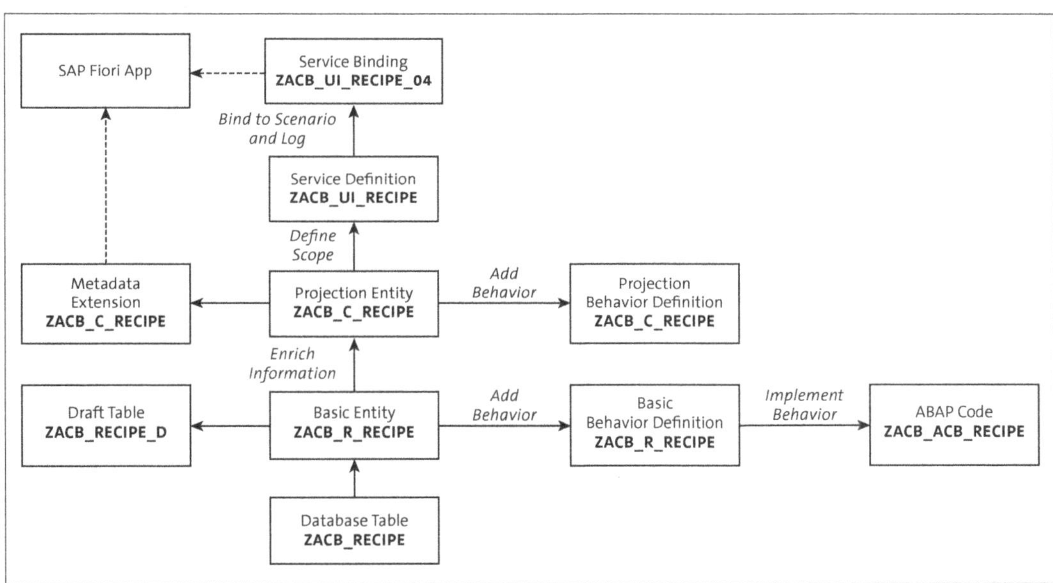

Figure 2.19 CDS Entity Names of the Generated Objects and Relationship Diagram

Base entity

You can now view the individual CDS entities and expand them as required. We want to start with the ZACB_R_RECIPE base entity (see Listing 2.4). This *CDS view* defines the Recipe business object. A CDS view is a virtual structure. It can be used to combine complex and calculated data from different sources and merge them into a single view.

```
@AccessControl.authorizationCheck: #CHECK
@EndUserText.label: 'Base entity recipe'
```

```
define root view entity ZACB_R_Recipe
  as select from zacb_recipe

{
  key recipe_id as RecipeId,
  recipe_name as RecipeName,
  recipe_text as RecipeText,
  @Semantics.user.createdBy: true
  created_by as CreatedBy,
  @Semantics.systemDateTime.createdAt: true
  created_at as CreatedAt,
  @Semantics.user.localInstanceLastChangedBy: true
  local_last_changed_by as LocalLastChangedBy,
  @Semantics.systemDateTime.localInstanceLastChangedAt: true
  local_last_changed_at as LocalLastChangedAt,
  @Semantics.systemDateTime.lastChangedAt: true
  last_changed_at as LastChangedAt,
  @Semantics.user.lastChangedBy: true
  last_changed_by as LastChangedBy,
}
```

Listing 2.4 Base Entity ZACB_R_RECIPE

The CDS code contains the following sections: First, the access control to the CDS view is described, and the label is specified. This is followed by the definition of a `root view entity` and access to a database table from which the data is to be imported (`as select from`). The last part describes the available data of the base entity. Here, it's specified which fields are `key` fields, and aliases are assigned for the field identifiers.

Behavior definition of the base entity

Let's now take a look at the behavior definition for the base entity (see Listing 2.5). Behavior definitions are a crucial part of CDS data models because they determine what can be done with our data.

```
managed implementation in class zbp_acb_r_recipe unique;
strict ( 2 );
with draft;

define behavior for ZACB_R_Recipe alias Recipe
persistent table zacb_recipe
draft table zacb_recipe_d
etag master LocalLastChangedAt
total etag LastChangedAt
authorization master ( global )
{
```

```
  field ( mandatory : create )
  RecipeId;

  field ( readonly )
  CreatedAt,
  CreatedBy,
  LastChangedAt,
  LocalLastChangedAt,
  LocalLastChangedBy;

  field ( readonly : update )
  RecipeID;

  create;
  update;
  delete;

  draft action Edit;
  draft action Activate optimized;
  draft action Discard;
  draft action Resume;
  draft determine action Prepare;

  mapping for zacb_recipe
    {
      RecipeId           = recipe_id;
      RecipeName         = recipe_name;
      RecipeText         = recipe_text;
      CreatedBy          = created_by;
      CreatedAt          = created_at;
      LocalLastChangedBy = local_last_changed_by;
      LocalLastChangedAt = local_last_changed_at;
      LastChangedAt      = last_changed_at;
      LastChangedBy      = last_changed_by;
    }
}
```

Listing 2.5 Behavior Definition ZACB_R_RECIPE for the Base Entity

Components of the behavior definition

In the behavior definition, the behavior is defined using the *Behavior Definition Language* (BDL). The behavior definition consists of the following information:

- **Type of scenario**
  The specification of the scenario determines the automatic availability

of operations. In the managed scenario, the method behavior is determined by the framework and can be supplemented by a custom logic. In the unmanaged scenario, you must implement the method.

- **Strict mode**
  In strict mode, additional syntax checks are applied to behavior definitions. This means that obsolete syntax can't be used, and implicit operations must be declared explicitly. The latest mode is always recommended (currently this is `strict(2)`).
- **Alias**
  An *alias* specifies a descriptive name for the business object so that the CDS view name doesn't always have to be used.
- **Draft mode (`with draft`)**
  The draft mode is a function that allows end users to start and pause their work on the entities and resume it later without having to save this data directly in the database. Following are the important objects for this mode:
  - `DraftTable`: Name of the draft table.
  - `Totaletag`: Timestamp for the draft.
  - `Draftaction`: Standard actions for the draft.
- **Lock**
  The `lock master` information is used to lock the instance during a changing action.
- **ETag**
  ETag prevents accidental overwriting when an object is edited concurrently. The `etag master <field name>` specification defines the field in which a timestamp is to be saved.
- **Authorization checks**
  The `authorization master ( instance )` definition calls the corresponding method in the implementation of the behavior class in which an authorization check is carried out.
- **Field properties (`field`)**
  The field properties define the behavior of a field. The following statements are used for this purpose:
  - `Field(readonly)`: Display only
  - `Field(readonly:update)`: Only display, except during creation
  - `Field(mandatory)`: Required field
  - `Field(suppress)`: Don't display metadata
  - `Field(numbering:managed)`: Automatic numbering

- **Standard actions**
  The standard actions to be implemented are specified. Possible actions are listed here:
  - CREATE: Adding a data record
  - UPDATE: Updating a data record
  - DELETE: Deleting a data record
- **Mapping**
  As part of the mapping action, the field names of the database table are assigned to those with field names in the CDS entity.

Draft table

Next, take a look at the draft table (see Listing 2.6). This is a separate database table for data in the design stage with the structure of the respective entity, that is, the field names correspond to those from the respective CDS entity.

```
@EndUserText.label : 'Draft recipe'
@AbapCatalog.enhancement.category : #EXTENSIBLE_ANY
@AbapCatalog.tableCategory : #TRANSPARENT
@AbapCatalog.deliveryClass : #A
@AbapCatalog.dataMaintenance : #RESTRICTED
define table zacb_recipe_d {

  key mandt           : mandt not null;
  key recipeid        : zacb_recipe_id not null;
  recipename          : zacb_recipe_name;
  recipetext          : zacb_recipe_text;
  createdby           : abp_creation_user;
  createdat           : abp_creation_tstmpl;
  locallastchangedby  : abp_locinst_lastchange_user;
  locallastchangedat  : abp_locinst_lastchange_tstmpl;
  lastchangedat       : abp_lastchange_tstmpl;
  lastchangedby       : abp_locinst_lastchange_user;
  "%admin"            : include sych_bdl_draft_admin_inc;
}
```

Listing 2.6 Draft Table ZACB_RECIPE_D

In addition to the fields of the original database table, the draft table also contains the administration fields of the SYCH_BDL_DRAFT_ADMIN_INC structure.

Projection entity

Furthermore, the ZACB_C_RECIPE *projection view* was created (see Listing 2.7). This projection view serves as a link between the ZACB_R_RECIPE base entity and the application. With the help of such projection entities, the fields and functions of applications can be further restricted because the full range of functions of the base entities may not be required for the application. Various projection entities can therefore be created from a base entity and made available to the application level. Compared to the base entity, our projection entity only contains the following addition:

```
provider contract transactional_query
as projection on ZACB_R_Recipe
```

This command is used to indicate that it's a projection of the ZACB_R_Recipe base entity. In addition, the behavior of the projection and the behavior of various functions are defined (provider contract, see Listing 2.7).

```
@AccessControl.authorizationCheck: #CHECK
@Metadata.allowExtensions: true
@EndUserText.label: 'Projection entity recipe'
@ObjectModel.semanticKey: [ 'RecipeID' ]
define root view entity ZACB_C_Recipe
  provider contract transactional_query
  as projection on ZACB_R_Recipe
{
  key RecipeId,
  RecipeName,
  RecipeText,
  LocalLastChangedAt,
}
```

Listing 2.7 Projection Entity ZACB_C_Recipe

Behavior definition of the projection entity

In addition to the base entity with its behavior definition, a ZACB_C_Recipe behavior definition was also created for the projection entity (see Listing 2.8).

```
projection;
strict ( 2 );
use draft;
define behavior for ZACB_C_Recipe alias Recipe
use etag

{
  use create;
  use update;
  use delete;
  use action Edit;
  use action Activate;
  use action Discard;
  use action Resume;
```

```
  use action Prepare;
  use association _Ingredient { create; with draft; }
  use association _Review { create; with draft; }
}
```

Listing 2.8 Behavior Definition of the Projection Entity

Metadata definition

In addition to the base and projection entities and their behavior definitions, a metadata extension was generated. A metadata extension uses annotations to map the graphical UI. Among other things, the following *UI annotations* can be found here:

- *Facets* are areas of the UI that are defined using the `@UI.facet` annotation. Listing 2.9 shows an example of such an annotation.

  ```
  @UI.facet: [ {
   id: 'idIdentification',
   type: #IDENTIFICATION_REFERENCE,
   label: 'Recipe',
   position: 10
  } ]
  ```

  Listing 2.9 Annotation for a Facet

  A special facet is the one for the identification area, which is defined using the `@UI.identification` annotation. Fields in this area are intended to identify the object. This annotation (see Listing 2.10) must be assigned to all fields that are supposed to be displayed here.

  ```
  @UI.identification: [ {
    position: 10 ,
    label: ''
  } ]
  RecipeId;
  ```

  Listing 2.10 Annotation for the Identification Area

- *General field annotations* can be used in different places. The following annotations are possible here:
  - `Hidden`: The field isn't displayed and can't be added during personalization.
  - `Position`: This specifies the order of the fields.
  - `Label`: This specifies the name of the field in the UI.

Service definition

The service definition (see Listing 2.11), which is also generated, describes which CDS entities of a data model are to be published. With the help of an alias name, the technical names for the service can be replaced by more descriptive names.

```
@EndUserText.label: 'Recipe'
define service ZACB_UI_RECIPE {
  expose ZACB_C_Recipe      as Recipe;
}
```

Listing 2.11 Service Definition

Service binding

The service binding is generated based on the service definition. You can also take a closer look at this. Service bindings assign a protocol to the previously defined services. They can be published directly locally. The following binding types can be selected:

- **OData 2.0 (V2) or OData 4.0 (V4)**
  SAP recommends using the OData V4 version, which supports all functions such as the draft function.
- **UI**
  This transfers the UI annotations.
- **Web API**
  This creates a web service without UI annotations.
- **InA – UI**
  This is used for analytical data models.
- **SQL – Web API**
  This provides access with ABAP SQL.

2.3.3 Publishing the OData Service

Manual or automatic publication

An OData V4 service is made available with the generated service binding. If it has the **Unpublished** status, there are two ways to publish it:

- **Button click**
  You can click on the **Publish** button to publish the service automatically.
- **Manual activation via Transaction /IWFND/V4_ADMIN**
  Depending on the SAP S/4HANA version, automatic publishing may not yet be supported. If so, a manual activation via Transaction /IWFND/V4_ADMIN is necessary.

[»]

OData

Open Data Protocol (OData) is an HTTP-based protocol for the exchange of data between systems. OData allows you to request and write data to resources using familiar operations such as GET, POST, PUT, DELETE, and PATCH.

Publishing in Transaction /IWFND/V4_ADMIN

To publish the OData service manually, follow these steps:

1. Log on to the SAP system, and call Transaction /IWFND/V4_ADMIN. This transaction enables OData V4 services to be published.
2. Click on the **Publish Service Groups** button.
3. Search for the service to be published using the **System Alias** and **Service Group ID** fields. You can also use wildcards. Click on the **Get Service Groups** button to start the search (see Figure 2.20).

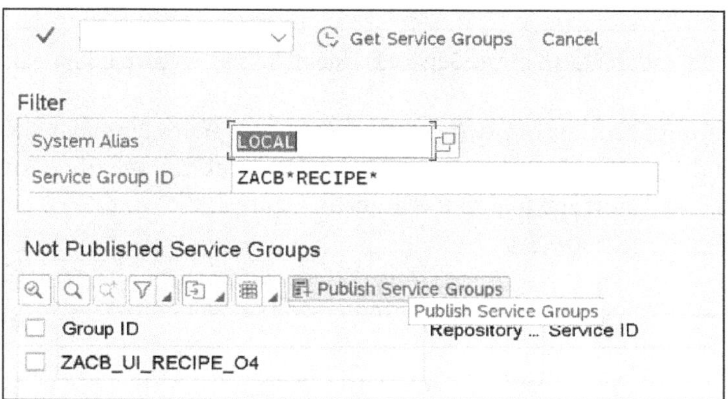

Figure 2.20 Search Filters and List of All Service Groups Found

4. As a result, all service groups matching the search criteria will be displayed. In our example, this list only contains one entry.
5. Select the entry with the appropriate **Group ID**, and click on the **Publish Service Groups** button.
6. You can change the description of the service group in the dialog box that opens.
7. In the information window that opens next, click on ☑ (**Execute**), and then click on the **Back** button.

At this point, the service has been published in a service group and can be used.

2.3.4 Testing the Application

"Published" status

Using the preview function of the ABAP development tools, you can now check directly in the development environment whether the publication of the OData UI service was successful. To do this, you need to call the service binding. If the **Published** status is displayed here in the **Local Service Endpoint** field (see Figure 2.21), you can open a preview of the SAP Fiori app UI.

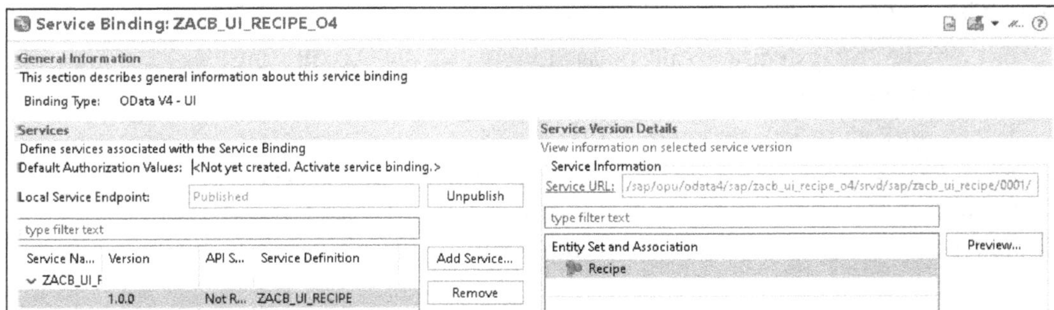

Figure 2.21 Publication Status of the Service Endpoint in the Service Binding

Preview of the SAP Fiori app Now select the appropriate entity (here, **Recipe**) in the **Service Binding** view and click on the **Preview** button to open the preview. You can see the UI of the generated SAP Fiori app in preview mode in Figure 2.22.

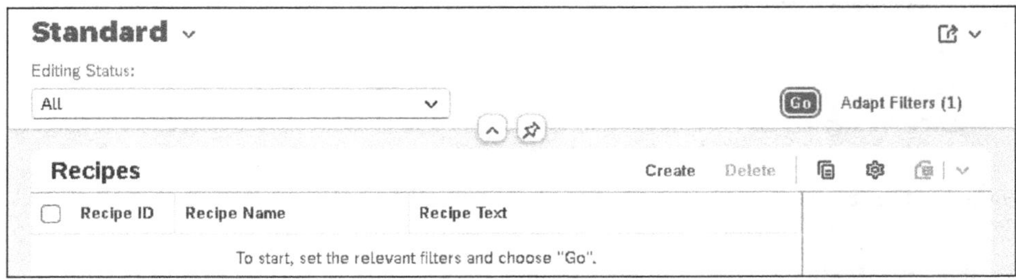

Figure 2.22 Preview Mode of the SAP Fiori App

2.4 Creating Associations

Adding more entities Before we look at the functions of the generated app from the user's perspective in the following section, we need to add the Ingredient and Review entities and link them to the Recipe root entity. In this section, we'll show you how to create an *association* for the root entity using the Ingredient entity. You can proceed in the same way for all other associated entities, in our example for the Review entity.

Creating a data definition First, you must create a base entity for the associated entity. To do so, follow these steps:

1. Open the context menu of the **Data Definitions** folder (see Figure 2.23), and select the **New Data Definition** item.

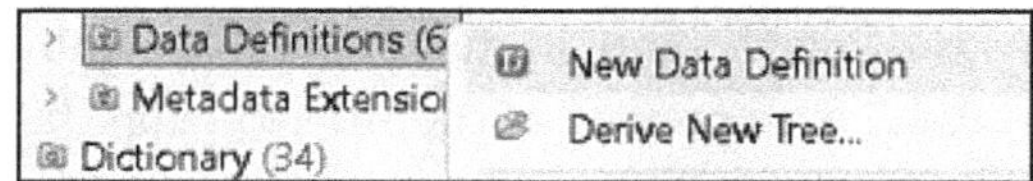

Figure 2.23 Context Menu Item: "New Data Definition"

2. In the dialog that opens, enter a name and description for the data definition. For our example, enter "ZACB_R_INGREDIENT" as the name. In the **Referenced Object** field, you must also enter a referenced object; in our example, this is the database table for the **Ingredient** object (see Figure 2.24). Then, click the **Next** button.

Name: * ZACB_R_INGREDIENT
Description: * Ingredient
Original Language: EN
Referenced Object: ZACB_INGREDIENT Browse...
< Back Next > Finish Cancel

Figure 2.24 Creating the Data Definition

3. In the subsequent step, select a transport request and confirm this by clicking the **Next** button.
4. Select a template. In our case, we want to create our own entity and therefore select **defineViewEntity** as the template (see Figure 2.25). This template generates the most important fields for the data definition.
5. Confirm the selection of the template by clicking the **Finish** button.

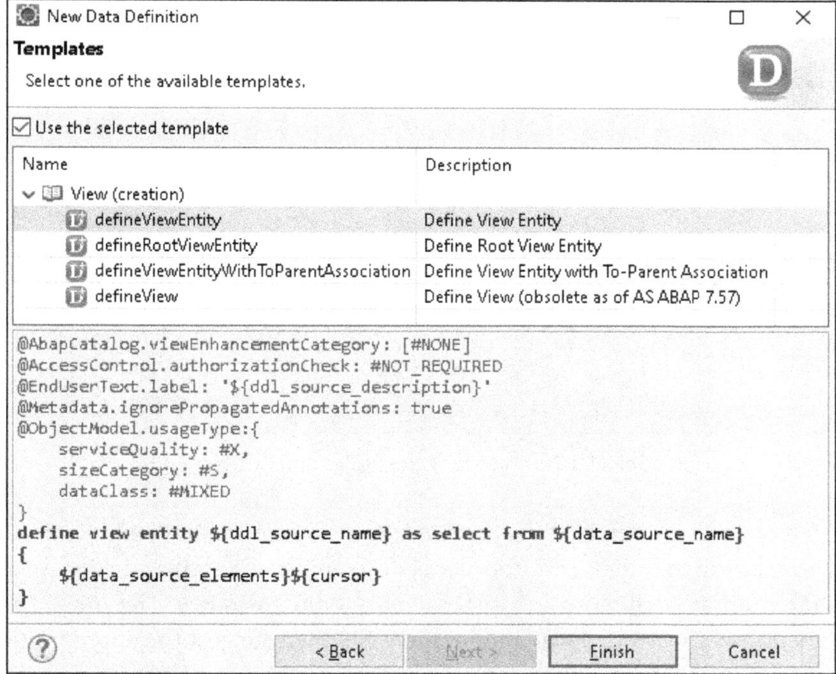

Figure 2.25 Selecting a Template for the Data Definition

The CDS coding will then be generated from Listing 2.12 based on the template and the referenced object.

```
define view entity ZACB_R_Ingredient select from zacb_ingredient
{
    key recipe_id as RecipeId,
    key ingredient_id as IngredientId,
    name as Name,
    quantity as Quantity,
    unit as Unit,
    created_by as CreatedBy,
    created_at as CreatedAt,
    local_last_changed_by as LocalLastChangedBy,
    local_last_changed_at as LocalLastChangedAt,
    last_changed_at as LastChangedAt,
    last_changed_by as LastChangedBy
}
```

Listing 2.12 Generated Base Entity ZACB_R_Ingredient

Creating a projection entity

Now you need to repeat steps 1–4 to create a projection entity. However, this time you don't select a template (see Figure 2.26), but click directly on the **Finish** button.

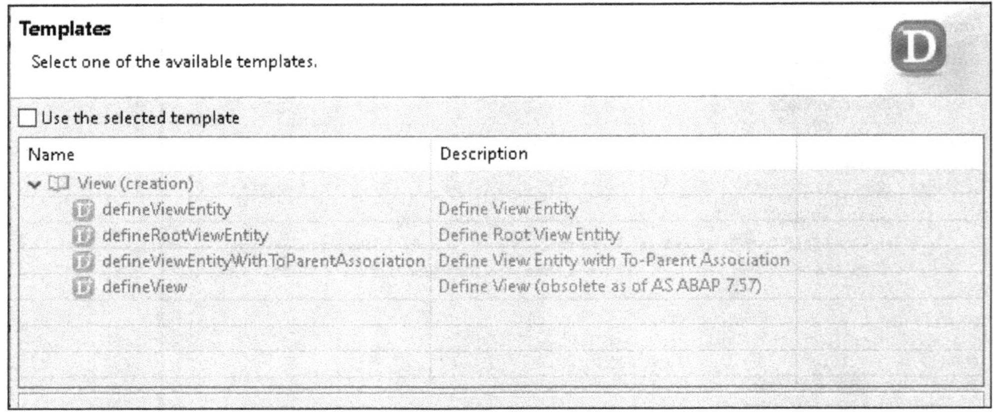

Figure 2.26 Don't Select a Template for the Projection Entity

An empty data definition gets created. You can use the **defineProjectionView** pattern to trigger an automatic code generation. For this purpose, you should enter "defineProje..." in the empty editor window. The quick fix function of the ABAP development tools will then suggest the appropriate template (see Figure 2.27).

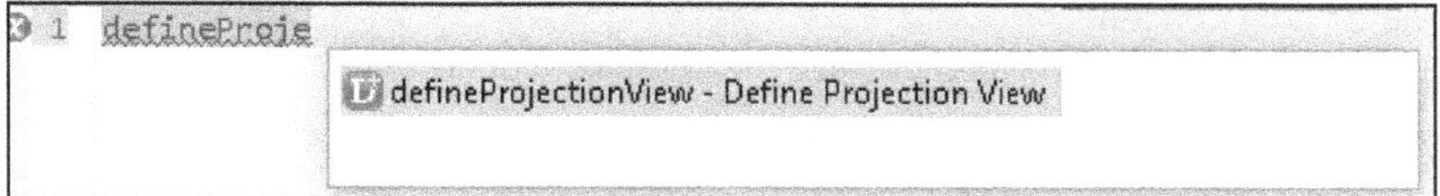

Figure 2.27 Selecting a Code Template for a Projection View

When you select the template, the code will automatically be generated from Listing 2.13.

```
@AccessControl.authorizationCheck: #NOT_REQUIRED
@EndUserText.label: 'Ingredient'
@Metadata.ignorePropagatedAnnotations: true
define view entity ZACB_C_Ingredient as projection on
  data_source_name
{

}
```

Listing 2.13 Generated Code for the Projection View

Adding entity names and fields

Add the name of the base entity here as data_source_name. You can also specify the fields of the base entity and use key to define which fields are key fields (see Listing 2.14).

```
define view entity ZACB_C_Ingredient
  as projection on ZACB_R_Ingredient as Ingredient
{
  key RecipeId,
  key IngredientId,
      Name,
      Quantity,
      Unit,
      LocalLastChangedAt,
      LocalLastChangedBy,
      LastChangedAt,
      LastChangedBy,
}
```

Listing 2.14 Completed Code of the Projection Entity

Linking the base entities

Now, the two base entities must first be linked to each other via an association. To do this, we first create a *composition* from the root entity in the direction of the linked entity (see Listing 2.15). A composition defines an existential dependency between two entities, whereby the child entity can't exist without the parent entity.

```
define root view entity ZACB_R_Recipe
  as select from zacb_recipe
  composition [0..*] of ZACB_R_Ingredient as _Ingredient
{
  key recipe_id              as RecipeId,
      recipe_name            as RecipeName,
...
```

Listing 2.15 Specifying a Composition Link

At the level of the associated entity, you must use `association to parent` to add an association to the root entity. This creates an upward relationship with the root entity (see Listing 2.16).

```
@EndUserText.label: 'CDS entity Ingredient'
define view entity ZACB_R_Ingredient
  as select from zacb_ingredient
  association to parent ZACB_R_Recipe as _Recipe
    on $projection.RecipeId = _Recipe.RecipeID
{
  key recipe_id     as RecipeId,
  key ingredient_id as IngredientId,
...
```

Listing 2.16 Adding "association to parent"

Linking the projection entities

A link between the two entities must also be created at the level of the projection entities. In the root entity, you must add the `redirected to composition child` expression (see Listing 2.17).

```
...
define root view entity ZACB_C_Recipe
  provider contract transactional_query
  as projection on ZACB_R_Recipe
{
  ...
      _Ingredient : redirected to composition child
        ZACB_C_Ingredient,
}
```

Listing 2.17 Adding the Link to the Child Entity

A similar link is made from the associated entity to the root entity. Here, you need to enter the `redirected to parent` addition (see Listing 2.18). The relationship defined in this way is used to navigate between the entities in the service binding.

```
...
define view entity ZACB_C_Ingredient
  as projection on ZACB_R_Ingredient as Ingredient
{
...
      _Recipe : redirected to parent ZACB_C_Recipe
}
```

Listing 2.18 Redirected to Parent

Extending the behavior definition

You must define a behavior for each entity. To do this, you want to extend the behavior definition as shown in Listing 2.19:

- Associations are published in the entities.
- Data records of associated entities such as the ingredient are created via the root entity.
- The behavior of locks and authorizations takes place via the root entity.

```
managed implementation in class zbp_acb_r_recipe unique;
strict ( 2 );
with draft;

define behavior for ZACB_R_Recipe alias Recipe
...
  association _Ingredient { create; with draft; }
...
define behavior for ZACB_R_Ingredient alias Ingredient
persistent table zacb_ingredient
lock dependent by _Recipe
authorization dependent by _Recipe
draft table zacb_ingredien_d
{
  update;
  delete;
  field ( readonly ) RecipeId;
  field ( readonly ) IngredientId;

  association _Recipe { with draft; }
  mapping for zacb_ingredient
    {
   ...
    }
}
```

Listing 2.19 Link in the Behavior Definition of the Base Entities

These associations must also be defined in the behavior definition of the projection entities (see Listing 2.20).

```
projection;
strict ( 2 );
use draft;
define behavior for ZACB_C_Recipe alias Recipe
...
  use association _Ingredient { create; with draft; }
...
define behavior for ZACB_C_Ingredient alias Ingredients
...
  use association _Recipe { with draft; }
```

Listing 2.20 Link in the Behavior Definition of the Projection Entities

Extending the metadata definition

To ensure that the linked entities are displayed in the SAP Fiori app, both the metadata extension of the root entity must be extended and a separate metadata extension must be created for the child entities. A facet must be added to the metadata extension of the root entity (see Listing 2.21).

```
{
id          : 'controlSection',
type        : #LINEITEM_REFERENCE,
position    : 20,
targetElement: '_Ingredient'
},
```

Listing 2.21 Metadata Extension of the Root Entity

The targetElement annotation is used to indicate that the metadata extension of the associated entity is supposed to be called and displayed in this section.

Extending the service definition

Finally, the association must be specified in the service definition (see Listing 2.22).

```
@EndUserText.label: 'Recipe'
define service ZACB_UI_RECIPE {
  expose ZACB_C_Recipe as Recipe;
  expose ZACB_C_Ingredient as Ingredient;
}
```

Listing 2.22 Extending the Service Definition

Once all created and modified development artifacts have been successfully activated, the link between the two entities can be seen in the service binding (see Figure 2.28).

Service Version Details
View information on selected service version
Service Information
Service URL: /sap/opu/odata4/sap/zacb_ui_recipe_o4/srvd/sap/zacb_ui_recipe/0001/
type filter text
Entity Set and Association
Preview...
Ingredient
_Recipe
Recipe
_Ingredient

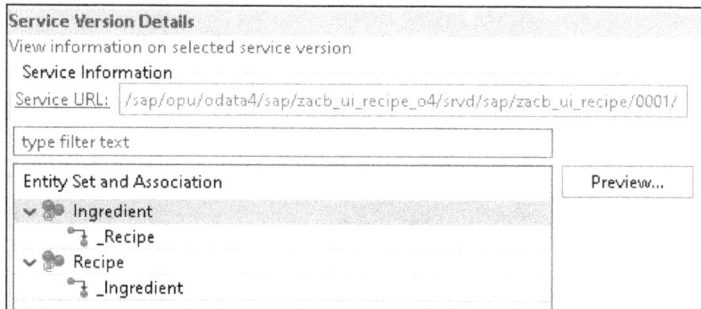

Figure 2.28 Service Binding with Associations

Repeat the steps shown here for all associated entities.

2.5 Application Scenario from a User's Perspective

We now want to explain the basic functions of our SAP Fiori app. This app serves as the basis for all the functions we'll successively add to the application in this book. For this purpose, you should display the application using the **Preview** function of the service binding.

List view of the recipes

The recipe portal is displayed as an SAP Fiori app in a browser. The individual recipes are displayed in a list view. Our application doesn't yet contain any recipes (see Figure 2.29).

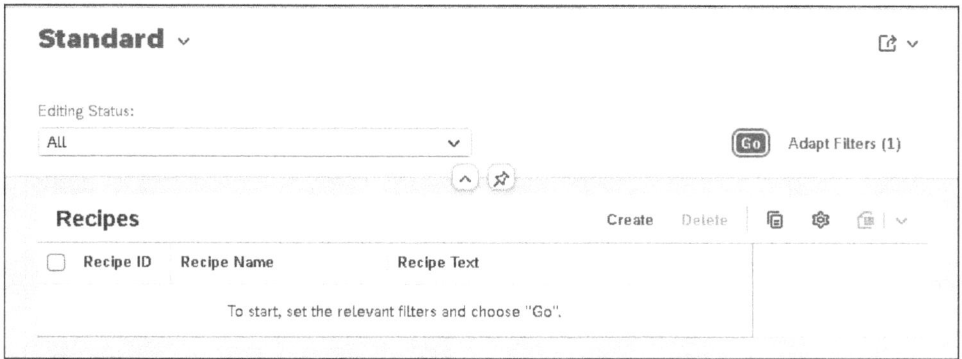

Figure 2.29 Homepage of the Recipe Portal

The **Go** button is used to import all recipes in the recipe portal from the SAP backend and display them in the list (see Figure 2.30).

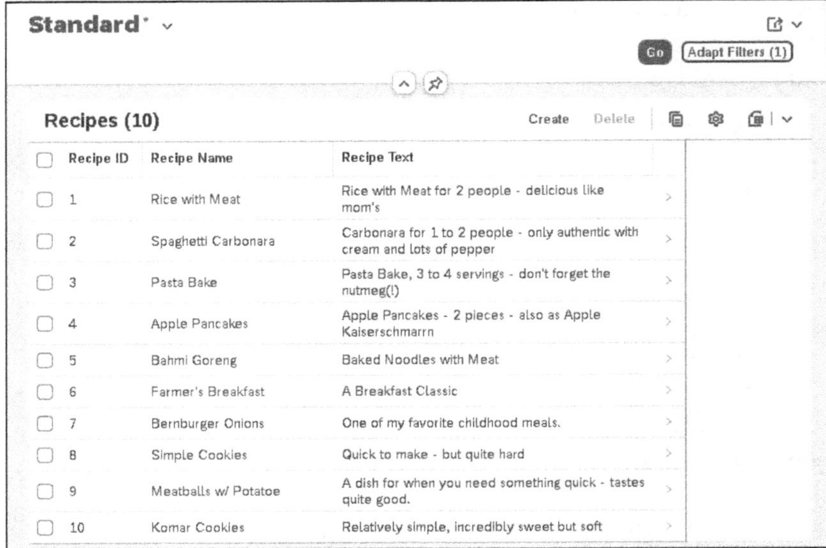

Figure 2.30 Display of All Recipes in the SAP Fiori App

Single view of a recipe

If a specific recipe is selected, the user is taken to the individual view of this recipe (see Figure 2.31). Information such as the recipe name and the recipe text as well as all ingredients and reviews are displayed here in table format.

Figure 2.31 Single View of a Recipe

Adding a recipe

To add a new recipe, the user must click on the **Create** button in the initial view (refer to Figure 2.30). A popup window appears in which a recipe ID of up to five digits must be entered (see Figure 2.32).

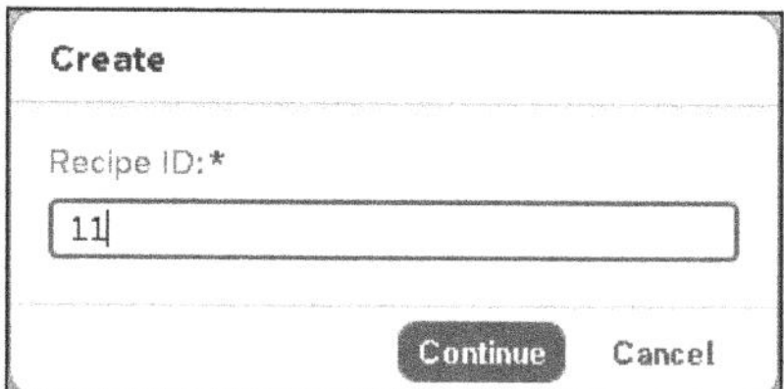

Figure 2.32 Specifying a Recipe ID

A name (**Recipe Name** field) and preparation instructions (**Recipe Text** field) must then be entered (see Figure 2.33).

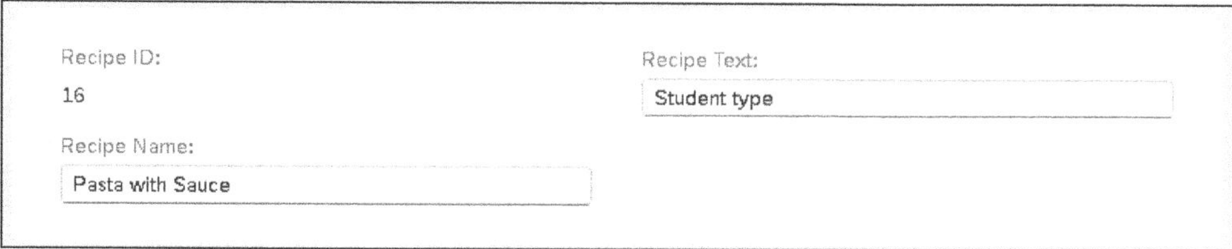

Figure 2.33 Entering the Recipe Data

[«]

Numbering

An explicit numbering by the user is required in the generated application. In Chapter 9, we show you how you can improve this using internal numbering.

Adding an ingredient

Next, you can add the ingredients for the recipe. To do this, click on the **Create** button. This opens the creation view for an ingredient (see Figure 2.34). There you can enter the required information for the ingredient, such as the name, quantity, and unit.

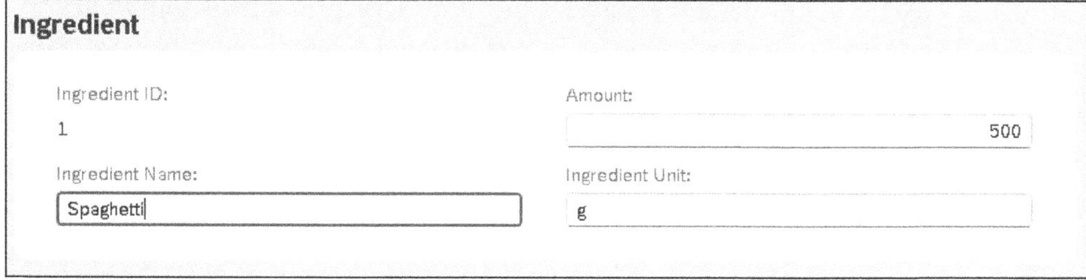

Figure 2.34 Entering an Ingredient

After confirming your entries by clicking the **Apply** button, the ingredient will be added to the recipe but not yet saved in the database. The data gets saved when you click on the **Create** button in the **Recipe** view (see Figure 2.35).

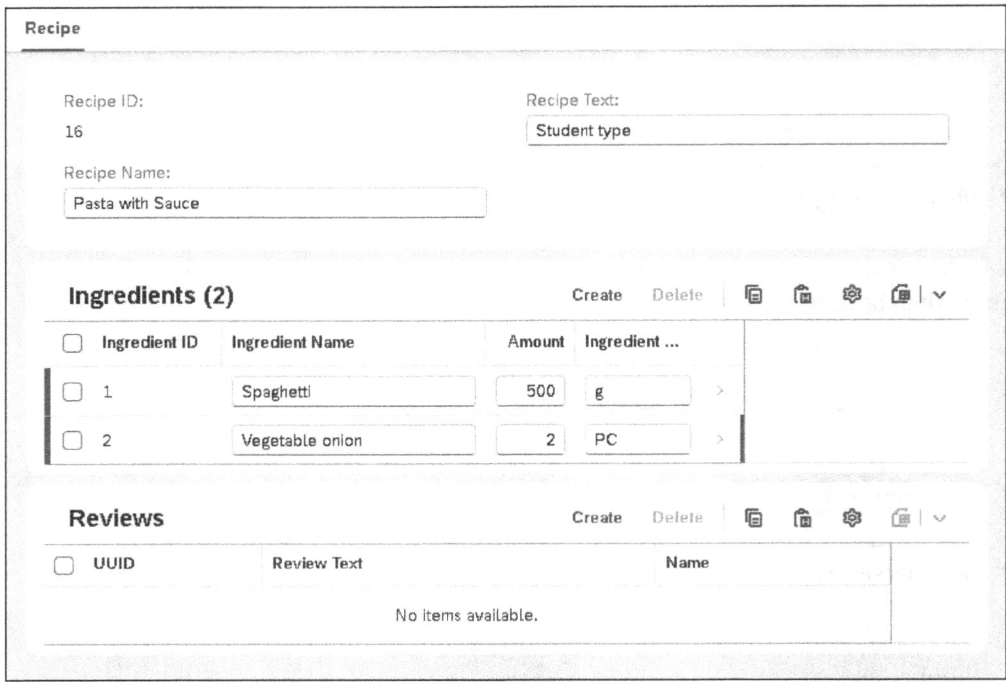

Figure 2.35 Recipe View with Details of All Ingredients

2.6 Summary

In this chapter, we've presented the sample application of a recipe portal that we'll use for the examples in this book. We derived the technical concepts of the application from the use case. The repository objects domain, data element, and database table were then created.

Based on the database table for the recipes, we generated an ABAP RESTful application programming model application and then described it step-by-step. Among other things, we discussed the difference between CDS base entities and projection entities. We've also introduced important components of the ABAP RESTful application programming model such as the behavior and metadata definition, the service definition, and the service binding.

For a recipe portal, it doesn't suffice to enter recipe names and recipe texts because it must also be possible to maintain ingredients. To do this, we've manually extended the automatically generated application. It's now also possible to add reviews to a recipe. We've shown you how to build relationships between multiple associated CDS entities and their behavior definitions.

Chapter 3
Handling System Fields and Runtime Information

3

To determine runtime information, there are some new access options in ABAP Cloud, some of which replace the system fields. In this chapter, you'll learn how to use these new APIs and find out when it makes sense to use a specific API.

In your custom applications, you very often need access to information that is only available at runtime and relates to the environment in which your program is currently being executed. It can therefore also be referred to as *runtime environment information*. The ABAP runtime environment deals with the execution of programs, with the processing blocks and with the implementation of the ABAP statements in the associated ABAP kernel.

Information on the ABAP runtime environment

The information you want to receive from the ABAP runtime environment usually consists of answers to the following questions:

- What is the ID of the user currently logged in?
- What is the date and time now?
- Was the last statement executed correctly?
- Which message was provided by the last statement?
- What is the name of the current program/transaction/system ID/client?
- Which ABAP release is installed in the system?

Global "sy" variable

Such questions could previously be answered using the predefined global `sy` variable, which provides various runtime information via its components in the `SYST` structure. The individual components are also referred to as *system fields*. The global `sy` variable continues to provide most of this information—`sy-subrc` and other system fields are therefore still available in ABAP Cloud. However, there are new access options to a lot of information with an extended range of functions or independent of a global variable. These are also implemented in a contemporary manner so that development teams in which different languages are spoken no longer have to learn the German terms `DATUM` and `UZEIT` (date and time).

Structure of this chapter

Section 3.1 deals with system fields and the way they behave in ABAP Cloud. Section 3.2 provides an overview of the available application programming interfaces (APIs) for runtime information that we'll use in this chapter.

Sections Section 3.3 to Section 3.7 deal with specific use cases. In Section 3.3, we start with the retrieval of time information. Section 3.4 deals with the handling of user data. Section 3.5 describes how you can access technical runtime information. In Section 3.6, you'll learn how to deal with messages. Section 3.7 deals with access to system information.

Runtime information in the sample application

With regard to our sample application, in this chapter, we develop an application that provides debug information on the recipe portal. As part of an error analysis, this application should first make it possible to find out what the system status looks like to then be able to delve deeper. We try out various concepts for determining runtime information and output this information via the console in an ABAP console application.

[»]

Using the Sample Application

You can run the application to output debug information yourself using the enclosed code and view it in the debugger. You'll find this application in the RUNTIME subpackage in the download material for this book. To execute it, open the ZCL_ACB_DEBUG_INFO class in the ABAP development tools for Eclipse, and run it as an ABAP console application via the F9 key.

3.1 System Fields in ABAP Cloud

Permitted access to system fields

System fields are still available in ABAP Cloud and can be accessed via the predefined global sy variable. Syntactically, read access to all components is possible, albeit with warnings in some cases. For example, the number of lines processed by an ABAP SQL statement can still be queried via the sy-dbcnt system field:

```
SELECT FROM zacb_recipe FIELDS COUNT(*).
out->write(
  |Number of recipes: { sy-dbcnt NUMBER = USER }| ).
```

Restricted access to system fields

However, some of the system fields available in classic ABAP aren't released in ABAP Cloud. Their use therefore results in a syntax warning. For this reason, you should look for an alternative at this point. The system fields that are no longer released and possible alternatives are grouped thematically in the following sections. For example, read access to the system date via the sy-datum system field isn't released in ABAP Cloud, as shown in Figure 3.1. Alternative ways to access time information can be found in Section 3.3.

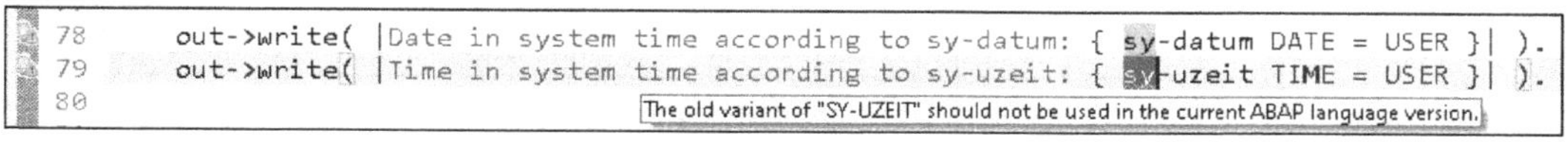

```
out->write( |Date in system time according to sy-datum: { sy-datum DATE = USER }| ).
out->write( |Time in system time according to sy-uzeit: { sy-uzeit TIME = USER }| ).
The old variant of "SY-UZEIT" should not be used in the current ABAP language version.
```

Figure 3.1 Syntax Warning for Read Access to System Fields Not Released in ABAP Cloud

Prohibited write access to system fields

Write access to system fields is completely prohibited in ABAP Cloud and leads to a syntax error. However, write access to system fields was already prohibited by programming guidelines in classic ABAP. Nevertheless, there were use cases, for example, in older enhancement technologies, in which communication with the standard SAP system took place via system fields. These are no longer relevant in ABAP Cloud, which is why write access could be syntactically prohibited.

In Figure 3.2, you can see an attempted write access to the sy-datum system field and the resulting syntax error. If you attempt a dynamic write access, it's prevented with the runtime error of type MOVE_TO_LIT_NOTALLOWED_NODATA.

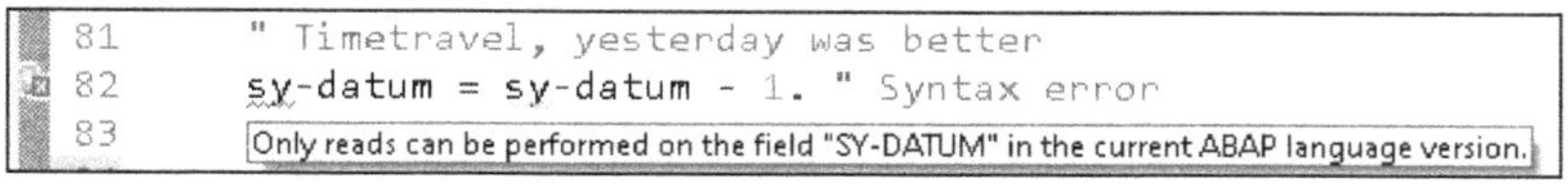

```
" Timetravel, yesterday was better
sy-datum = sy-datum - 1. " Syntax error
Only reads can be performed on the field "SY-DATUM" in the current ABAP language version.
```

Figure 3.2 Syntax Error After a Write Access to a System Field

[«]

Write Access to System Fields in the Debugger

Surprisingly, write access to system fields in the debugger is still permitted in SAP S/4HANA 2023 with the appropriate authorization. If you want to use this tool to analyze program behavior, it's still possible in this release. In the SAP BTP ABAP environment and in SAP S/4HANA Cloud Public Edition; however, you'll receive an error message.

System fields released in ABAP Cloud

The system fields listed in Table 3.1 can be used in ABAP Cloud without warnings or errors. For some of them, you may still want to use other access options that have a wider range of functions or are easier to replace in the testing context.

System Field	Meaning	Note
BATCH	Indicator for batch processing	–
DBCNT	Number of table rows processed after ABAP SQL statement	–

Table 3.1 Released System Fields in ABAP Cloud

System Field	Meaning	Note
FDPOS	Location for operations with character or byte-type data types	–
INDEX	Loop index	–
LANGU	Language of the text environment	Alternatives in Section 3.5
MANDT	Client ID of the current client	Explanation in Section 3.5
MSGID	Message class of the last message	Alternatives in Section 3.6
MSGNO	Message number of the last message	
MSGTY	Message type of the last message	
MSGV1, MSGV2, MSGV3, MSGV4	Message variables of the last message	
SUBRC	Return value of ABAP statements	–
SYSID	System identifier of the current system	Explanation in Section 3.5
TABIX	Row number of the last row addressed in an internal table by primary or secondary index	–
UNAME	User name of the current user session	Alternatives in Section 3.4

Table 3.1 Released System Fields in ABAP Cloud (Cont.)

If the system field you require isn't included in Table 3.1 or there's no reference to alternatives, you can find ABAP Cloud-compliant access options in the following sections.

3.2 Overview of the Available APIs

Various access options to runtime information are provided specifically for ABAP Cloud in the form of methods of global classes. These have been released and are therefore upgrade stable. We show you how to find APIs that you can use in ABAP Cloud in general in Chapter 17.

The following classes are particularly relevant for us in the context of runtime information:

- The CL_ABAP_CONTEXT_INFO class
- Classes and interfaces of the XCO library (extension components)
- Built-in functions such as utclong_current

The "CL_ABAP_SYST" class in ABAP Cloud

If you're familiar with the CL_ABAP_SYST class as an alternative to direct access to the system fields, you can no longer use this in ABAP Cloud because it hasn't been released. The CL_ABAP_CONTEXT_INFO class is to be regarded as a cloud-compliant successor.

The XCO library

The *XCO library* is a very extensive collection of standard functions of the ABAP platform, which are provided via a standardized object-oriented abstraction. These classes and interfaces are released in the Cloud Platform Edition of the XCO library and can therefore be used in ABAP Cloud. The library can be considered as one of the first points of contact for common problems. It's divided into two categories:

- The ABAP repository APIs enable a program-based interaction with workbench objects and the transport system.
- The second category is the standard library, which covers various technical topics.

We want to use the XCO library in this chapter to access runtime information. The sy method of the XCO_CP class provides a central point of entry for the components of the XCO library that affect system fields.

3.3 Access to Time Information

When implementing transactional applications, you often need access to administrative data, such as the user name of the currently logged in user or information on the current time in the form of date, time, or timestamp. In this section, we start with this time information. With regard to our sample application, the recipe portal requires the time of creation and the last change of data records for the application tables, such as the table for recipes and reviews. This allows the first table to be sorted by the most recent recipes, for example, or the recipe that hasn't been updated for the longest time can be found.

Determining the system date and time

As described in Section 3.1, the usual candidates sy-datum and sy-uzeit can be used in ABAP Cloud, but their use is no longer recommended. The situation is similar with sy-datlo and sy-timlo. Their use generates a syntax warning.

UTC times

The system time, for example, depends on a configuration setting of the SAP system. In the SAP BTP ABAP environment and in SAP S/4HANA Cloud

Public Edition, you have no influence on this configuration. You should therefore consider using UTC times (Coordinated Universal Time) for new applications and converting this to the user's time zone when displaying it in the frontend.

SAP provides the CL_ABAP_CONTEXT_INFO class as a replacement for the sy-datum and sy-uzeit system fields. You can use the get_system_date and get_system_time methods of this class to obtain the current date and time. However, these times refer to the UTC time zone.

[!]

Time Zones in the "CL_ABAP_CONTEXT_INFO" Class

Despite the method identifiers get_system_date and get_system_time and the data types syst_datum and syst_uzeit, the methods of the CL_ABAP_CONTEXT_INFO class don't return any time information in system time. The results always refer to the UTC time zone. This unusual behavior is described in the method documentation, an excerpt of which is shown in Figure 3.3. However, the documentation is easy to overlook.

get_system_date
returning value(**rv_date**) type cl_abap_context_info=>ty_system_date

Documentation

Gets the current date in UTC.

Important note: Note that, despite its name, this method will always calculate and return the current date with respect to UTC.

Parameters

rv_date The current date in UTC

Figure 3.3 Method Documentation for get_system_date

In cloud systems and corresponding new developments, this behavior makes sense because the system time is a configuration detail over which you have no control. When using ABAP Cloud in SAP S/4HANA and SAP S/4HANA Cloud Private Edition, however, incompatibilities may occur if your system doesn't use UTC as the system time zone. For this reason, you shouldn't replace sy-datum and sy-uzeit in existing applications or in integrations with the method calls without having checked the time zone first.

From a technical perspective, the time information is determined using a timestamp. Unlike the system fields, it's therefore not necessary to call the GET TIME statement beforehand to obtain current data. You can see a sample call in Listing 3.1.

```
out->write(
  |Date in UTC according to cl_abap_context_info: | &&
  |{ cl_abap_context_info=>get_system_date( )
```

```
    DATE = USER }|
).
out->write(
  |Time in UTC according to cl_abap_context_info: | &&
  |{ cl_abap_context_info=>get_system_time( )
    TIME = USER }|
).
" Date in UTC according to cl_abap_context_info:
" 02.03.2025
" Time in UTC according to cl_abap_context_info: 13:45:49
```

Listing 3.1 Using the CL_ABAP_CONTEXT_INFO Class to Determine the Date and Time

Determining timestamps

Depending on your application, you may need a combined field in the form of a timestamp instead of separate fields for date and time. ABAP recognizes several timestamp formats, so you should pay close attention to what kind of timestamp you need. In the ABAP RESTful application programming model, you can choose between timestamps with packed numbers with the TZNTSTMPL domain of type DEC(21,7) and the built-in utclong type. The packed numbers, for example, for the timestamp for the last change with data element ABP_LASTCHANGE_TSTMPL, correspond to the format of the TIMESTAMPL data element. You can also fill this data element in ABAP Cloud in the classic way using the GET TIME STAMP statement (see Listing 3.2).

```
DATA now TYPE timestampl.
GET TIME STAMP FIELD now.
out->write(
  |Timestamp in DEC(21,7) according to | &&
  |GET TIME STAMP: | &&
  |{ now TIMESTAMP = USER TIMEZONE = 'UTC' }| ).
" Timestamp in DEC(21,7) according to GET TIME STAMP:
" 02.03.2025 13:45:49,9519390
```

Listing 3.2 Timestamp Determination via GET TIME STAMP

Precision differences in timestamps

Note that this statement doesn't return the timestamp in the system time zone, but always in UTC. However, applications that use timestamps usually work with UTC anyway, so that no conversion logic is usually required. However, you should pay attention to which of the two timestamp formats is used for packed numbers. In addition to the TIMESTAMPL data element of type DEC(21,7), which is accurate to 100 nanoseconds, there's also the TIMESTAMP data element of type DEC(15,0). This data element only works with a precision down to the second. This affects us here in that the GET TIME STAMP statement uses the less precise type when using inline declarations with the FIELD DATA(timestamp) addition. If you subsequently want to use this value to assign a more precise timestamp, data would be lost unnecessarily.

Timestamp with "utclong_current"

Another disadvantage of the statement-based determination of a timestamp is that these statements can't be used in expressions. For a one-liner, you can use the built-in utclong_current function in combination with a data type conversion instead. utclong_current returns a current timestamp with the built-in utclong type (see Listing 3.3). This natively represents a timestamp in ABAP instead of using packed numbers or character-type data types. You can use the CL_ABAP_TSTMP class to convert the timestamp to the DEC(21,7) type, for example, if required. The ABAP RESTful application programming model also supports utclong directly via reuse data elements with the UTCL suffix, such as ABP_LASTCHANGE_UTCL.

```
DATA(now_one_line) =
  cl_abap_tstmp=>utclong2tstmp( utclong_current( ) ).
out->write(
  |Timestamp in DEC(21,7) according to | &&
  |utclong_current: | &&
  |{ now_one_line TIMESTAMP = USER TIMEZONE = 'UTC' }| ).
" Timestamp in DEC(21,7) according to utclong_current:
" 02.03.2025 13:45:49,9520500
```

Listing 3.3 Timestamp Determination via utclong_current

The XCO standard library

The XCO standard library has a module that provides functions for time data. This includes date and time information as well as timestamps. Using this API is somewhat more demanding. However, its range of functions is much more extensive.

The XCO library provides its functions via *API classes*. Their identifiers don't start with the usual CL prefix for classes, but with XCO. A subsequent CP stands for Cloud Platform. The classes that contain the abbreviation CP in their name are wrapper classes that are released in ABAP Cloud and form the *Cloud Platform Edition* of the XCO library. There's also a *Key User Edition*, whose classes contain the name component KU. XCO classes are available for standard ABAP without an additional naming convention. Using the *code completion* of the ABAP development tools, you can easily find the classes required for the desired function in the XCO library and open up the object-oriented designed API.

System fields and time specifications in the XCO library

A central point of entry for the XCO library from the system field perspective is the static sy attribute of the global XCO_CP class. It enables direct access to modules of the library that are relevant for system fields. You can access the API step-by-step via code completion and *ABAP element info*, which you can call in the ABAP development tools using the [F2] keyboard shortcut. Access is shown in Figure 3.4.

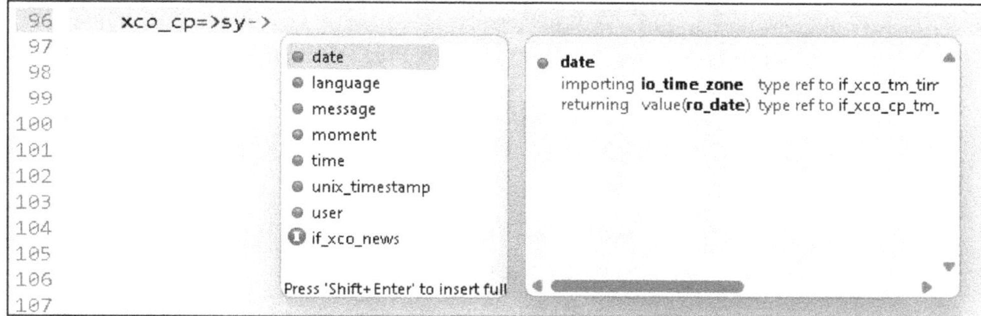

Figure 3.4 Code Completion for the Modules of the XCO Standard Library Relevant for System Fields

You can obtain the current date by calling the date method. This contains an optional parameter for the time zone. If you don't provide this parameter, the user time zone is used by default. In the following, we provide it to obtain the date in the UTC time zone:

```
DATA(system_date) = xco_cp=>sy->date(
  xco_cp_time=>time_zone->utc ).
```

The available time zones are provided via attributes of a factory class. In the XCO library, this is the usual way to provide fixed values in a type-safe manner and with additional functions. You can find these time zones in the API class of the respective module; in our case, it's the XCO_CP_TIME class.

Using date objects

The generated object, which we address via the IF_XCO_CP_TM_DATE interface, enables further manipulation of the date through methods, for example, via the add or subtract method. The object doesn't get changed directly during such operations. It's immutable, but a copy of the object with a changed state gets returned.

The object can also be converted into a target format using the as method. Similar to the time zones, the formats are provided via attributes of a factory class. In Listing 3.4, a day is first added to the generated date and then the date is output in different formats. The ABAP format corresponds to the familiar format of the built-in d type with YYYYMMDD as the sequence of year, month, and day.

```
DATA(tomorrow) = system_date->add( iv_day = 1 ).
out->write( |Tomorrow in UTC according to XCO in...| ).
out->write( |  string: | &&
  tomorrow->as( xco_cp_time=>format->abap )->value ).
```

```
out->write( |  ISO 8601: | &&
  tomorrow->as( xco_cp_time=>format->iso_8601_basic
    )->value ).
out->write( |  ISO 8601 extended: | &&
  tomorrow->as( xco_cp_time=>format->iso_8601_extended
    )->value ).
" Tomorrow in UTC according to XCO in ...
"   string: 20250303
"   ISO 8601: 20250303
"   ISO 8601 extended: 2025-03-03
```

Listing 3.4 Using XCO Date Objects

Type conversion in the XCO library

Note that the as method always returns a string instance that implements the IF_XCO_STRING interface. This can be useful if you subsequently want to use the value as text because this interface provides various methods for text operations. To work directly with the value, you must first access the value attribute, which itself is of the built-in string type. If you subsequently want to use statements that require the d type, you must also perform a type conversion, for example, using the EXACT expression.

In Listing 3.5, the d type is required to use user-specific date formatting in a string template.

```
DATA(system_date_d) = EXACT d(
  system_date->as( xco_cp_time=>format->abap )->value ).
out->write( |Date in UTC according to XCO (d): | &&
  |{ system_date_d DATE = USER }| ).
" Date in UTC according to XCO (d): 02.03.2025
```

Listing 3.5 Type Conversion of an XCO Date for the d Type

Times in the XCO library

Working with times is possible in a similar way. The xco_cp=>sy->time method gives you a time object that represents the current time—here, too, in the user time zone by default. Operations for manipulation and conversion or formatting are provided in the same way as for dates.

Timestamp in the XCO library

The combination of both elements of time information—date and time—in the XCO library is interesting. It summarizes this timestamp via a *moment*, which is represented by the IF_XCO_CP_TM_MOMENT interface. Technically, this is simply a combination of associations of the already known date and time objects with additional functions. In Listing 3.6, an instance is created at the current time in the UTC time zone and output in various formats.

```
DATA(moment) = xco_cp=>sy->moment(
  xco_cp_time=>time_zone->utc ).
out->write(
  |Timestamp in UTC according to XCO (string): | &&
  moment->as( xco_cp_time=>format->abap )->value ).

DATA(now_xco) = EXACT timestamp(
  moment->as( xco_cp_time=>format->abap )->value ).
out->write(
  |Timestamp in UTC according to XCO (timestamp): | &&
  |{ now_xco TIMESTAMP = USER TIMEZONE = 'UTC' }| ).
out->write(
  |Timestamp in UTC according to XCO (ISO 8601): | &&
  moment->as( xco_cp_time=>format->iso_8601_basic
    )->value ).
out->write(
  |Time stamp in UTC according to XCO | &&
  |(ISO 8601 extended): | &&
  moment->as( xco_cp_time=>format->iso_8601_extended
    )->value ).
" Timestamp in UTC according to XCO (string):
" 20250302134549
" Timestamp in UTC according to XCO (timestamp):
" 02.03.2025 13:45:49
" Timestamp in UTC according to XCO (ISO 8601):
" 20250302T134549
" Timestamp in UTC according to XCO (ISO 8601 extended):
" 2025-03-02T13:45:49
```

Listing 3.6 Handling Timestamps via Moments in the XCO Library

Here, too, you can see that to use timestamp-specific formatting options in string templates, a type conversion to a specific timestamp data type must first be carried out in ABAP. In this case, the `xco_cp_time=>format->abap` formatting option returns a string that is intended for the packed number format of the `TIMESTAMP` data element. As the times in the XCO library are based on the composition of a date and a time object, they are at most accurate to the second. For this reason, the `EXACT` constructor expression with the `TIMESTAMP` data type was used in the type conversion instead of `TIMESTAMPL`. This is where a disadvantage of this timestamp representation lies. If you need more precise timestamps, you should choose one of the techniques described previously, such as the built-in `utclong_current` function or the `GET TIME STAMP` statement.

Advantages of timestamps in the XCO library

However, the moments of the XCO library also provide some advantages that the other techniques don't directly provide. You can use the get_moment method to generate a time from a date or time object in the XCO library, for example. You then only need to add the missing information.

Compared to the date and time object, the timestamp object provides a significantly greater number of methods. In addition to the known manipulation options, comparison methods are provided, as well as methods for intervals. In Listing 3.7, a time is first generated from a date and then checked to see whether tomorrow at 3 pm is more or less than 24 hours in the future.

```
DATA(tomorrow_3pm) = tomorrow->get_moment(
  iv_hour   =  '15'
  iv_minute =  '00'
  iv_second =  '00' ).
DATA(in_24_hours) = moment->add( iv_hour = '24' ).
DATA(interval) = moment->interval_to( tomorrow_3pm ).
IF interval->contains( in_24_hours ).
  out->write(
    |There are more than 24 hours until 3pm tomorrow| ).
ELSE.
  out->write(
    |There are less than 24 hours until 3pm tomorrow| ).
ENDIF.
```

Listing 3.7 Extended Functions of Points in Time

Objects independent of the current time

If you want to create specific timestamp, time, or date objects without reference to the current system time, you can use the corresponding static methods of the XCO_CP_TIME class.

3.4 Accessing User Data

In addition to information on when changes are made or new data records are created in transactional applications, you usually also need information on which user is operating the application. In combination, this data forms the basis for change documents and administrative fields in data records. We discussed time information in the previous section. This is now about determining the information about the registered user.

System field "sy-uname"

In Section 3.1, we described the system fields and their behavior in ABAP Cloud. The system field relevant for user information is sy-uname. As mentioned previously, it's released for ABAP Cloud and can be used in the usual way and directly without generating syntax errors or a syntax warning.

"CL_ABAP_ CONTEXT_INFO" class

You can also use the get_user_technical_name method of the CL_ABAP_CONTEXT_INFO class to obtain the user ID of the user currently logged in. Its use doesn't provide any direct technical added value. If you use it, this runtime information is accessed in the same way as the system date and time. Further information on the user master record can be accessed via the code completion of the class, which you'll see in the next section. In addition, the name of the get_user_technical_name method is more meaningful than that of the uname system field. The call of the corresponding method is shown in Listing 3.8.

```
out->write(
  |Logged-in user according to sy-uname: { sy-uname }| ).
out->write(
  |Logged-in user according to cl_abap_context_info: | &&
  cl_abap_context_info=>get_user_technical_name( ) ).
```

Listing 3.8 User Name Determination via CL_ABAP_CONTEXT_INFO

Additional user information

You can also use the class to obtain additional information about the relevant user account, such as the alias, the formatted name, the time zone, and the business partner ID, as shown in Listing 3.9.

```
out->write( |Alias: | &&
  cl_abap_context_info=>get_user_alias( ) ).
out->write( |Description: | &&
  cl_abap_context_info=>get_user_description( ) ).
out->write( |Formatted name: | &&
  cl_abap_context_info=>get_user_formatted_name( ) ).
out->write( |Timezone: | &&
  cl_abap_context_info=>get_user_time_zone( ) ).
out->write( |Business partner: | &&
  cl_abap_context_info=>get_user_business_partner_id( ) ).
```

Listing 3.9 Querying Additional User Information via the CL_ABAP_CONTEXT_INFO Class

When using this class, however, you should note that the get_user_description and get_user_formatted_name methods don't determine the first name and last name of a user, but only the description of a technical user.

User name via the XCO library

Alternatively, you can also use the XCO library. The static sy attribute of the xco_cp class provides you with a central point of entry to relevant components of the library, which provide similar functions for the user information as for the system fields. The user method returns an instance of a class that represents the user who is currently logged in. Only one attribute is currently available in the relevant IF_XCO_CP_USER interface: the user ID. In our application with debug information, we want to output the ID of the user currently logged in. The call reads as follows:

```
out->write( |Logged-in user according to XCO: | &&
  xco_cp=>sy->user( )->name ).
```

Information about other user accounts

Both APIs, CL_ABAP_CONTEXT_INFO and the XCO library, also provide options for determining information about user accounts that aren't currently logged in. Listing 3.10 shows a corresponding request.

```
out->write( |Specific language: | &&
  cl_abap_context_info=>get_user_language_abap_format(
    sy-uname ) ).
out->write( |User ID of specific user XCO: | &&
  xco_cp_system=>user( sy-uname )->name ).
```

Listing 3.10 Querying User Information with a Specific User ID

3.5 Access to Technical Information on the Current Program Execution

In your application, you usually want to retrieve environment data about the current program execution or the call location for logging purposes. This environment data can make it much easier to eliminate errors.

System ID and client ID

You would receive the login client and the system ID in the classic way via the sy-mandt and sy-sysid system fields. And, surprisingly, this is still the approach in ABAP Cloud. The system fields mentioned have been released and are largely without alternative. These two types of data are less relevant in cloud systems.

Current program

You can also still obtain the identifier of the currently running program via the sy-repid system field. Technically, this is a predefined constant to which you still have read access.

Callstack

To analyze errors based on logs, you may have already retrieved the *callstack* in a program. The SYSTEM_CALLSTACK function module is available in standard ABAP for this purpose. This function module isn't released in ABAP Cloud and leads to a syntax error. However, the XCO library provides object-oriented modeled access to the same data.

Here again, the XCO_CP class represents the starting point. This time we use the current attribute to obtain information about the current ABAP session. The callstack is then determined using the call_stack method.

You have the option of querying this callstack completely or from a specific level onward. You can then search within the callstack or filter it using the from and to methods. You can also have the result prepared in ADT format. The ADT format is a text-based representation of the callstack, which is based on the view in the ABAP debugger of the ABAP development tools.

In Listing 3.11, we filter for entries that contain our class names used in the application example. The converted lines are returned as an object that implements the IF_XCO_STRINGS interface. The lines can be concatenated into a string using the join method.

```
DATA(callstack) = xco_cp=>current->call_stack->full( ).
callstack = callstack->to->last_occurrence_of(
  xco_cp_call_stack=>line_pattern->method(
    )->where_class_name_matches( 'Z(CL|IF|BP)_ACB_.*' ) ).
out->write(
  callstack->as_text( xco_cp_call_stack=>format->adt( )
    )->get_lines( )->join( |\n| )->value ).
```

Listing 3.11 Determining the Callstack Using the XCO Library

Logon language

To implement application logic that implements user interface elements, you want to obtain the logon language of the current user so that you can then format texts depending on the language. You can also continue to use the sy-langu system field released in ABAP Cloud for this purpose. However, there's an alternative available that has more functions: You can use the XCO library to obtain an object that represents the logon language. To do this, you need to call the xco_cp=>sy->language method. The return value is typed with the IF_XCO_LANGUAGE interface, which in addition to the technical language key from sy-langu also returns a translated name of the language and the language key in ISO 639 format. The access is shown as an example in Listing 3.12.

```
DATA(language) = xco_cp=>sy->language( ).
out->write(
  |Language (XCO): { language->get_name( ) } | &&
  |({ language->value }/| &&
  |{ language->as( xco_cp_language=>format->iso_639 ) })|
).
```

Listing 3.12 Querying the Logon Language Using the System Field and the XCO Library

You can also create specific instances for a specific language using the language method of the global XCO_CP class.

3.6 Accessing Messages

Message classes in ABAP Cloud

Message classes continue to be used in ABAP Cloud to format texts. There's a native editor for message classes available in the ABAP development tools. Message classes enable cross-object text maintenance, are integrated into the translation concept, and allow the use of variables. Unfortunately, the editor in the ABAP development tools doesn't yet support long texts. In systems in which SAP GUI is available, SAP GUI–based maintenance opens for this purpose.

Messages in ABAP code

Messages from message classes are integrated into ABAP using the MESSAGE statement. In addition, function modules and methods can use messages to enrich classic exceptions. They can also be used in class-based exceptions. Access varies depending on the technology, but it's often via the system fields for messages: sy-msgid, sy-msgno, sy-msgty, and sy-msgv1 to sy-msgv4. As described in Section 3.1, these fields are also available in ABAP Cloud. However, there's an alternative in the XCO library to view the fields bundled together.

Messages in the XCO library

You can use the xco_cp=>sy->message method to create a message object based on the message currently stored in the system fields. Access is defined via the IF_XCO_MESSAGE interface, which makes some methods for handling the message available. Our application with debug information checks whether the database tables are filled with recipes. If this isn't the case, a classic exception with message reference is triggered via MESSAGE ... RAISING. In Listing 3.13, this exception was handled using the technique presented. For this purpose, we first create a message object and write the formatted text and type into the log. In this case, the "Table ZACB_RECIPE is empty (tip: run generator ZCL_ACB_DEMO_GENERATOR)" message is output.

```
check_tables_not_empty(
  EXCEPTIONS
    data_not_generated = 1
    OTHERS             = 2 ).
IF sy-subrc <> 0.
  DATA(message) = xco_cp=>sy->message( ).
  out->write(
    |{ message->get_type( )->value }: | &&
    message->get_text( ) ).
ENDIF.
```

Listing 3.13 Retrieving Messages Using the XCO Library

[!]

The Language of Messages

In the example shown in Listing 3.13, a message is converted into a text and written to the console output. Note that this should only be done in user-oriented scenarios, as otherwise you'll lose the link to the message and therefore also to the translations. For example, if you create a log that is to be displayed by different users in their respective logon language, the message itself must be logged. If required, you can also access the relevant system fields indirectly via the message object. They are provided via the value attribute of the symsg type.

3.7 Accessing System Data

To conclude this chapter, let's take a look at accessing data about the SAP system and its installation details. This involves the installed software components, their releases and patch levels, languages, and information on the tenant in cloud systems. This data is important in relation to our application with debug information, as it can be used to analyze whether used ABAP functions or a library may not yet be available in the specific system, which would cause problems during the operation of the recipe portal.

Tenant information about the XCO library

The current attribute of the global XCO_CP class provides information on the callstack as well as data on the tenant if it's a cloud system. If this is the case, the tenant method returns an object that provides information about getter methods, such as the URL for the system's SAP Fiori launchpad. Otherwise, a null pointer will be returned. In Listing 3.14, you can see an example of the output of this data.

```
DATA(tenant) = xco_cp=>current->tenant( ).
IF tenant IS BOUND.
  DATA(ui_url) = tenant->get_url(
    xco_cp_tenant=>url_type->ui ).
  IF ui_url IS BOUND.
    out->write( |  UI URL: | &&
      |{ ui_url->get_protocol( ) }://| &&
      |{ ui_url->get_host( ) }:| &&
      |{ ui_url->get_port( ) }| ).
  ELSE.
    out->write( |  UI URL: not available| ).
  ENDIF.
  out->write( |  ID: | &&
    tenant->get_id( ) ).
  out->write( |  Subaccount ID: | &&
```

```
      tenant->get_subaccount_id( )->as_string( ) ).
    out->write( |  Global Account ID: | &&
      tenant->get_global_account_id( )->as_string( ) ).
    out->write( |  GUID: | &&
      tenant->get_guid( )->as(
        xco_cp_uuid=>format->c36 )->value ).
ELSE.
  out->write( |  System without tenant| ).
ENDIF.
```

Listing 3.14 Querying Tenant Data Using the XCO Library

Software components in the XCO library

Depending on your application, you may want to run a program-based query as to whether a certain software component is installed in the system and whether it can be extended. This information can also be found via the XCO library. The API class for software components is called XCO_CP_SYSTEM. It provides a method for accessing data on the installed software components by name. The extensibility of the SAP_BASIS software component is determined in Listing 3.15.

```
DATA(basis) =
  xco_cp_system=>software_component->for_name(
    'SAP_BASIS' ).
IF basis->get_extendability( )
   = xco_cp_software_component=>extendability->extendable.
  out->write(
    |Software component { base->name } is extensible| ).
ELSE.
  out->write(
  |Software component { base->name } is not extensible| ).
ENDIF.
```

Listing 3.15 Querying the Extensibility of the SAP_BASIS Software Component Using the XCO Library

Unfortunately, you can't currently find out the release and patch level of a software component in this way. Specifically for SAP_BASIS, you could switch to the nonreleased sy-saprl system field or write a wrapper around classic APIs in SAP S/4HANA and SAP S/4HANA Cloud Private Edition.

Determining installed languages

In addition to the software components, you can also query the languages installed in the system. Again, the XCO library can be used for this purpose. The point of entry is the languages method in the familiar XCO_CP_SYSTEM class. The languages installed in the system are determined and displayed in Listing 3.16.

```
LOOP AT xco_cp_system=>languages->installed->get( )
     INTO DATA(language).
  out->write(
 | { language->as( xco_cp_language=>format->iso_639 ) } |
    && language->get_long_text_description( ) ).
ENDLOOP.
```

Listing 3.16 Querying Installed Languages Using the XCO Library

[«]

Additional Query Options Using the XCO_CP_SYSTEM Class

In addition to the methods for determining software components and languages, the XCO_CP_SYSTEM class also provides methods for application components and namespaces, which aren't addressed individually here.

Determining all languages via a CDS view

If you're more interested in all language abbreviations, including those that aren't installed or can't be installed, you can alternatively use the shared CDS view I_Language. In Listing 3.17, the output of all languages available in the system is based on this view as the data source. The association to the text table is resolved directly in the ABAP SQL command to obtain the language-dependent name of the language in the logon language.

```
SELECT FROM I_Language
  FIELDS
    LanguageISOCode,
    \_Text[ ONE TO ONE WHERE Language = @sy-langu
          ]-LanguageName
  ORDER BY LanguageISOCode
  INTO TABLE @DATA(languages).
LOOP AT languages
     ASSIGNING FIELD-SYMBOL(<language>).
  out->write(
    | { <language>-LanguageISOCode }: | &&
    |{ <language>-LanguageName }| ).
ENDLOOP.
```

Listing 3.17 Querying All Languages Using CDS View I_Language

3.8 Summary

In this chapter, you've learned how to handle runtime information in ABAP Cloud. Many of the techniques known from classic ABAP still work here, some of them trigger syntax warnings, and others are no longer available at all. Alternatively, a range of APIs can already be used in ABAP Cloud to access the required information.

From the perspective of SAP S/4HANA and SAP S/4HANA Cloud Private Edition, however, some of the APIs are apparently designed more for the SAP BTP ABAP environment and SAP S/4HANA Cloud Public Edition. The fact that the system ID and the client ID haven't been given any methods in the CL_ABAP_CONTEXT_INFO class or that this class doesn't return the system time indicates that the APIs were originally designed for these cloud environments only. Such limitations don't apply to older approaches, such as the CL_ABAP_SYST class, but they aren't available in ABAP Cloud.

This chapter is the first in this book where you have come into contact with the XCO library. It's the new standard library and you'll therefore encounter it again and again in the following chapters. Using the XCO library requires some familiarization, as the API is very much object-oriented. Even strings are represented by instances of a class. From an ABAP developer's point of view, such abstractions at the lowest level are rather unusual.

Chapter 4
Table Analysis

How can you view the data records of a database table in ABAP Cloud without using SAP GUI? This chapter gives you the answer.

Previously, the analysis of customer data in the SAP backend was generally performed using the Transaction SE16 family. These transactions provide the usual selection screens of a report with fields for entering the selection options and deliver the usual table output based on the *SAP List Viewer* (ALV; see Figure 4.1).

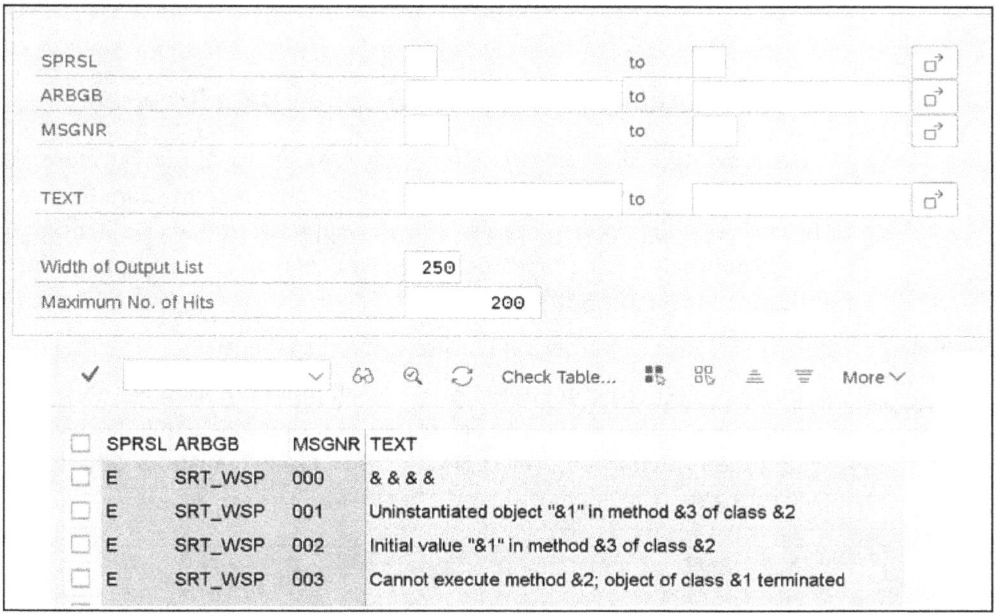
SPRSL
ARBGB
MSGNR
TEXT
to
Width of Output List
250
Maximum No. of Hits
200
Check Table...
More
SPRSL ARBGB MSGNR TEXT
E SRT_WSP 000 & & & &
E SRT_WSP 001 Uninstantiated object "&1" in method &3 of class &2
E SRT_WSP 002 Initial value "&1" in method &3 of class &2
E SRT_WSP 003 Cannot execute method &2; object of class &1 terminated

Figure 4.1 A Selection Screen of the Transaction SE16 Family

Conversion to SAP Fiori UIs

However, SAP is currently moving away from SAP GUI transactions and converting its products to SAP Fiori interfaces. Only SAP Fiori interfaces can be used in SAP S/4HANA Cloud Public Edition. The ABAP development tools for Eclipse development environment and the ABAP for cloud development language version setting therefore provide new options for customer data analysis:

- **Customer data browser objects (see Section 4.1)**
  This option is primarily intended for end users for data analysis.
- **Data preview in the ABAP development tools (see Section 4.2)**
  This option is primarily aimed at development teams for use in the development process.

Table analysis in our sample application

In this chapter, we create the customer data browser object named ZACB_I_RECIPE_CDBO on the basis of core data services (CDS) view ZACB_I_RECIPE_CDBO, which is based on database table ZACB_RECIPE with the recipe data. We use this object to explain the functions of the customer data browser. Database table ZACB_RECIPE also serves as the basis for presenting the data preview in the ABAP development tools.

Using the Sample Application

The objects created in this chapter are located in the TABLEDISPLAY subpackage in our sample application.

4.1 Table Analysis Using the Customer Data Browser

A *customer data browser object* makes it possible to display a reference table or a data definition in a framework called the *customer data browser*. The customer data browser is a self-service application that can be used to display data belonging to specific CDS views users can access based on their authorizations in the SAP system.

Creating a customer data browser object

You can create a customer data browser object as follows:

1. In the project structure in the ABAP development tools, select the package in which the object is to be created, and right-click on it. In the context menu that opens, select **New • Other ABAP Repository Object** (see Figure 4.2).

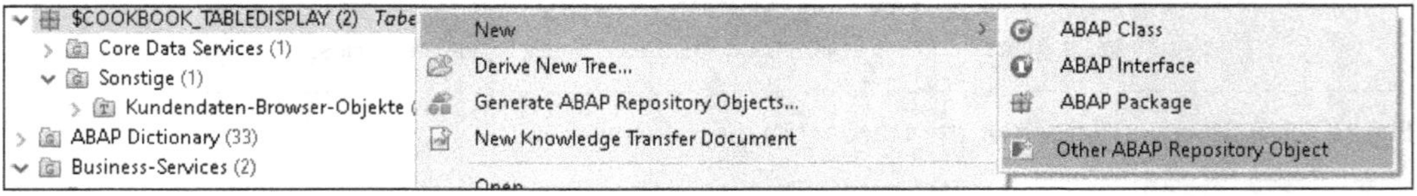

Figure 4.2 Context Menu Item for Creating Another ABAP Repository Object

2. In the repository browser that opens, select the **Customer Data Browser Object** item (see Figure 4.3). Confirm your selection by clicking on the **Next** button.

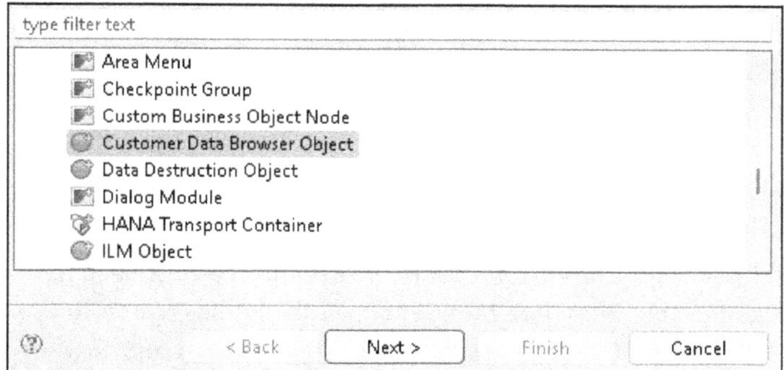

Figure 4.3 Creating a Customer Data Browser Object

3. Enter the following information on the new object in the next screen, as shown in Figure 4.4:
   - **Name**: The name for the customer data browser object must correspond to that of the referenced object, in this case our CDS view.
   - **Description**: Enter a suitable description for the customer data browser object here.
   - **Object Type**: The object type can be specified in this dropdown menu. A distinction is made between the following options:
     - **Table**: A database table can be specified.
     - **Data Definition**: A CDS view can be specified.
   - **Object Name**: You can enter the object name directly or use the **Browse** button to search for a CDS view or a database table in the system.

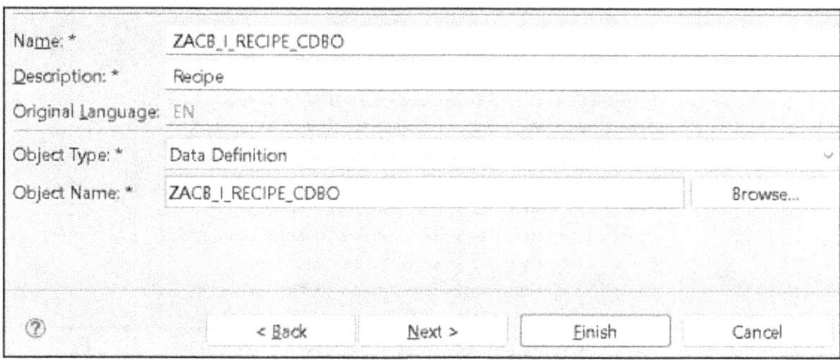

Figure 4.4 Parameters of a Customer Data Browser Object

[«]

Table versus Data Definition

Even if it's possible to select the **Table** object type at this point, the **Data Definition** object type should always be used. SAP recommends the use of data definitions to access database tables. If no data definition exists for a standard SAP table, you can develop your own data definition for this standard table. If you have your own database tables, they should be accessed via your own data definitions.

[!]

Package Assignment

The object referenced by the customer data browser object must be in the same package. If a customer data browser object has to be created for a standard data definition, a separate data definition with access to the standard data definition must be developed beforehand. This separate data definition must then have the same package assignment as the customer data browser object.

4. Once all the information has been entered, you can go to the next step of the wizard by clicking the **Next** button. Enter a transport request here, and end the wizard by clicking the **Finish** button.

This will create the customer data browser object (see Figure 4.5). Activate the created object to be able to use it.

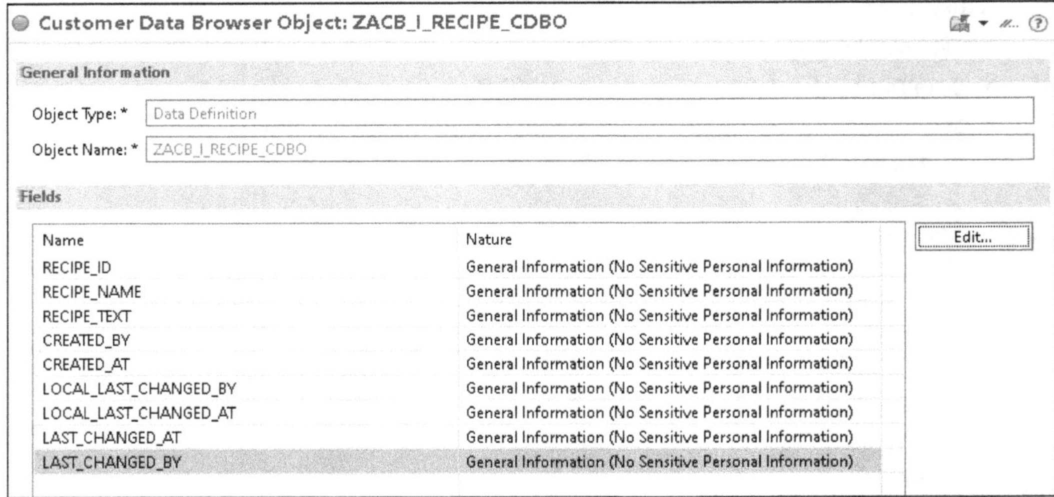

Figure 4.5 Created Customer Data Browser Object

Changing a field property

You can now adapt the object created by the wizard according to your requirements. In our example, we'll provide the recipe data with the **Recipe ID**, **Recipe Name**, and **Recipe Text** columns:

1. Select the field to be modified—in our example, this is the **CREATED_BY** field—and click on the **Edit** button (refer to Figure 4.5).

2. The **Edit Field Details** dialog box opens (see Figure 4.6).

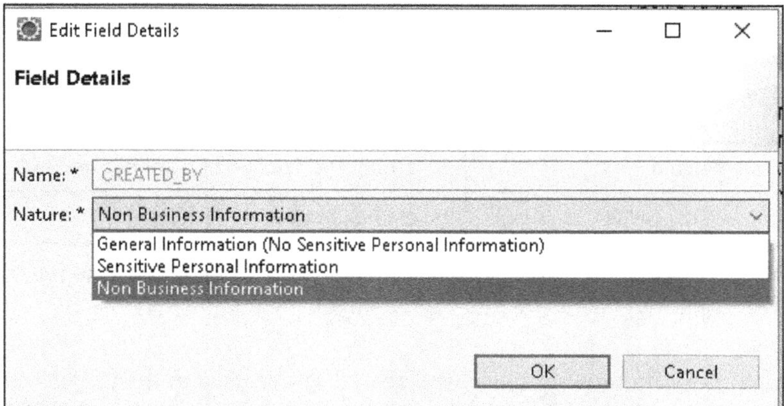

Figure 4.6 Edit Field Details Dialog Box

3. Select a property for the **CREATED_BY** field in the **Nature** dropdown menu. The following properties are available for selection here (see Figure 4.6):
   - **General Information (No Sensitive Personal Information)**: The data in this field doesn't contain any sensitive personal information.
   - **Sensitive Personal Information**: The data in this field contains sensitive personal information.
   - **Non Business Information**: The data of such fields doesn't contain any important information and its display can be suppressed.

   In our example, we enter the date and personal information such as CREATED_BY or CREATED_AT as **Non Business Information**. This means it won't be displayed in the output.
4. Confirm your selection by clicking the **OK** button.

The changes have been applied to the display of the customer data browser object (see Figure 4.7). Activate the object again so that the changes are applied.

RECIPE_TEXT	General Information (No Sensitive Personal Information)
CREATED_BY	Non Business Information
CREATED_AT	General Information (No Sensitive Personal Information)

Figure 4.7 Transferring the Field Property

Changes to the data definition

Table 4.1 shows how the behavior of the customer data browser object changes when changes are made to the data definition.

Changes to the Data Definition	Behavior of the Customer Data Browser Object
Fields are added.	Automatic transfer of the added fields from the data definition occurs. However, the fields still need to be classified.
Fields are deleted.	No automatic deletion of fields occurs.
The order of the fields is changed.	No change to the order is made.

Table 4.1 Behavior of the Customer Data Browser Object After Changes to the Data Definition

The customer data browser object we created can be used in the Customer Data Browser app (app ID F5746).

[»]

Customer Data Browser App

All information on the Customer Data Browser app can be found in the SAP Fiori apps reference library at *https://fioriappslibrary.hana.ondemand.com/*. Select **All apps**, and enter the ID "F5746" in the search bar. Click on Customer Data Browser when it appears in the list. The **IMPLEMENTATION INFORMATION** tab describes all the requirements for making the app available in your system. Although the app is currently only available for SAP S/4HANA Cloud Public Edition according to the SAP website, it's also available for SAP S/4HANA.

Displaying a customer data browser object

If your system and your user account meet the technical requirements, you can access the app. To do this, start the SAP Fiori launchpad, and click on the **Customer Data Browser** tile (see Figure 4.8).

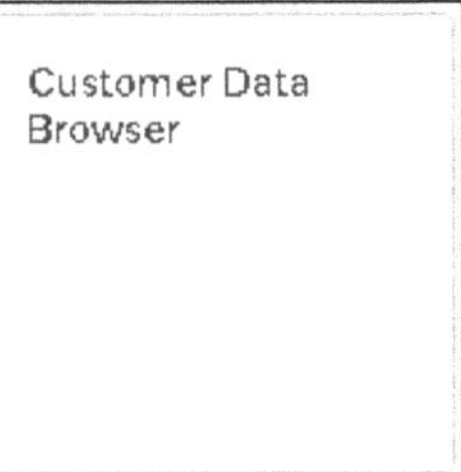

Figure 4.8 Selecting the Customer Data Browser App Tile

You can confirm the note that sensitive data is being accessed and that this must comply with your company's data protection guidelines by clicking **OK**. Then, the customer data browser objects available with your user authorizations will get displayed (see Figure 4.9).

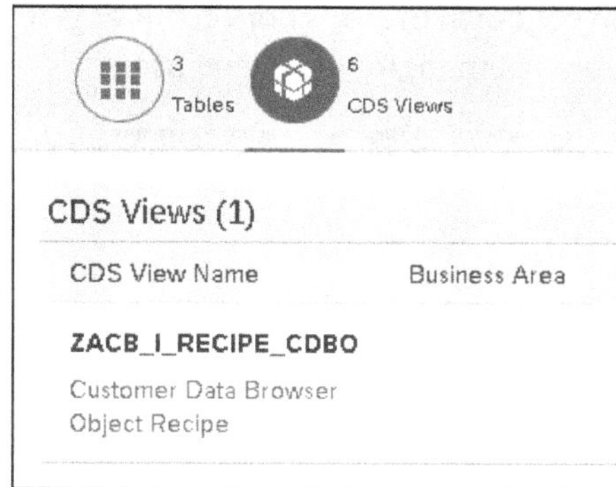

Figure 4.9 Display in the Customer Data Browser

There, you can click the customer data browser object **ZACB_I_RECIPE_CDBO** to select it. The data of the customer data browser object is then displayed (see Figure 4.10).

Functions of a customer data browser object

You can use the following functions within the Customer Data Browser app:

- **Filtering**
  Various filters can be used to restrict the columns displayed and the number of data records displayed.
- **Search**
  The search line can be used to search for a specific term to limit the display. A cross-column search is performed.

Figure 4.10 Display of the Customer Data Browser Object

- **Column properties**
  Within the app, you can specify which columns are to be displayed in which order. To do this, click on the **Settings** button. In the window that opens next, you can select the columns to be displayed and adjust their order using the arrow icons (see Figure 4.11). Click **OK** to apply the

changes and apply them directly to your own user. It's not yet possible to save the configuration in variant management for other end users.

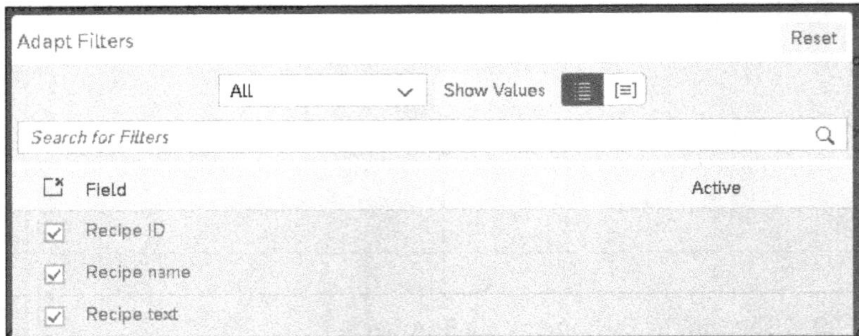

Figure 4.11 Defining the Column Properties

- **Displaying names**
  You can use this function to switch between displaying the technical identifiers and the business field names. To do this, click on the **View technical names** button (not shown; found at the top of the screen displayed in Figure 4.10). If the technical field names are currently displayed as column headings, this button is called **View business names**.
- **Total number of data records**
  By clicking on the **Total Records** button (not shown; found at the top of the screen displayed in Figure 4.10), the number of data records from the data source is determined and displayed.
- **Data export**
  The data records displayed can also be exported via the **Export** button, either as a Microsoft Excel file or as a CSV export. Figure 4.12 shows an Excel file with the exported data created in this way.

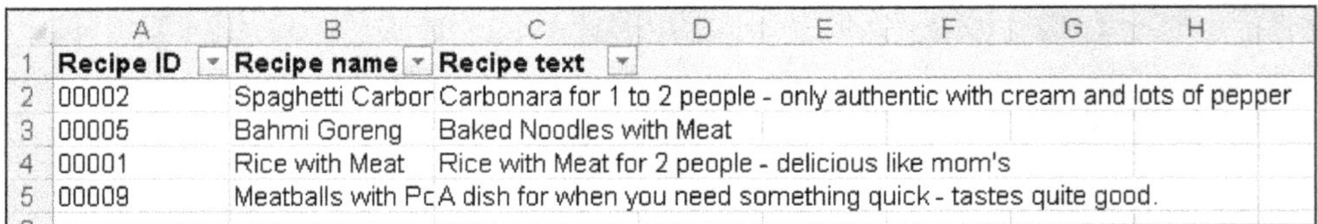

	A	B	C	D	E	F	G	H
1	Recipe ID	Recipe name	Recipe text					
2	00002	Spaghetti Carbor	Carbonara for 1 to 2 people - only authentic with cream and lots of pepper					
3	00005	Bahmi Goreng	Baked Noodles with Meat					
4	00001	Rice with Meat	Rice with Meat for 2 people - delicious like mom's					
5	00009	Meatballs with Pc	A dish for when you need something quick - tastes quite good.					

Figure 4.12 File Export as Excel Table

4.2 Table Analysis Using ABAP Development Tools

The table analysis in the ABAP development tools has similar functions to those of the Transaction SE16 family. It displays ABAP Dictionary tables, views, and CDS data definitions. In addition, it provides functions such as sorting and filtering entries as well as downloading the displayed data.

Display of a database table

In this section, we'll show you how to display a database table in the ABAP development tools. To search for such a data source in the ABAP repository, press the keyboard shortcut [Ctrl]+[Shift]+[A] or [Alt]+[F8], or click on the **Open ABAP Development Object** icon. You can then search for the name of the data source in the window that opens (see Figure 4.13). Then, select the database table in the results list and click **OK**, or select the table by double-clicking on it.

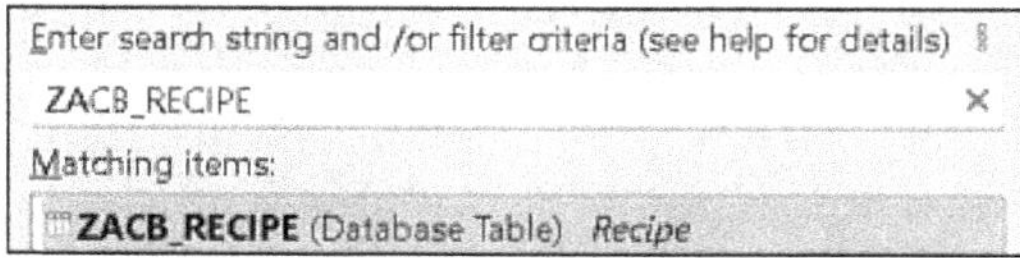

Figure 4.13 Searching for a Database Table

Calling the data preview

In the database editor that opens, right-click on the table name, and select **Open With • Data Preview** in the context menu (see Figure 4.14), or press the [F8] function key. The contents of the database table are then displayed as shown in Figure 4.15.

```
@EndUserText.label : 'Recipe'
@AbapCatalog.enhancement.category : #NOT_EXTENSIBLE
@AbapCatalog.tableCategory : #TRANSPARENT
@AbapCatalog.deliveryClass : #A
@AbapCatalog.dataMaintenance : #RESTRICTED
define table zacb_recipe {

  key client              : abap.clnt not null;
  key recipe_id           : zacb_recipe_id not null;
  recipe_name             : zacb_recipe_name;
  recipe_text             : zacb_recipe_text;
  published               : zacb_published;
  created_by              : abp_creation_user;
  created_at              : abp_creation_tstmpl
  local_last_changed_by   : abp_locinst_lastcha
  local_last_changed_at   : abp_locinst_lastcha
  last_changed_at         : abp_lastchange_tstn
  last_changed_by         : abp_lastchange_user

}
```

Undo Ctrl+Z
Revert File
Save Ctrl+S
ABAP Type Hierarchy F4
Quick Type Hierarchy Ctrl+T
Navigate To F3
Open in Project Ctrl+Alt+P >
Open With >
Open Others >
Show In Alt+Shift+W >
Quick Outline Ctrl+O

Data Preview
Data Preview...
SQL Console
SAP GUI
Dictionary Log

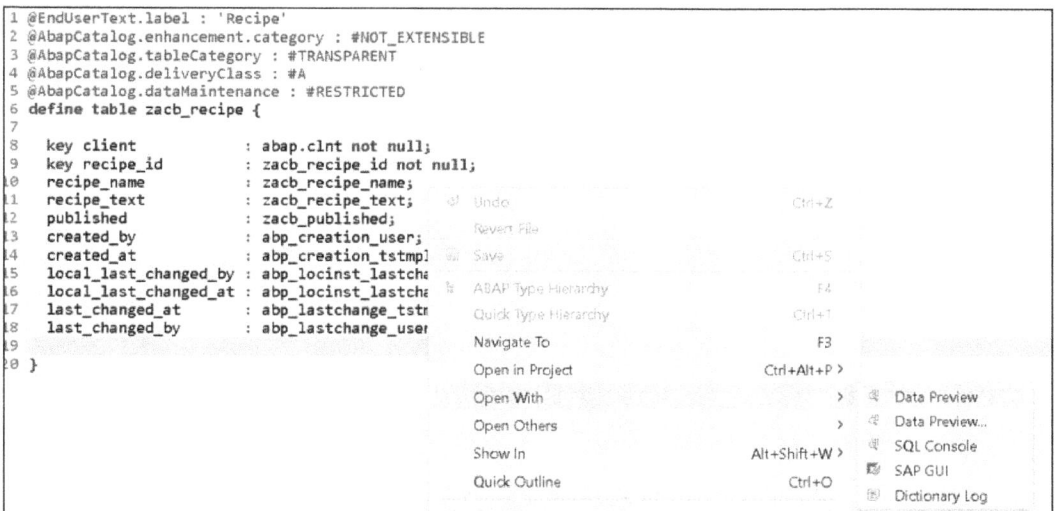

Figure 4.14 Opening the Data Preview via the Context Menu

Data Preview

find pattern | 10 rows retrieved - 23 ms

CLIENT	RECIPE_ID	RECIPE_NAME	RECIPE_TEXT
100	00001	Rice with Meat	Rice with Meat f...
100	00002	Spaghetti Carbon...	Carbonara for 1...

Figure 4.15 Table Display in the Data Preview of the ABAP Development Tools

Data preview functions

There are various functions for using the data in the top right-hand menu:

- **Number of Entries**
  This function can be used to determine the number of entries.
- **Select Columns**
  This function allows you to show and hide columns. All columns in the table are available (see Figure 4.16).

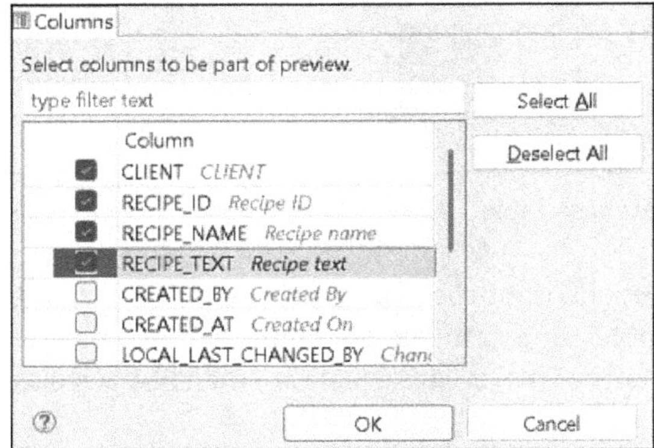

Figure 4.16 Select Columns Function

- **Add Filter**
  This function allows you to restrict the display to specific data records. In Figure 4.17, we filter for recipes whose title contains the "* with*" character string.

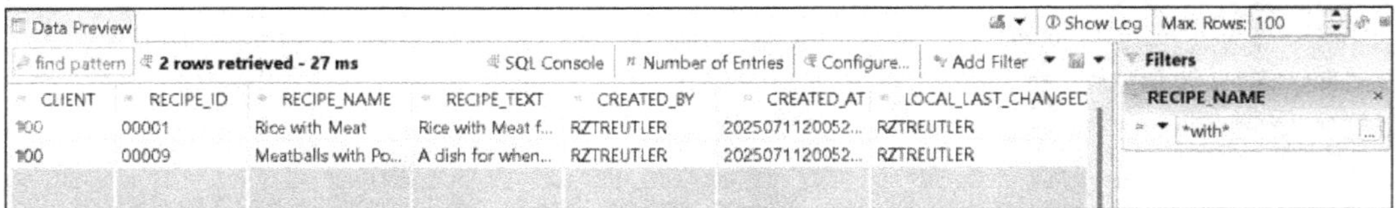

Figure 4.17 Add Filter Function

SQL console

In addition to these table display functions, the ABAP development tools also include the *SQL console*, which enables you to access a database table with SQL statements. You can call the SQL console via the context menu of the database table using the path **Open With • SQL Console** (see Figure 4.18).

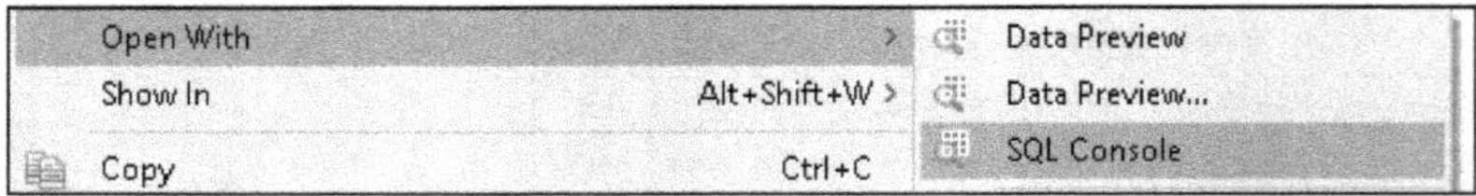

Figure 4.18 Context Menu of the SQL Console

Alternatively, you'll find the **SQL Console** button in the data preview, which you can use to open the console. The advantage of this procedure is that all settings are adopted from the data preview (see Figure 4.19).

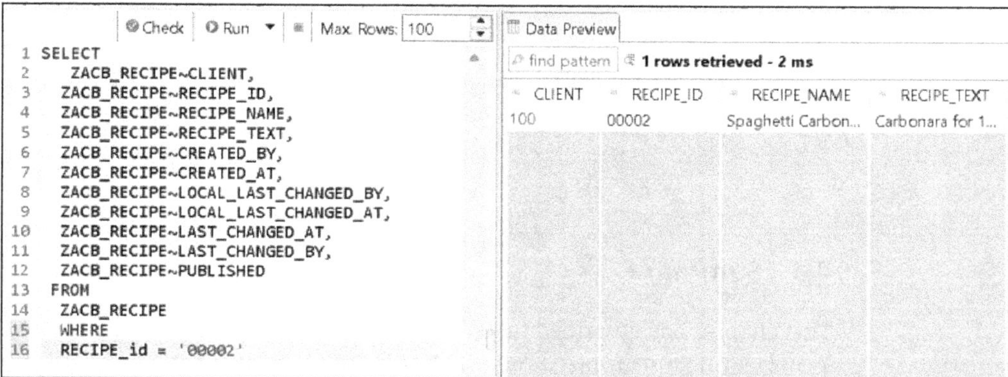

Figure 4.19 Adopting the Settings from the Data Preview

Any ABAP SQL commands can be entered in the SQL console. In contrast to other database applications, for example, Transaction DBACOCKPIT, you can't use native SQL here.

Exporting ABAP source code

If you right-click in the table in the data preview, the table content can be copied as an ABAP statement. To do this, select **Copy All Rows As • ABAP Value Statement** from the context menu (see Figure 4.20).

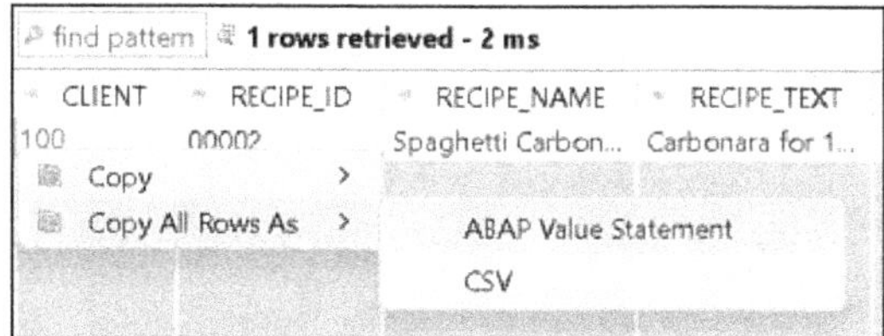

Figure 4.20 Copying the ABAP Statement

You can then add the exported source code as a VALUE statement to an existing ABAP program. Listing 4.1 shows the copied ABAP statement.

```
VALUE #(
( RECIPE_ID = '00005' RECIPE_NAME = 'Spaghetti Carbonara' RECIPE_TEXT
= 'Carbonara for 1 to 2 people - only authentic with cream and lots
of pepper' )
```

Listing 4.1 Copied ABAP Statement

For data analysis and testing SQL statements, the data preview in the ABAP development tools is faster and more effective than the customer data browser. If a connection to the production systems has been set up in the ABAP development tools, data can also be analyzed here using ABAP SQL in the SQL console.

[!]

Authorizations in the Production System

You must take the authorization concept of the production system into account in your analyses and handle the S_TABU_DIS authorization object with care.

4.3 Summary

This chapter has described table analyses in ABAP Cloud. In the future, the Transaction SE16 family should no longer be used. We've shown two alternatives to table analysis. The first alternative is the new *customer data browser object*. This repository object enables key users to display the data of a CDS data definition in the Customer Data Browser app.

The second alternative for displaying the data is table analysis in the ABAP development tools. The data can even be displayed in the development environment during the development process, which is why this type of display is primarily intended for developers. In addition to table analysis, it's also possible to try out SQL statements in an SQL console.

Chapter 5
Table Maintenance Using Business Configuration Maintenance Objects

5

With the help of business configuration maintenance objects, you can easily provide Customizing settings for your own applications. The ABAP RESTful application programming model and SAP Fiori elements are used in the background.

Even at a low level of complexity, the question of setting options arises for almost every business application. The application should act differently depending on the options selected, without requiring any development work. In the SAP environment, we refer to this as *Customizing*. Customizing is also often provided for in-house developments to make program sequences more dynamic or to outsource the responsibility of configuring from developer teams to application support, consulting, or specialist departments that don't need to have development skills.

Customizing tools in the development environment

Because Customizing is so highly relevant, corresponding options for ABAP development were integrated into the development environment at an early stage. These included Customizing database tables, maintenance views and view clusters, Implementation Guide (IMG) and Customizing activities, change logging, and also buffered access to table contents. In this context, the *table maintenance generator* should be mentioned in particular. However, many of these tools are based on SAP GUI or aren't compatible with cloud concepts. For this reason, there's a modern successor concept in ABAP Cloud that enables you to make setting options for your application available quickly and easily via a *business configuration app* based on SAP Fiori. The setting options are implemented via the *business configuration maintenance objects*, which you'll learn about in this chapter.

Structure of this chapter

Section 5.1 provides an overview of the concept of business configuration and the relevant objects. We then create the relevant objects for our sample application step-by-step in Section 5.2 and Section 5.3. In Section 5.4, we take a look at assigning authorizations before discussing the actual handling from the user's perspective in Section 5.5. The subsequent sections go deeper into some of the details, such as the setting options in a business configuration maintenance object (Section 5.6), lifecycle management

(Section 5.7), and the documentation of the configuration options provided (Section 5.8). This extensive chapter covers many of the technical options provided here because Customizing continues to be highly relevant in everyday development with ABAP Cloud.

Settings options in the application scenario

The recipe portal is supposed to provide the option of maintaining labels for recipes. Any number of these labels can be assigned to a recipe. Recipes can then be searched and filtered using the labels. Examples of such labels are vegan, inexpensive, or quick meal. The available labels are firmly defined and should be maintained by the application support team in a Customizing table.

Procedure for implementation

The data model must be extended to implement these settings options. New database tables are required. The ZACB_LABEL and ZACB_LABELT Customizing tables contain the definitions of labels, including translatable texts. The two tables are to be maintained using Customizing. This Customizing option should be implemented in accordance with ABAP Cloud, which means using a business configuration maintenance object.

Using the Sample Application

The fully implemented business configuration app can be found in the enclosed coding in the CONFIG subpackage. After the installation, you must make sure that the ZACB_UI_LABEL_04 service binding is published in your system. The other steps needed for use are described in Section 5.4 and Section 5.5.

5.1 Overview of the New Table Maintenance Concept

Custom Business Configurations app

The ABAP RESTful application programming model and SAP Fiori elements are used to generate a business configuration app for the business configuration maintenance objects to provide the maintenance option for Customizing tables. Users can then access these via the Custom Business Configurations app (app ID F4579) provided in the SAP Fiori launchpad.

From a developer's point of view, the entire development takes place in the ABAP stack. Although an SAP Fiori app is created, there's no need to switch to SAP Business Application Studio because the generated app is automatically available in the system. There's also no dependency on the SAP IMG via Transaction SPRO, the Implementation Activities app (app ID SIMG), or SAP Central Business Configuration. However, you can optionally add a jump to the IMG. Regardless of the specific SAP system and its runtime environment, access to the business configuration and the technology are always the same, which is exactly the idea behind ABAP Cloud.

ABAP RESTful application programming model objects in the business configuration scenario

As is usual with the ABAP RESTful application programming model, various objects of different object types are required to model and implement the application logic. In addition, there are specific objects required for the Customizing use case. The objects needed to implement the label maintenance requirement outlined in the introduction to this chapter and how these objects relate to each other are shown in Figure 5.1.

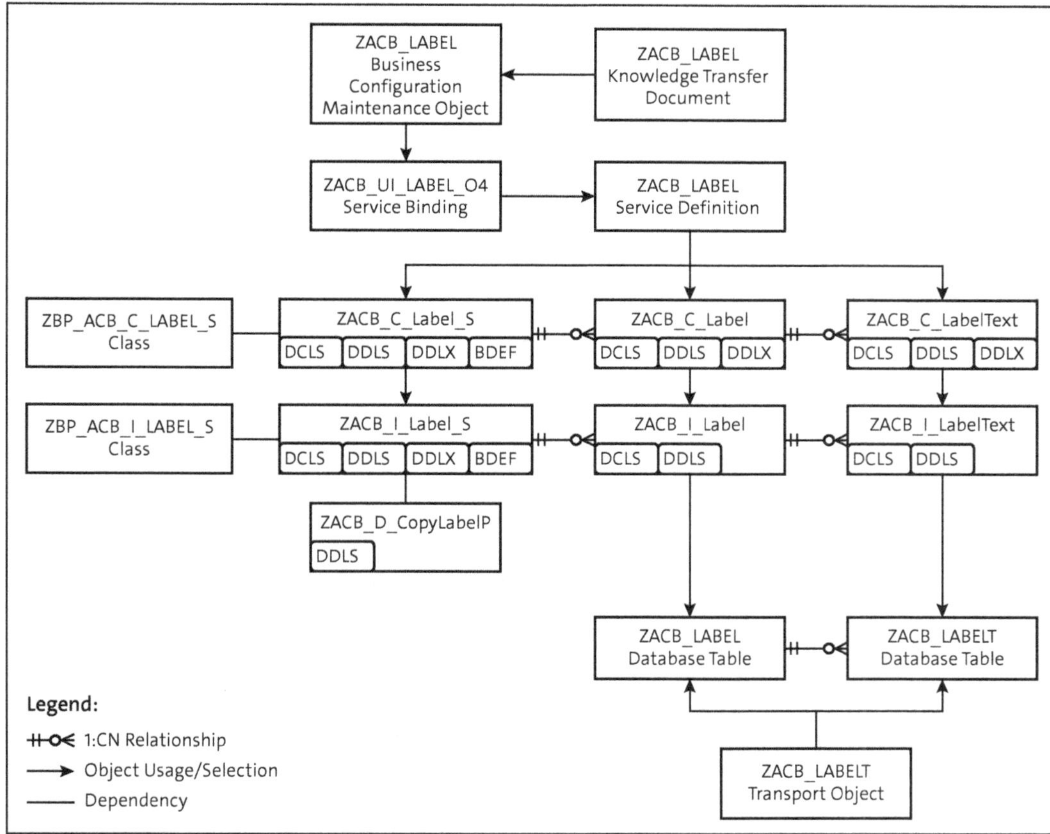

Figure 5.1 Overview of Objects and Dependencies in Label Maintenance

The graphic may seem surprisingly complex for the maintenance of two Customizing tables. However, the construct of a business configuration app is very similar to that of normal ABAP RESTful application programming model applications for application data with transactional behavior. You'll see that you only need to create or manually modify very few of the objects shown because an ABAP repository object generator will do the hard work for you. However, you *can* modify all the objects. These are objects that are generated once, are then your area of responsibility, and can be completely changed as required. This has the advantage that you have maximum flexibility—but it also has the disadvantage that you're on your own after generating them.

Specific object types

The following object types are new in this context:

- The business configuration maintenance object
- The transport object
- Optionally, a knowledge transfer document

You may also have noticed that additional view entities with an S suffix are created that have the behavior definitions and associated classes of the behavior implementation. The S stands for the *singleton* design pattern. The singleton objects are used as brackets to provide a central point of entry into possibly numerous Customizing tables and to implement transport integration.

5.2 Creating Customizing Tables

Two Customizing tables

To persist the Customizing settings, the Customizing tables must first be created. In our case, there are two tables:

- ZACB_LABEL
  This is the primary table for labels.
- ZACB_LABELT
  This is the text table for the language-dependent identifiers.

You can create Customizing tables in the ABAP development tools in the same way as normal database tables with application data. To do this, you must first select the **New • Other Repository Object** path from the context menu of the relevant package. Then, filter for the **Database Table** object type, or select it in the **ABAP Dictionary** folder. Assign an identifier and a short description in the original language. In our example, we entered the technical identifier "ZACB_LABEL" and the short description "Labels". Finally, you need to enter a transport request and close the creation wizard by clicking the **Finish** button.

Customization of the table definition via dictionary DDL

After specifying the transport request, the text-based maintenance interface for database tables opens. We'll now adapt the definition of the fields using the Data Definition Language (DDL) syntax for the ABAP Dictionary, which we'll refer to as *dictionary DDL syntax*. We add the LABEL_ID key field with the ZACB_LABEL_ID type and the LABEL_COLOR field with the ZACB_LABEL_COLOR type. We also add the administrative fields LAST_CHANGED_AT with the ABP_LASTCHANGE_TSTMPL type and LOCAL_LAST_CHANGED_AT with the ABP_LOCINST_LASTCHANGE_TSTMPL type, as well as the CONFIGDEPRECATIONCODE field with the CONFIG_DEPRECATION_CODE. At this point, the source code for defining our database table looks like Figure 5.2. This state is still unfinished, and we'll continue to adapt the table before it can be used as a Customizing table.

```
[S4P] ZACB_LABEL ×
@EndUserText.label : 'Labels'
@AbapCatalog.enhancement.category : #NOT_EXTENSIBLE
@AbapCatalog.tableCategory : #TRANSPARENT
@AbapCatalog.deliveryClass : #A
@AbapCatalog.dataMaintenance : #RESTRICTED
define table zacb_label {

  key client                 : abap.clnt not null;
  key label_id               : zacb_label_id not null;
  label_color                : zacb_label_color;
  configdeprecationcode      : config_deprecation_code;
  last_changed_at            : abp_lastchange_tstmpl;
  local_last_changed_at      : abp_locinst_lastchange_tstmpl;

}
```

Figure 5.2 Intermediate Status of Text-Based Maintenance for the Definition of Database Table ZACB_LABEL

Timestamp fields

To be able to use the ABAP repository object generator, the table must meet specific technical requirements. The two timestamp fields will be required by the ABAP RESTful application programming model business object, which we'll generate later. The LOCAL_LAST_CHANGED_AT timestamp field is supposed to be included in every associated Customizing table. The LAST_CHANGED_AT field is to be contained in the primary table, in this case, table ZACB_LABEL. Both fields are required for the implementation of the optimistic lock behavior. The CONFIGDEPRECATIONCODE field is optional. It's required for the lifecycle management of entries, which we'll return to in Section 5.7.

Creating data elements

The ZACB_LABEL_ID and ZACB_LABEL_COLOR types don't yet exist and are therefore underlined in red in the editor (see Figure 5.2). Using forward navigation, we now create a data element for each of these types. To do this, position the cursor on the underlined code element and open *Quick Assists* using the keyboard shortcut [Ctrl] + [1] (see Figure 5.3).

```
key client                 : abap.clnt not null;
key label_id               : zacb_label_id not null;
label_color                : zacb_la   [Create data element zacb_label_id]
configdeprecationcode      : config_
last_changed_at            : abp_las
local_last_changed_at      : abp_loc
```

Figure 5.3 Quick Assist for Creating a Data Element

Select the **Create data element zacb_label_id** option and follow the steps of the creation wizard. You must specify an identifier, the package, and a transport request. After that, you can confirm the creation. This opens the data element maintenance. Figure 5.4 shows this maintenance screen for the ZACB_LABEL_ID data element.

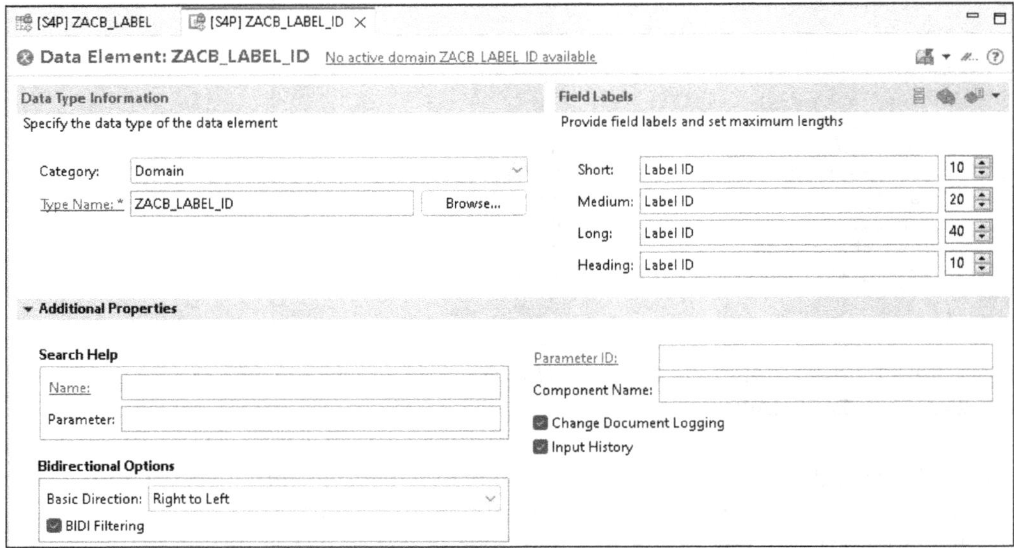

Figure 5.4 ZACB_LABEL_ID Data Element Settings

Enter the settings from Table 5.1 here and also for the ZACB_LABEL_COLOR data element.

Property	ZACB_LABEL_ID Data Element Values	ZACB_LABEL_COLOR Data Element Values
Category	Domain	Domain
Type Name	ZACB_LABEL_ID	ZACB_LABEL_COLOR
Field Labels	Label ID	Label color
Change Document Logging	Enabled	Enabled
Input History	Enabled	Enabled

Table 5.1 Properties of the Data Elements Created for Label ID and Color

[+]

Change Logging

You should deactivate the **Change Document Logging** option for data elements only in exceptional cases. In this specific example, activation means that it's possible to subsequently trace which labels were assigned to which recipe and when, as well as who changed these assignments. Details on writing change documents with ABAP Cloud can be found in Chapter 7.

Creating domains

The two data elements refer to the domains that also don't yet exist. By clicking on **Type Name**, you can also create these via the forward navigation. Assign the properties listed in Table 5.2 for our example.

Property	ZACB_LABEL_ID Domain Value	ZACB_LABEL_COLOR Domain Value
Data Type	CHAR	CHAR
Length	15	1
Output Length	0 (filled automatically)	0 (filled automatically)
Conversion Routine	Blank	Blank
Case Sensitive	No	No
Fixed Values	None	0 = neutral 1 = red 2 = orange 3 = green
Value Table	ZACB_LABEL	(blank)

Table 5.2 Properties of the Label Domains Created

Figure 5.5 shows an example of maintenance for the ZACB_LABEL_ID domain.

The ZACB_LABEL_ID domain is supposed to contain technical, text-type keys, such as VEGAN or QUICK_AND_EASY. As such, they aren't language-dependent and are shown in English by default. The language-dependent, translatable label text will be stored later in the text table. The ZACB_LABEL_COLOR domain is intended to define the color design of a label and uses the color values that are available in SAPUI5 for status colors. These color values can also be used in SAP Fiori elements via the `@UI.lineItem.criticality` annotation.

[«]

Colors in SAP Fiori Elements

The use case for the colors here is that the labels assigned to the recipes stand out visually. In SAP Fiori elements, there's currently no concept for directly influencing the color scheme. This would partially contradict the concept of SAP Fiori elements because compliance with the SAP Fiori design guidelines and the feasibility of custom themes for all SAP Fiori apps couldn't be guaranteed. In this specific case, we therefore use the colors as a workaround, with which a criticality can be clarified, to at least be able to set which labels have the same color, even if the specific coloring can't be guaranteed.

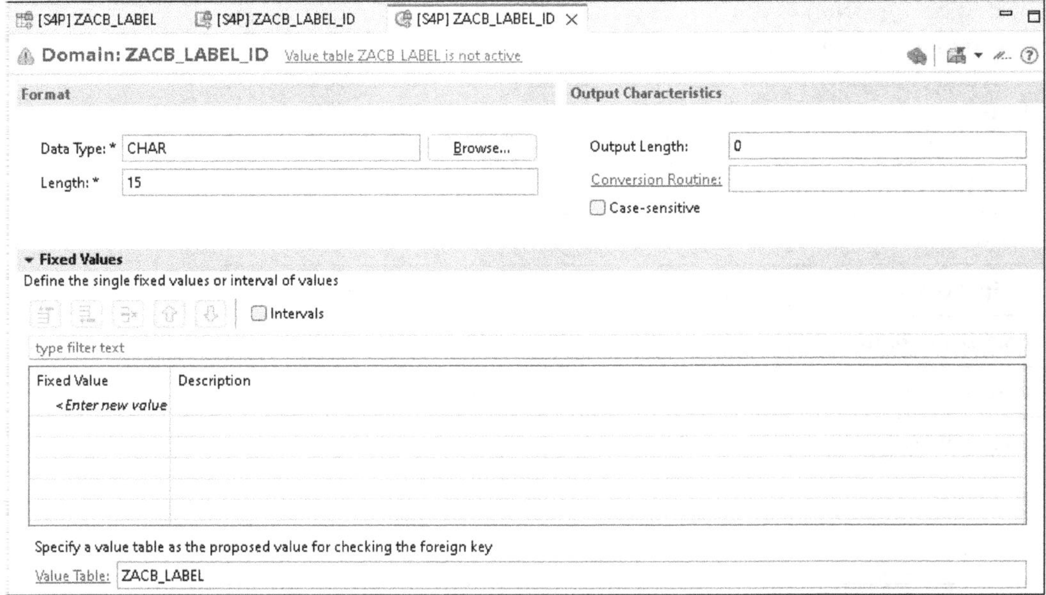

Figure 5.5 ZACB_LABEL_ID Domain Settings

Annotations for Customizing tables

By default, the database table was created as an application table with delivery class A. However, our table is supposed to be a Customizing table. For this purpose, you want to change the value of the `@AbapCatalog.deliveryClass` annotation in the source code editor to `C`. Also make sure that displaying and maintaining the table with standard tools is permitted. To do this, the `@AbapCatalog.dataMaintenance` annotation must have the `#ALLOWED` value.

Display and Maintenance Indicators

Setting the table maintenance indicator to `#ALLOWED` (**Display/Maintenance allowed** in Transaction SE11) may seem unusual at first. In the previous concept of maintenance views, we would have selected the `#RESTRICTED` option (**Display/Maintenance restricted** in Transaction SE11) so that the Customizing table can only be maintained via Transaction SM30 and only in combination with the maintenance view. However, the generator for business configuration maintenance objects requires the

#ALLOWED value, whereas the generated business configuration app doesn't. We therefore recommend that you change the value back to #RESTRICTED after generating the app. Otherwise, the unpleasant effect in on-premise systems is that, with the appropriate authorizations, it's also possible to maintain the table directly via Transactions SE16 and SE16N—bypassing the consistency checks and the transport obligation.

Listing 5.1 contains the finished table definition.

```
@EndUserText.label : 'Labels'
@AbapCatalog.enhancement.category : #NOT_EXTENSIBLE
@AbapCatalog.tableCategory : #TRANSPARENT
@AbapCatalog.deliveryClass : #C
@AbapCatalog.dataMaintenance : #ALLOWED
define table zacb_label {
  key client             : abap.clnt not null;
  key label_id           : zacb_label_id not null;
  label_color            : zacb_label_color;
  configdeprecationcode  : config_deprecation_code;
  last_changed_at        : abp_lastchange_tstmpl;
  local_last_changed_at  : abp_locinst_lastchange_tstmpl;
}
```

Listing 5.1 Definition of Table ZACB_LABEL

Technical settings for the database table

Finally, the technical settings need to be modified. In the ABAP development tools, you can access these either in the context menu of the source code editor (**Open Others • Technical Settings** option) or by opening the database table in the repository explorer and double-clicking on the name of the database table under **Technical Table Settings**.

In our example, we change the data class to APPL2 (Organization and Customizing). The default size category 0 is already suitable for the expected number of entries. Because this is a Customizing table, change logging should be activated here. Otherwise, you'll receive a warning in combination with the delivery class or even have to explain with a short text why logging hasn't been activated.

Buffering settings

Finally, the buffering settings must be checked. As we expect few write accesses but many read accesses, we decide to buffer the table in its entirety. The complete technical settings are shown in Figure 5.6. In the final step, you must activate the table.

Creating text tables

Repeat the steps shown for the ZACB_LABELT database table. Because it contains the language-dependent label texts, it's a text table. In this context, there are a few differences compared to the creation of the primary ZACB_LABEL table:

- We need to add a key field to create language dependency.
- We need to add a field for the language-dependent label identifier.
- We need to define the foreign key relationship to the primary table.
- We need to set up a language-dependent buffering of the table entries.

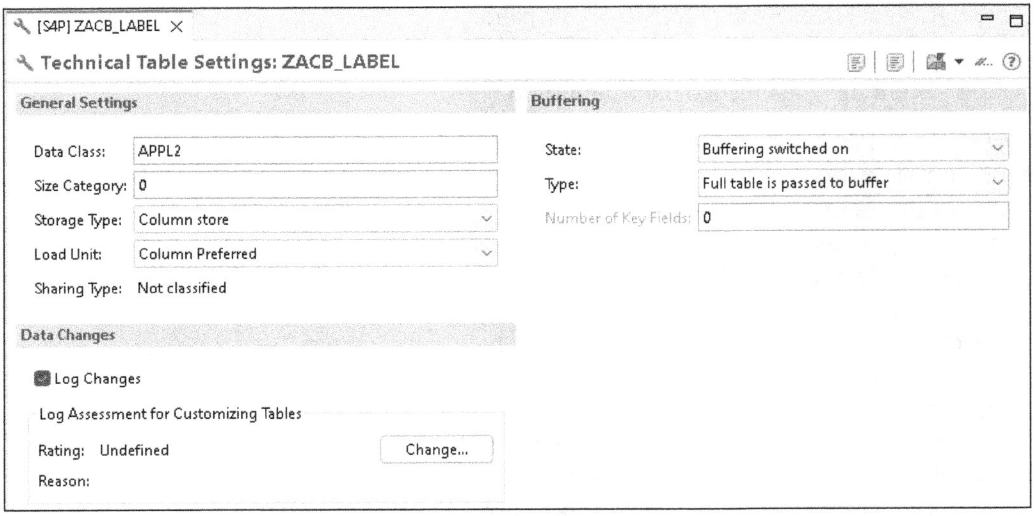

Figure 5.6 Technical Settings of Table ZACB_LABEL

Language-dependent primary key

As is usual with text tables, we adopt all primary key components of the primary table, in this case CLIENT and LABEL_ID, and add an additional language key (LANGUAGE) with the LANG dictionary type. We deliberately add this in the second position directly after the client field CLIENT. This enables us to create language-dependent buffers later on.

We also add the LABEL_TEXT field, which contains the language-dependent identifier for a label. To enable the concurrency control, we add the LOCAL_LAST_CHANGED_AT timestamp field with the ABP_LOCINST_LASTCHANGE_TSTMPL type as in table ZACB_LABEL. We don't need to include the LAST_CHANGED_AT field, which is only required in the primary table.

Text table relationship by foreign key

To create the text table relationship, you need to use the with foreign key addition for a field definition in the CDS source code. We use this addition to set the LABEL_ID field in the text table in relation to the field with the same name in the primary table. The relevant section is shown in Listing 5.2.

```
key label_id : zacb_label_id not null
  with foreign key [0..*,1] zacb_label
    where label_id = zacb_labelt.label_id;
```

Listing 5.2 Definition of a Foreign Key Relationship with Dictionary DDL Syntax

The cardinality is specified as [0..*,1]. This must be read as [n,m], where n stands for the foreign key table (here, the ZACB_LABELT text table), and m stands for the related check table (here, ZACB_LABEL), which is specified after the square brackets. Possible values for n, the foreign key table, are listed in Table 5.3.

Cardinality of foreign keys

Value	Meaning
1	For each entry in the check table, there's exactly one corresponding entry in the foreign key table.
0..1	For each entry in the check table, there's a maximum of one corresponding entry in the foreign key table.
1..*	For each entry in the check table, there's at least one corresponding entry in the foreign key table.
0..*	For each entry in the check table, there are any number of corresponding entries in the foreign key table.

Table 5.3 Possible Cardinality Values in a Foreign Key Relationship for the Foreign Key Table

Possible values for m, the check table, are listed in Table 5.4.

Value	Meaning
1	For each entry in the foreign key table, there's exactly one corresponding entry in the check table.
0..1	For each entry in the foreign key table, there's a maximum of one corresponding entry in the check table.

Table 5.4 Possible Cardinality Values in a Foreign Key Relationship for the Check Table

In our example, *each* entry in the ZACB_LABELT text table has *exactly one* corresponding entry in the ZACB_LABEL check table. Conversely, *each* entry in the check table has *any number of* corresponding entries in the text table. This results in the cardinality of [0..*,1].

[!]

Cardinality Compared to Transaction SE11

If you're used to maintaining foreign keys using the ABAP Workbench or Transaction SE11, there's a pitfall here: The specification of the cardinality in the with foreign key addition is exactly inverted to the specification in Transaction SE11. To maintain the foreign key definition from the example there, you would make the settings from Figure 5.7.

Foreign key field type	○ Not Specified
	○ Non-key fields/candidates
	○ Key fields/candidates
	◉ Key fields of a text table
Cardinality: 1 : CN	

Figure 5.7 Semantic Properties of a Foreign Key Relationship in Transaction SE11

In text-based maintenance with dictionary DDL syntax, on the other hand, we use the same foreign key definition with the information from Listing 5.3.

```
@AbapCatalog.foreignKey.keyType : #TEXT_KEY
@AbapCatalog.foreignKey.screenCheck : false
key label_id : zacb_label_id not null
  with foreign key [0..*,1] zacb_label
    where label_id = zacb_labelt.label_id
```

Listing 5.3 Definition of the Foreign Key Relationship with Dictionary DDL Syntax

The cardinality specifications `1:CN` from Transaction SE11 and `[0..*,1]` from the ABAP development tools lead to the same result. In the course of implementing the dictionary DDL syntax, SAP decided to change the cardinality specification to the order expected by many: foreign key table to check table. In Transaction SE11, this was always the other way around. In addition, the notation of the individual details has been changed, which may take some getting used to.

Type of foreign key and validity checks

We also add the `@AbapCatalog.foreignKey.keyType` annotation and set its value to `#TEXT_KEY` to specify the type of foreign key. Using the `@AbapCatalog.foreignKey.screenCheck` annotation and the `#FALSE` value, we deactivate screen-based validity checks. Unfortunately, this is necessary because the generator of business configuration maintenance objects with the **Add Data Consistency Check** option enabled would otherwise display an error message as only foreign keys for tables outside the generated ABAP RESTful application programming model construct can be checked. Technically, this is due to the activated draft function.

[!]

Effects of Foreign Key Definitions

The functionality of foreign key relationships hasn't changed due to the text-based editor and the dictionary DDL syntax. This means that the definition of such relationships primarily fulfills a documentation purpose.

In addition, some of them are analyzed by tools, as was traditionally the case with maintenance views or lock objects. However, there's no integration at the database level or in ABAP SQL.

The finished definition of the database table is shown in Listing 5.4.


```
@EndUserText.label : 'Labels Texts'
@AbapCatalog.enhancement.category : #NOT_EXTENSIBLE
@AbapCatalog.tableCategory : #TRANSPARENT
@AbapCatalog.deliveryClass : #C
@AbapCatalog.dataMaintenance : #ALLOWED
define table zacb_labelt {
  key client             : abap.clnt not null;
  key langu              : abap.lang not null;
  @AbapCatalog.foreignKey.keyType : #TEXT_KEY
  @AbapCatalog.foreignKey.screenCheck : false
  key label_id           : zacb_label_id not null
    with foreign key [0..*,1] zacb_label
      where client = zacb_labelt.client
        and label_id = zacb_labelt.label_id;
  label_text             : zacb_label_text;
  local_last_changed_at : abp_locinst_lastchange_tstmpl;
}
```

Listing 5.4 Definition of the ZACB_LABELT Text Table

Technical settings for text tables

We also make the technical settings. ZACB_LABELT is also a table of data class APPL2, and the size category 0 fits as well. Activate change logging and also buffering. As this is a text table, we opt for generic buffering with two key fields. This results in language-dependent buffering. If a read access is made to the database table, all entries with a suitable combination of clients and language are loaded into the table buffer.

To be able to use this function, the key field for the language must be entered directly after the client. You'll now also find an option for the translation process in the technical settings of the table. This should already be automatically set to the correct **Translation using standard translation procedure** value for the standard translation procedure. The settings made are shown in Figure 5.8. Finally, you need to activate the table.

In this section, we've created two Customizing tables using the ABAP development tools. We used the text-based editor with dictionary DDL syntax for this purpose. The result is the same as with the classic creation of Customizing tables via the ABAP Workbench. In the following section, we can now continue with the creation of the business configuration maintenance object.

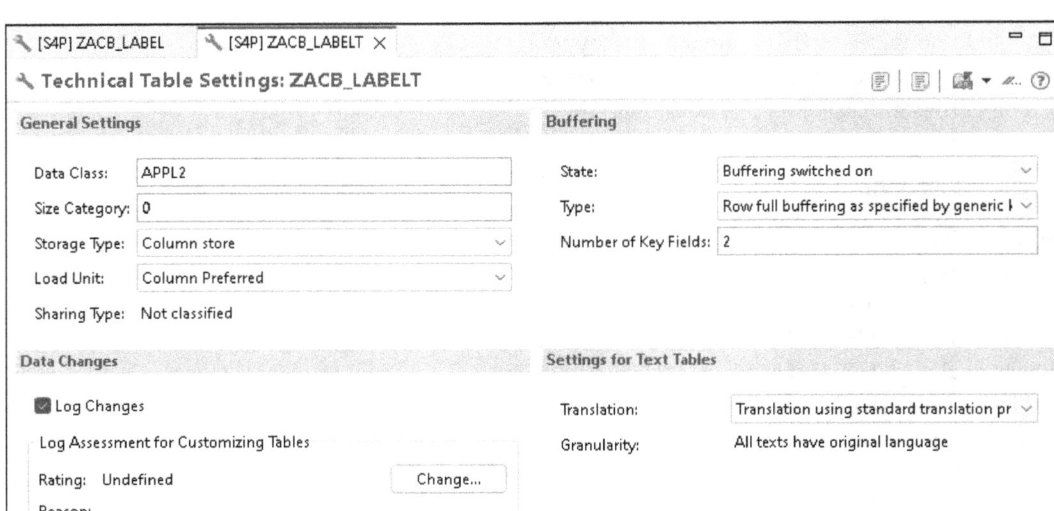

Figure 5.8 Technical Settings of Table ZACB_LABELT

5.3 Generating the Business Configuration App

The ZACB_LABEL and ZACB_LABELT Customizing tables were created in the previous section. Based on this, we now want to create the objects required to provide the maintenance interface. Theoretically, you could create these manually. In real life, however, these objects must follow a strict structure and even contain sections of code that you probably can't type from your memory. For this reason, the use of the generator is more or less mandatory.

Generating a business configuration app

The ABAP repository object generator for maintenance objects generates an application implemented with the ABAP RESTful application programming model, including the SAP Fiori interface, for you, which automatically integrates into the Custom Business Configurations app. Various development objects are generated by the system for this purpose, which you can then modify as required. The generator uses an architecture for the ABAP RESTful application programming model objects that enables the editing of multiple root object instances in a table. For this purpose, a singleton object is created, which represents the central point of entry.

Creating the Underlying Objects

You should create the Customizing tables as completely as possible in advance and only use the generator at the end. There's no function available to "regenerate" the application again following changes to the tables.

With good preparatory work, as described in the previous section, you give the generator the option to do most of the work for you.

Starting the ABAP repository object generator

To start the generator, select the primary database table in the project explorer, and then click on the **Generate ABAP Repository Objects** item in the context menu, as shown in Figure 5.9. In our example, we start the generator for the ZACB_LABEL table.

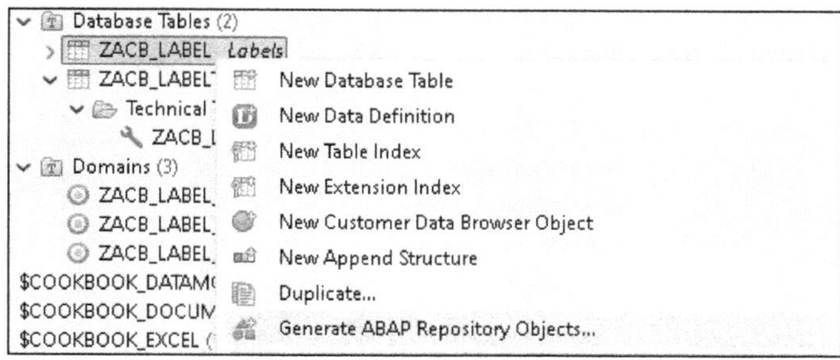

Figure 5.9 Generator in the Context Menu of the Project Explorer

In the dialog box that opens, you need to select the **Maintenance Object** generator in the **Business Configuration Management** folder and confirm your selection by clicking the **Next** button, as shown in Figure 5.10.

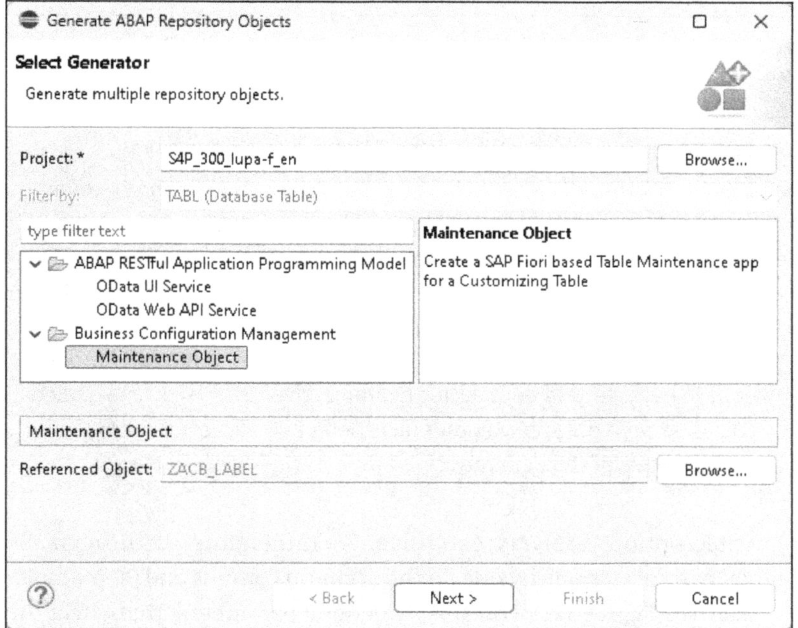

Figure 5.10 Generator Selection in the ABAP Development Tools

Generator parameters

In the next step, select the package in which the objects are to be created. After confirming again by clicking **Next**, you'll be asked for the parameters for the generation process. You can see the initial screen in Figure 5.11. The system proposes default values for the various parameters. In our example, the ZACB_LABELT text table was automatically recognized via the foreign key relationship.

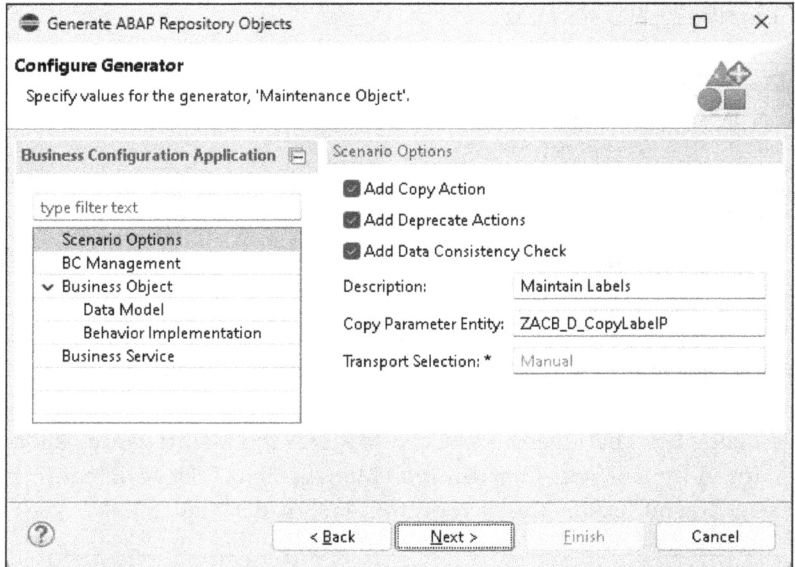

Figure 5.11 Parameterization in the Generator for Maintenance Objects

[»]

Observing Naming Conventions When Using the Generators

The parameter values suggested in the generation dialogs don't necessarily have to correspond to the development guidelines that apply to your specific case. They try to follow the naming convention for the ABAP RESTful application programming model as much as possible. However, this doesn't always work, and it may well happen that other conventions apply to you. For this reason, you should check the parameter values and modify them where necessary. In our specific example, the ACB prefix we intended for all development objects wouldn't have been taken into account in various default values of the generator.

The possible options are very extensive. We'll therefore go through the individual parameters one by one. In the **Scenario Options** and **BC Management** areas (see Figure 5.12), you specify general parameters that affect the basic generation. These include, in particular, the identifiers for the business configuration maintenance object and the transport object.

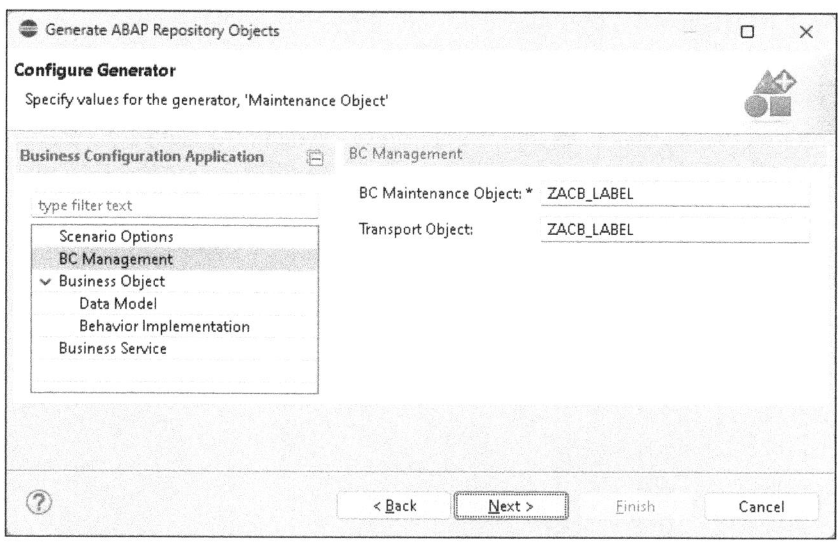

Figure 5.12 Scenario Options and BC Management Areas in the Generator Parameters

Transport object

It's not mandatory to specify a transport object. If no explicit specification is made, the Customizing settings are transported via the TABU object type. However, we recommend using your own transportation object. Such an object can be used, for example, to allow maintenance in production systems in the case of current settings. The identifier of the transport object created later doesn't correspond exactly to the parameter specified in the dialog; instead, a T is appended.

Functional scope of the business configuration app

The **Scenario Options** area provides additional options to extend the functional scope of the generated business configuration app:

- **Add Copy Action/Copy Parameter Entity**
  This option generates an action for duplicating existing entries and provides it in the app. Another entity is generated to display a popup window for specifying the key of the copy. The action itself is created as a factory action in the behavior definition of the singleton entity and used in the projection behavior definition.
- **Add Deprecate Action**
  This option can be used to manage the lifecycle of individual entries. This function is described in more detail in Section 5.7.
- **Add Data Consistency Check**
  This option can be used to generate consistency checks for fixed domain values and foreign key specifications. The foreign keys reference tables outside your own ABAP RESTful application programming model construct, such as units of measure or currency keys.

We activate all of the options mentioned for our example.

Parameters for the Customizing table

The parameters that affect the Customizing table (in our case, table ZACB_LABEL) are shown in Figure 5.13. As we've accessed the generator via the context menu of this table, its name is entered directly in the **Table Name** field. The value can't be changed. The generated ABAP RESTful application programming model construct is automatically draft enabled; that is, it supports the draft concept described in Chapter 2, Section 2.3.2. In the **Draft Table Name** field, you must specify the name of the draft table the generator will create. We entered "ZACB_LABEL_D" for this.

Customizing Table	
Table Name: *	ZACB_LABEL
Draft Table Name: *	ZACB_LABEL_D
Entity Name: *	Label
Entity Label: *	Label
Interface View: *	ZACB_I_Label
Projection View: *	ZACB_C_Label
ETag Master Field:	LOCAL_LAST_CHANGED_AT
Total ETag Field:	LAST_CHANGED_AT

Figure 5.13 Parameters of the Customizing Table

We adjust the other parameters in such a way that our prefix ACB is included in the respective identifiers for all objects in this book. In addition, we make sure that the objects are referred to in the singular and that the language-dependent texts are in English. The **ETag Master Field** was automatically set to the LOCAL_LAST_CHANGED_AT value. The LAST_CHANGED_AT value was also set in the **Total ETag Field**. The generator can automatically derive these fields from the data elements of the two fields in the table and preassign them correctly.

Parameters for the text table

Next, the parameters for the text table follow (see Figure 5.14). The generator has automatically determined the values shown here via the maintained foreign key relationship and preassigned the parameters. We adjust them on the basis of the same considerations as for the primary table.

[»]

Release Dependencies in the Generator

A newer version of the generator for business configuration maintenance objects is already available in the SAP BTP ABAP environment and in SAP S/4HANA Cloud Public Edition. This version allows you to specify more than one text table and one primary table.

Customizing Text Table

Text Table Name: *	ZACB_LABELT
Draft Table Name: *	ZACB_LABELT_D
Entity Name: *	LabelText
Entity Label: *	Label Text
Interface View: *	ZACB_I_LabelText
Projection View: *	ZACB_C_LabelText
ETag Master Field:	LOCAL_LAST_CHANGED_AT
Text Field: *	LABEL_TEXT

Figure 5.14 Parameters of the Text Table

Singleton entity

The information in the **Business Object • Data Model** section is concluded with some parameters for the singleton entity, which serves as a bracket for the objects of the business configuration app. The values we've selected here are shown in Figure 5.15.

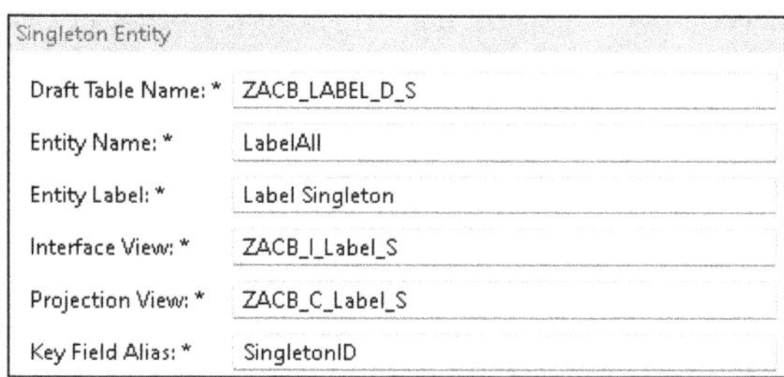
Singleton Entity

Draft Table Name: *	ZACB_LABEL_D_S
Entity Name: *	LabelAll
Entity Label: *	Label Singleton
Interface View: *	ZACB_I_Label_S
Projection View: *	ZACB_C_Label_S
Key Field Alias: *	SingletonID

Figure 5.15 Parameters of the Singleton Entity

Behavior implementation

The next section—**Business Object • Behavior Implementation**—contains the parameters for implementing the generated business object, that is, the business configuration app. There, you enter the names of the ABAP classes used for implementation, once for the behavior implementation in the **Implementation Class** field and once for the behavior implementation projection in the **Projection Implementation Class** field. You can see the class identifiers selected for our example in Figure 5.16.

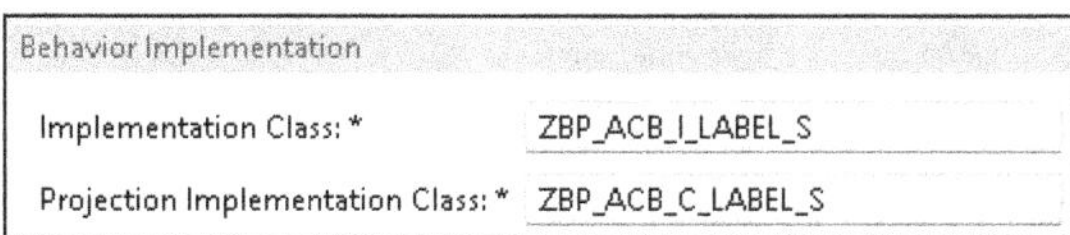
Behavior Implementation

Implementation Class: *	ZBP_ACB_I_LABEL_S
Projection Implementation Class: *	ZBP_ACB_C_LABEL_S

Figure 5.16 Parameters for Behavior Implementation in the Behavior Implementation Section

Business service

Finally, you must make the relevant settings in the **Business Service** area. For this purpose, you need to enter the names of the service definition and the service binding. As usual, we've slightly adapted the suggested names for our example (see Figure 5.17). The identifier of the service definition doesn't contain any special name components because the service definition should be reusable for different scenarios, such as APIs or user interfaces (UIs). The service binding, on the other hand, is specific to an OData V4–based UI service and therefore also contains the UI and 04 components in the identifier.

Figure 5.17 Parameters of the Business Service

Generation suggestion

You can now continue with the generation by clicking the **Next** button. Note that the objects can no longer be easily renamed once the generation has been completed. For this reason, you should check once again whether all identifiers have been entered correctly. You'll then receive a generation suggestion, as shown for our example in Figure 5.18.

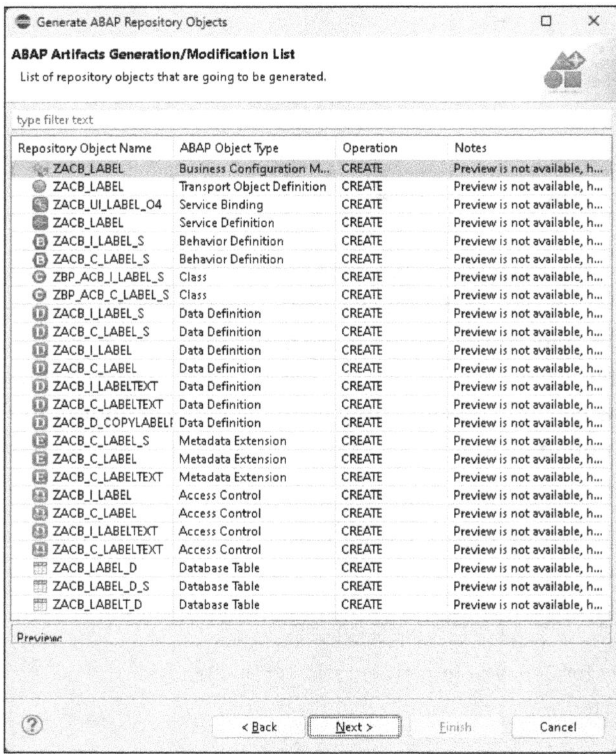

Figure 5.18 Generation Suggestion of the Generator for Business Configuration Maintenance Objects

The suggestion contains the various ABAP RESTful application programming model and CDS objects as well as the draft tables to provide the maintenance application, that is, the business configuration app. The output is also very similar to that of the regular generator of ABAP RESTful application programming model applications. Additional objects here are the business configuration maintenance object and the transport object, which are particularly relevant for the use case of maintaining Customizing tables.

Completion of the generation

After another click on the **Next** button, you need to enter a workbench transport with which the generated objects are to be transported and confirm the generation by clicking the **Finish** button. The system then creates the objects in the package selected at the beginning using the specified parameters. You'll then find the newly created objects in the project explorer and are then automatically directed to the business configuration maintenance object.

Before we look at the possible settings in Section 5.6, we'll first take a look at the provision and use of the generated application in the following paragraphs. To do this, the service binding must be published in the last step after the business configuration app has been generated.

Publishing the service binding

To publish the service binding, you must open it first. To do this, you can click on the **Service Binding** link in the automatically opened maintenance screen for the business configuration maintenance object. Alternatively, you can also find the service binding in the project explorer in the target package of the generation via the **Business Services • Service Bindings** path.

In our example, the binding is named ZACB_UI_LABEL_04. Click on the **Publish** button in the service binding maintenance. The value in the **Local Service Endpoint** field should then change from **Unpublished** to **Published**, and the endpoints of the service should be displayed under **Service Version Details**, as shown in Figure 5.19.

[«]

Publishing Service Bindings with the OData V4 Binding Type

Depending on your release and the system configuration, you may receive a generic error message in SAP S/4HANA or SAP S/4HANA Cloud Private Edition when publishing a service binding with the OData V4 binding type, as was done here by the generator. In this case, it's not possible to publish the service binding via the ABAP development tools. If that happens, use Transaction /IWFND/V4_ADMIN (see Chapter 2, Section 2.3.3).

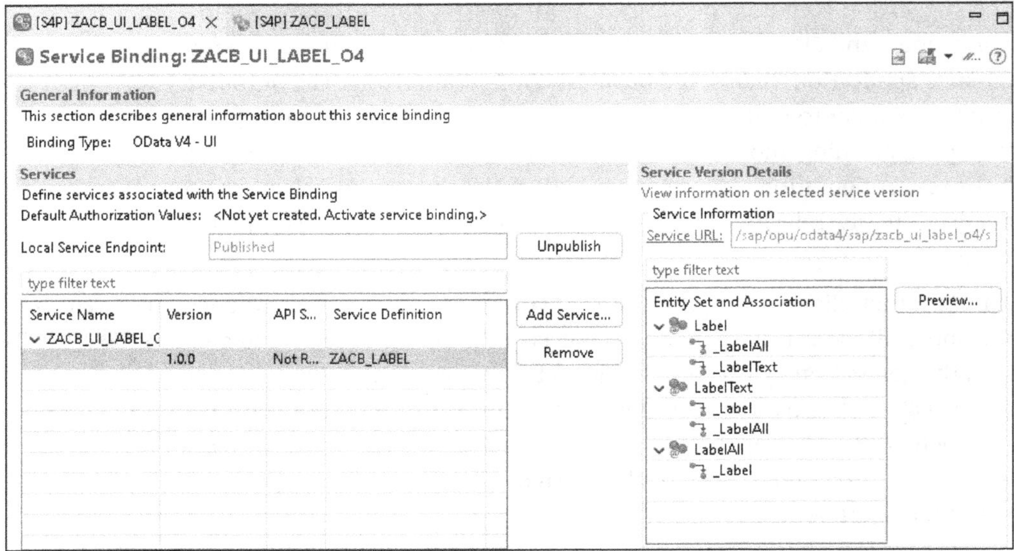

Figure 5.19 Maintenance of the Published Service Binding

Resetting the display and maintenance indicator

As described in Section 5.2, display and maintenance had to be permitted without restriction using the `@AbapCatalog.dataMaintenance` annotation and the `#ALLOWED` value to use the maintenance object generator. Because the generation of the app is now complete, you can now set the value back to `#RESTRICTED`.

We've now completed the minimum development steps required to maintain a Customizing table with an associated text table. In the following section, we'll discuss the assignment of authorizations so that the developed business configuration app can then be called.

5.4 Assigning Authorizations

Authorizations in on-premise solutions and in the cloud

To use the generated business configuration app, you or the users of the app require appropriate authorizations. At this point, there are some differences between SAP S/4HANA and SAP S/4HANA Cloud Private Edition on the one hand and SAP S/4HANA Cloud Public Edition and the SAP BTP ABAP environment on the other. In the public cloud systems, authorizations are assigned via Identity and Access Management (IAM), while on-premise solutions and the private cloud traditionally work with Transaction PFCG–based roles.

In this section, we describe the integration of the business configuration app into the SAP authorization concept from the perspective of the development team and for on-premise and private cloud systems.

Maintenance of default authorization values

First of all, it should be possible to assign authorizations for our business configuration app. To simplify this for those responsible for managing authorizations, you should maintain *default authorization values*. To do this, open Transaction SU24 and select the **Application types** item from the **Select By** dropdown list on the initial screen, and then the **SAP Gateway OData V4 Backend Service Group & Assignments** item under **Type of Application** (see Figure 5.20). In addition, you need to enter the technical identifier "ZACB_UI_LABEL_04" of our service binding in the **Object Name** field. Then, run the transaction.

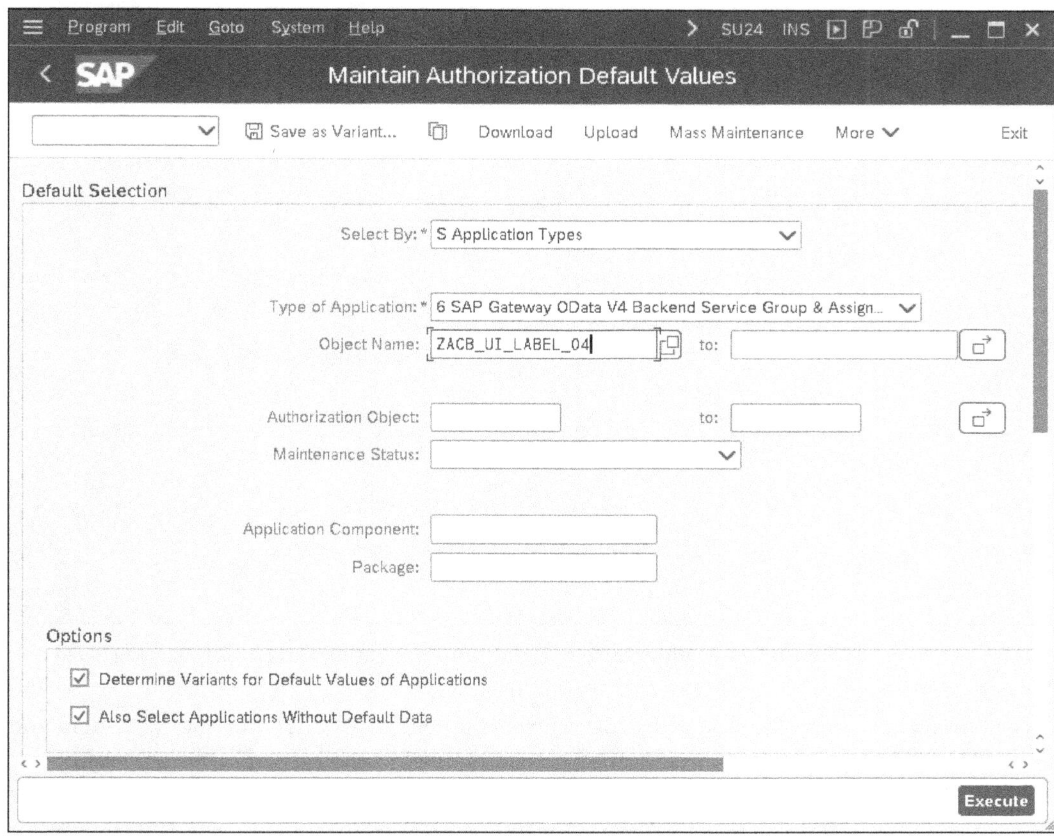

Figure 5.20 Selection Screen in Transaction SU24

You'll then find the selected service binding on the left-hand side of the screen and maintain the default authorization values on the right. There, you must add default values to the authorization objects used by our business configuration app. If you've implemented different authorization objects in the generated ABAP RESTful application programming model application, you should use those. Otherwise, the S_TABU_NAM authorization object will be used by default.

Switch to change mode, select **Object • Add Objects (Manually)**, and enter the authorization object. Confirm the dialog box and add the interface view of the primary table in the TABLE field in the default values of the added authorization object. In our example, this is interface view ZACB_I_LABEL. You also need to set the default values for the activity to 02 (Change), 03 (Display), and 08 (Display change documents). After saving, you should see the display shown in Figure 5.21.

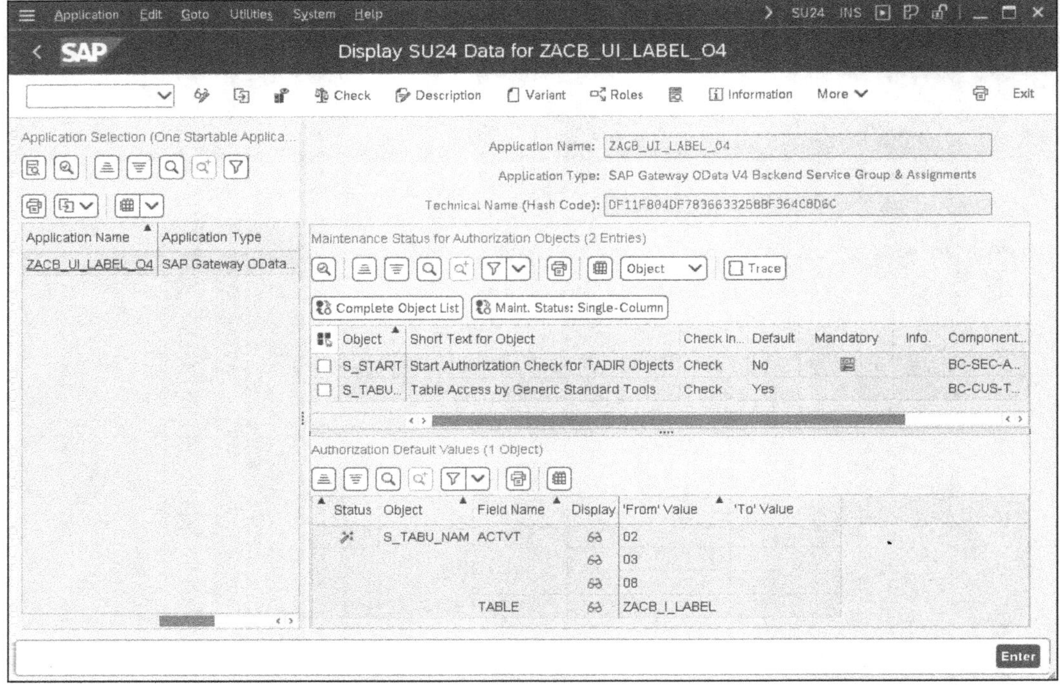

Figure 5.21 Maintenance of Default Authorization Data in Transaction SU24

Implementation details of the authorization check

You can find out which views you need to add here for the TABLE field and which authorization object is involved in the corresponding Data Control Language (DCL) sources. You can find these in your package under **Core Data Services • Access Controls**. In our example, only the DCL source for interface view ZACB_I_Label contains a pfcg_auth addition and therefore a reference to an authorization object; in this case, it's S_TABU_NAM. All other DCL sources use the principle of inheritance to adopt their authorization restrictions from interface view ZACB_I_Label, as you can see in Listing 5.5.

```
@MappingRole: true
define role ZACB_I_Label {
  grant select on ZACB_I_Label
  where ( ) = ASPECT pfcg_auth (
```

```
    'S_TABU_NAM', ACTVT = '03', TABLE = 'ZACB_I_LABEL' );
}
@MappingRole: true
define role ZACB_I_LabelText {
  grant select on ZACB_I_LabelText
  where INHERITING CONDITIONS FROM ENTITY ZACB_I_Label;
}
```

Listing 5.5 Authorization Restrictions in DCL Sources ZACB_I_Label and ZACB_I_LabelText

A second place where you can find the implementation of the authorization concept is the implementation class of the behavior definition. The get_global_authorizations method of the local handler class also checks the named authorization object, as shown in Listing 5.6.

```
CLASS lhc_zacb_i_label_s DEFINITION FINAL
  INHERITING FROM cl_abap_behavior_handler.
  PRIVATE SECTION.
    METHODS:
      ...
      get_global_authorizations FOR GLOBAL AUTHORIZATION
        IMPORTING
           REQUEST requested_authorizations FOR LabelAll
        RESULT result.
ENDCLASS.
CLASS lhc_zacb_i_label_s IMPLEMENTATION.
  METHOD get_global_authorizations.
    AUTHORITY-CHECK OBJECT 'S_TABU_NAM'
      ID 'TABLE' FIELD 'ZACB_I_LABEL'
      ID 'ACTVT' FIELD '02'.
    ...
  ENDMETHOD.
ENDCLASS.
```

Listing 5.6 get_global_authorizations Method in the Local Handler Class

These would therefore be the exact two places where you could implement different authorization logic.

Creating the role and associated objects

At this point, the responsibility of the development team often ends and the authorization administration takes over. When implementing an authorization concept, it often has to adhere to company-specific specifications and naming conventions. The subsequent steps are therefore not presented in detail here. The following tasks still need to be completed:

1. **Create or extend a role.**
   A new role must be created in Transaction PFCG or an existing role must be extended. This role references the service binding via a TADIR service of the **SAP Gateway OData V4 Backend Service Group & Assignments** type and adopts your maintained default values. If no other role assigns the relevant authorization, the role in question must also contain the authorization for the SAP Fiori–based apps for business configuration. This happens in the second step by adding a business catalog. The corresponding tiles should also be made available to users. It makes sense to include the delivered SAP_BASIS_SP_BPC space in the role.
2. **Create a business catalog.**
   In SAP S/4HANA 2023 FPS01, there's unfortunately no business catalog delivered by SAP for the purpose of business configuration. That is why a separate catalog must be created that refers to the technical catalog SAP_BASIS_TCR_T. The relevant apps can be found in the SAP_BASIS_SP_BPC space.
3. **Assign the role to the user.**
   Finally, the created role must be assigned to the users.

5.5 Configuration from a User's Perspective

Business configuration in the SAP Fiori launchpad

You can access your business configuration app with the authorizations described in Section 5.4. To do this, open the SAP Fiori launchpad of your system, and select the **Business Process Configuration** space. Alternatively, you can also access the apps contained in this space directly via the App Finder. The relevant space is shown in Figure 5.22.

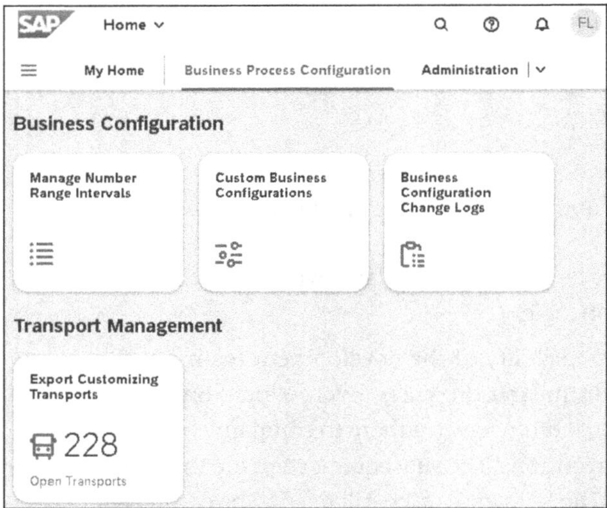

Figure 5.22 Business Process Configuration Space in the SAP Fiori Launchpad

In SAP S/4HANA 2023 FPS01, the space contains two sections:

Sections of the launchpad space

- **Business Configuration**
  This is primarily relevant to us and contains three tiles:
  - **Manage Number Range Intervals**: We can use this to define new intervals in number range objects or edit existing ones. Details on number range objects can be found in Chapter 9. The management of number range intervals is relevant in the context of business configuration in that you have to assign number range intervals in various Customizing settings.
  - **Custom Business Configurations**: This provides access to the individual business configuration apps. We'll deal with this app first in the following. You can find an alternative way to access this app in the SAP Reference IMG via Transaction SPRO or via SAP Fiori app Implementation Activities. Follow the **Custom Configuration • Custom Business Configurations** path.
  - **Business Configuration Change Logs**: You can use this to open the change logs and see who changed what in the settings and when.
- **Transport Management**
  This section contains the **Export Customizing Transports** tile. This tile makes it possible to create, edit and release Customizing transports and tasks via the SAP Fiori interface. To make these settings, you don't need to log on to SAP GUI or the ABAP development tools. The necessary activities, including transport creation and release, can all be carried out via the **Business Process Configuration** space in the SAP Fiori launchpad.

Then, open the Custom Business Configurations app.

[«]

Providing the Business Configuration via the Custom Business Configurations App

Because the business configuration app is an application that was implemented using the ABAP RESTful application programming model and SAP Fiori elements, you can theoretically also call it via the preview of the service binding in the ABAP development tools. Technically, however, this is neither necessary nor does it make much sense. The business configuration maintenance object automatically integrates the business configuration app into the Custom Business Configurations app and provides various configuration settings that you would otherwise only have found when calling SAP Business Application Studio. Your app is automatically available in the system and doesn't have to be deployed via SAP Business Application Studio.

Custom business configurations

After starting the app, you'll see the initial screen shown in Figure 5.23. You can select the desired business configuration here and go to the detail screen. If you can't directly find the entry you're looking for, you can display additional columns via the settings ⚙ or use the full-text search and filter options.

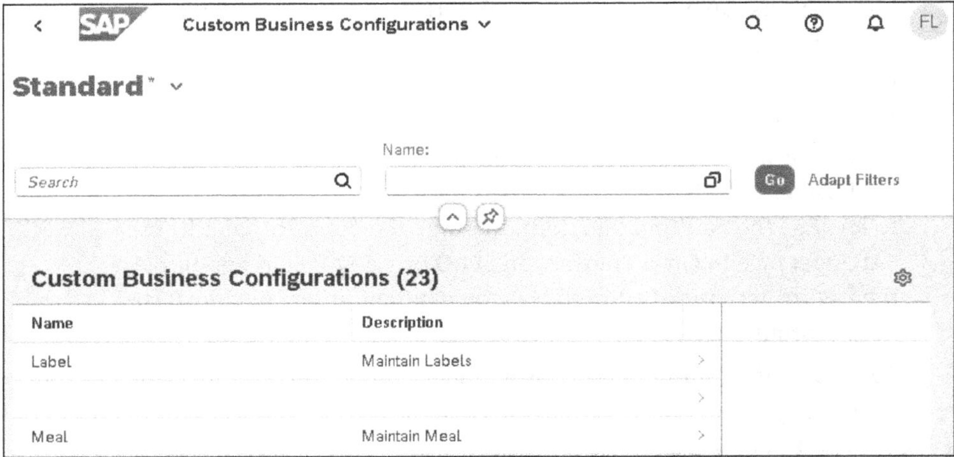

Figure 5.23 Initial Screen of the Custom Business Configurations App

Select our business configuration **Label** created in Section 5.3. This will redirect you to our business configuration app. It starts with a table containing a list of the labels, as shown in Figure 5.24. The floorplan list report from SAP Fiori elements is used as the layout for the initial screen.

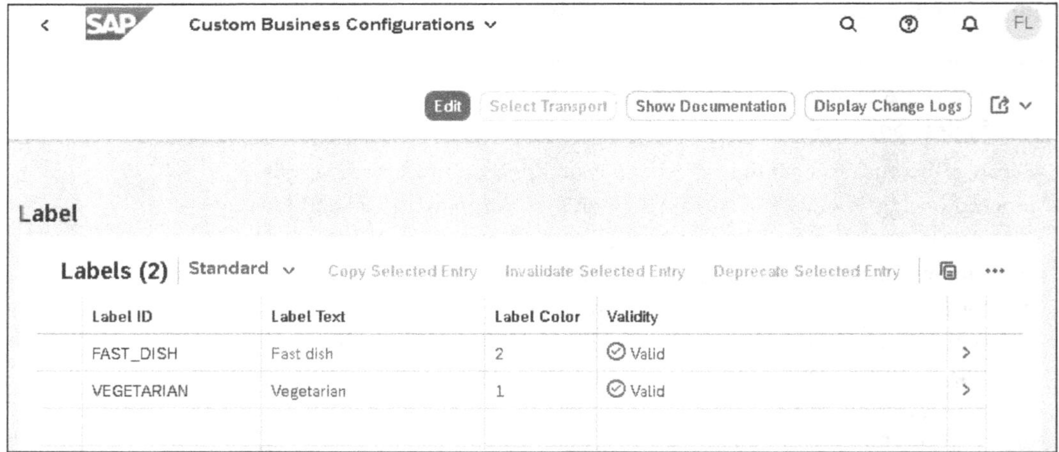

Figure 5.24 List Report of Our Business Configuration App

Two labels have already been maintained in Figure 5.24. These are displayed with their fields in a table. In the header toolbar above this table, you'll find the usual functions for switching to edit mode, marking the app as a favorite or sending a link to the app via email. The latter can be accessed via the share icon. What is new are the **Select Transport** and **Display Change Logs** functions, which we'll use immediately.

Creating a new data record

We now want to create another label called **Inexpensive**. To do this, you need to switch to edit mode by clicking the **Edit** button. The fields in the table are now ready for input, and you can add a new entry in the empty row. Then, confirm your entries by pressing the Enter key. The table then sorts itself automatically to take the new entry into account. This status is shown in Figure 5.25.

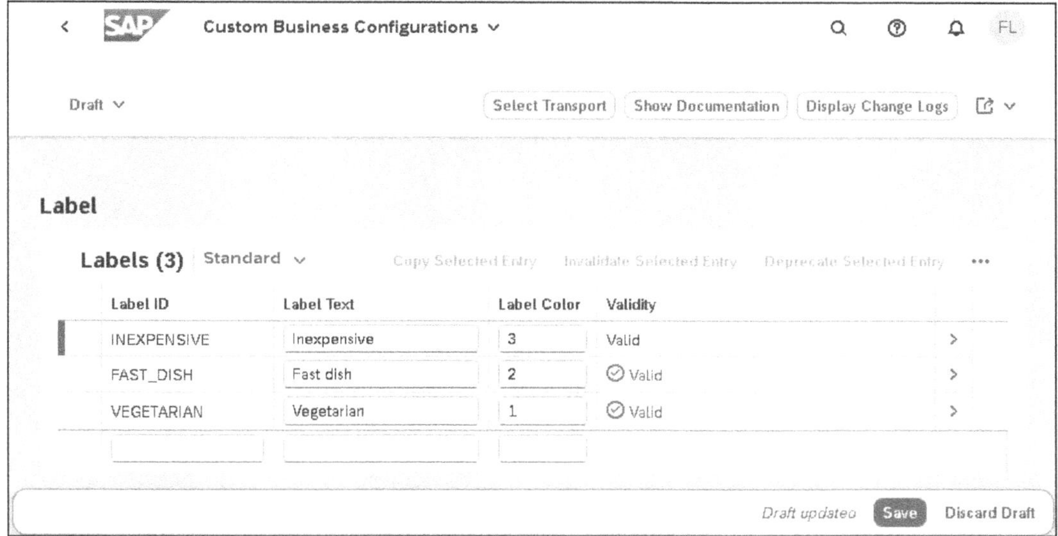

Figure 5.25 Change Mode in a Business Configuration App

At this point, you can also see that changes are automatically saved as a draft. The generated business configuration app is automatically *draft-enabled*; that is, it supports the draft concept of the ABAP RESTful application programming model.

Object page

In addition to the table overview, an object page is also available for maintaining individual records. You can use the arrow symbol next to an entry to navigate to this detailed view, which is shown in Figure 5.26. The object page also provides the function of maintaining the language-dependent texts, in this case, the label text, in languages other than the logon language.

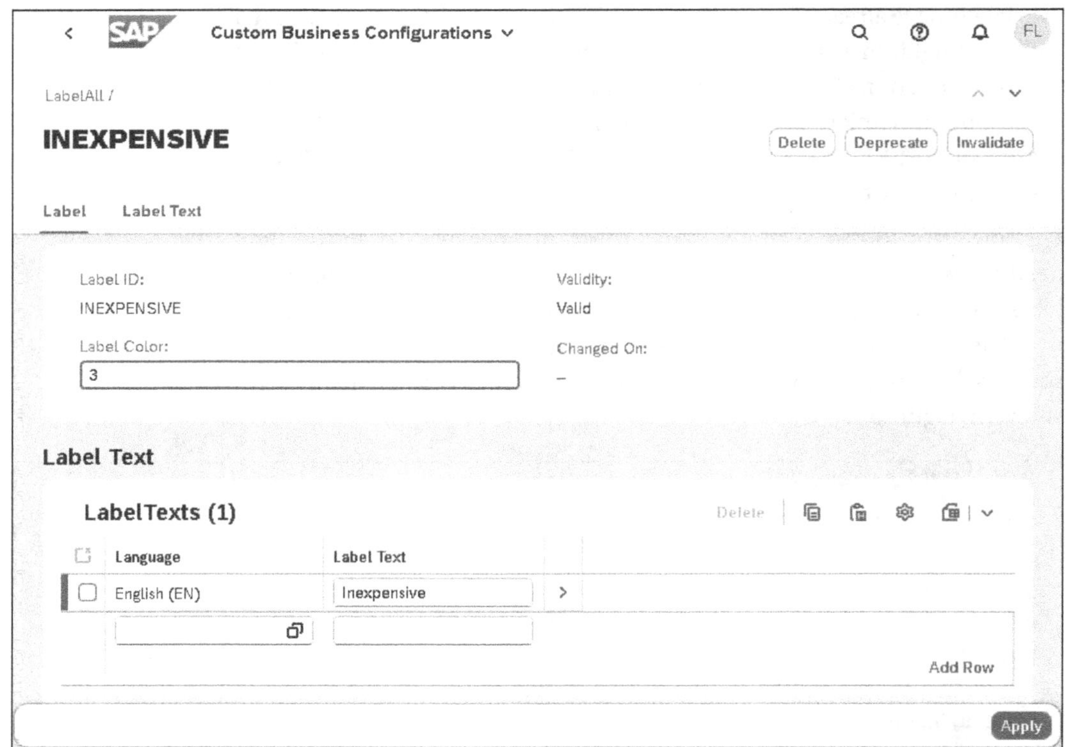

Figure 5.26 Object Page in a Business Configuration App

Transport link

To actively save the new entry, you must first select a Customizing transport. You can make this selection by clicking the **Select Transport** button. Note that you can only select existing Customizing transports in which you're involved. In contrast to maintenance with classic maintenance views, the Customizing transport must have been created previously. You can do this, for example, using the Export Customizing Transports app (app ID F5069), which is also provided in the **Business Process Configuration** space, and then select your transport in the popup window described here (see Figure 5.27).

Figure 5.27 Selecting the Customizing Transport Request

The selected transport request then gets displayed in the header area of the application. To save it, click on the **Save** button at the bottom right or use the keyboard shortcut Ctrl + S. The draft is then actively saved, the associated active database tables are updated, and the entries are removed from the draft tables. The entries are also transferred to the selected transport request so that they can be transported to subsequent systems. Finally, you'll see the initial screen of the app. You're back in display mode, and the table has been updated. Now you can release the Customizing transport via the Export Customizing Transports app.

[«]

Release Dependency of the Transport Selection

The exact process for selecting the Customizing transport in the business configuration app depends on the release. In the system used here, SAP S/4HANA 2023, the selection must be made via the **Select Transport** button before saving. Otherwise, you'll receive two error messages. In the public cloud environments, the process has already been further developed so that you're automatically redirected to the transport selection if you don't specify a transport.

Tracking changes

To see which settings were changed when and by whom, you can open the change log. At the top of your business configuration app, next to the transport selection and the function for switching between the display and change modes, you'll find the **Display Change Logs** button. After clicking on this button, the database tables that can be maintained in the current application will be displayed in a popup window. After selecting the table, you're redirected to the SAP Fiori app Business Configuration Change Logs (app ID F4327). The selected table is selected directly in this app. On the initial screen, you can see the number of available logs in a table (see Figure 5.28).

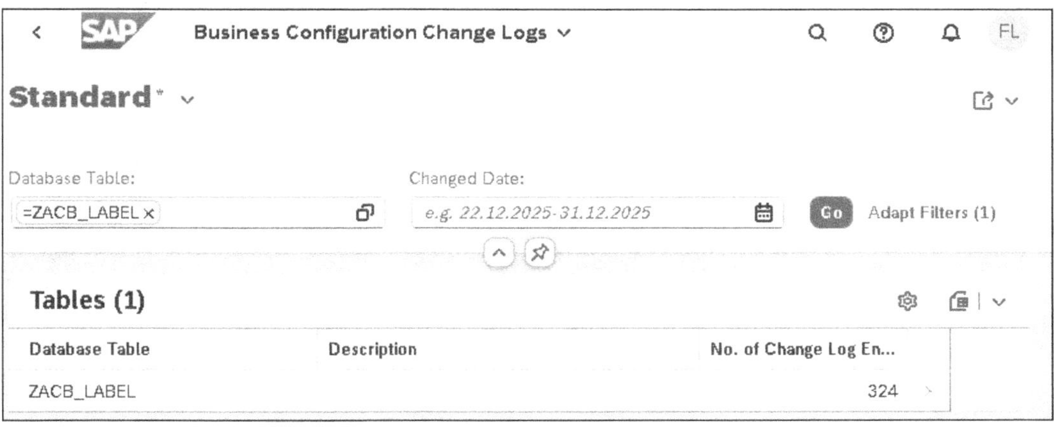

Figure 5.28 Initial Screen of the Change Log Display

Change log entries

After selecting the desired table, all changes to a table are displayed at the individual record and field level. For the text table of the labels, this change overview is shown in Figure 5.29.

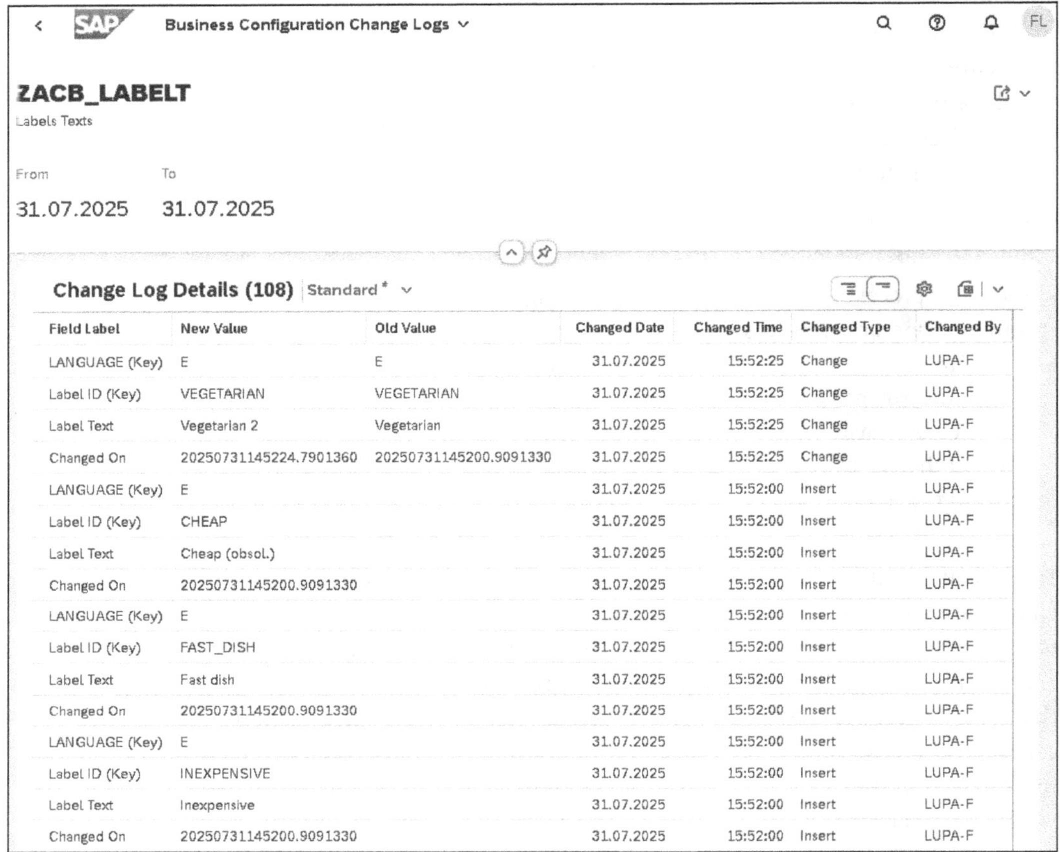

Field Label	New Value	Old Value	Changed Date	Changed Time	Changed Type	Changed By
LANGUAGE (Key)	E	E	31.07.2025	15:52:25	Change	LUPA-F
Label ID (Key)	VEGETARIAN	VEGETARIAN	31.07.2025	15:52:25	Change	LUPA-F
Label Text	Vegetarian 2	Vegetarian	31.07.2025	15:52:25	Change	LUPA-F
Changed On	20250731145224.7901360	20250731145200.9091330	31.07.2025	15:52:25	Change	LUPA-F
LANGUAGE (Key)	E		31.07.2025	15:52:00	Insert	LUPA-F
Label ID (Key)	CHEAP		31.07.2025	15:52:00	Insert	LUPA-F
Label Text	Cheap (obsol.)		31.07.2025	15:52:00	Insert	LUPA-F
Changed On	20250731145200.9091330		31.07.2025	15:52:00	Insert	LUPA-F
LANGUAGE (Key)	E		31.07.2025	15:52:00	Insert	LUPA-F
Label ID (Key)	FAST_DISH		31.07.2025	15:52:00	Insert	LUPA-F
Label Text	Fast dish		31.07.2025	15:52:00	Insert	LUPA-F
Changed On	20250731145200.9091330		31.07.2025	15:52:00	Insert	LUPA-F
LANGUAGE (Key)	E		31.07.2025	15:52:00	Insert	LUPA-F
Label ID (Key)	INEXPENSIVE		31.07.2025	15:52:00	Insert	LUPA-F
Label Text	Inexpensive		31.07.2025	15:52:00	Insert	LUPA-F
Changed On	20250731145200.9091330		31.07.2025	15:52:00	Insert	LUPA-F

Figure 5.29 Table Overview of Change Log Entries

In this specific case, you can see that the label text for the label with the technical name `VEGETARIAN` has been changed from **Vegetarian** to **Vegetarian 2**. Pay attention to the change type and the lines marked with **(Key)** in the field name. Groups can be identified in the table based on the change time.

As an alternative starting point, you can also open the Business Configuration Change Logs app directly and manually enter the database tables for which you want to view the logs. You can find this app in the **Business Process Configuration** space.

5.6 Settings in the Business Configuration Maintenance Object

In the preceding sections, we first created the Customizing tables with the associated data elements and domains. Various objects were then created in the system using the maintenance object generator to implement the business configuration app. We then assigned the authorizations for this app and finally tested it from a user's perspective.

Customizing the default settings

Up to this point, we've largely used the default settings everywhere, which have already returned an acceptable result. However, there are various places where these default settings can be modified. Many of these can be found in the maintenance dialog of the business configuration maintenance objects, which is shown in Figure 5.30. In this section, we'll discuss some of these settings and their effects.

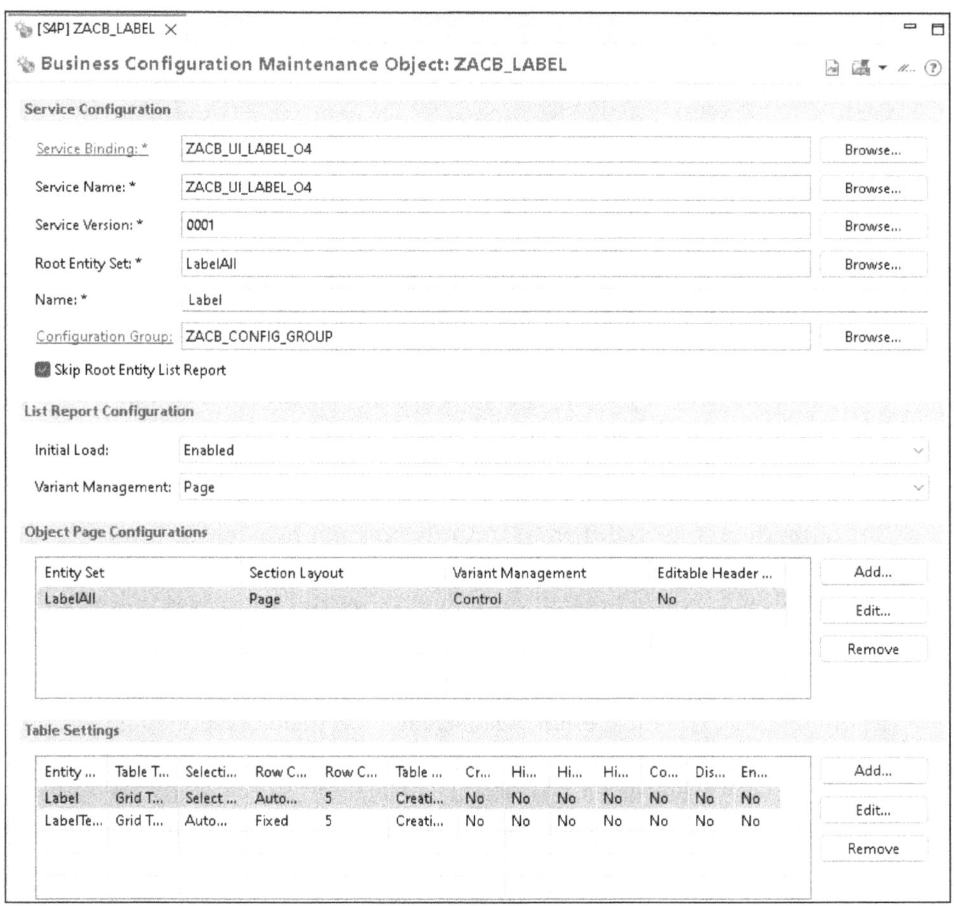

Figure 5.30 Maintenance Dialog of the Business Configuration Maintenance Object

Configuration groups

In the business configuration maintenance object settings, you'll find a **Configuration Group** field. You can use it to group related business configuration maintenance objects. This allows users to filter or sort according to their group.

In our example, we've defined and assigned a configuration group named **ABAP Cookbook**. The entries in the table for selecting the business configuration in the Custom Business Configurations app can thus be filtered according to this group, as shown in Figure 5.31.

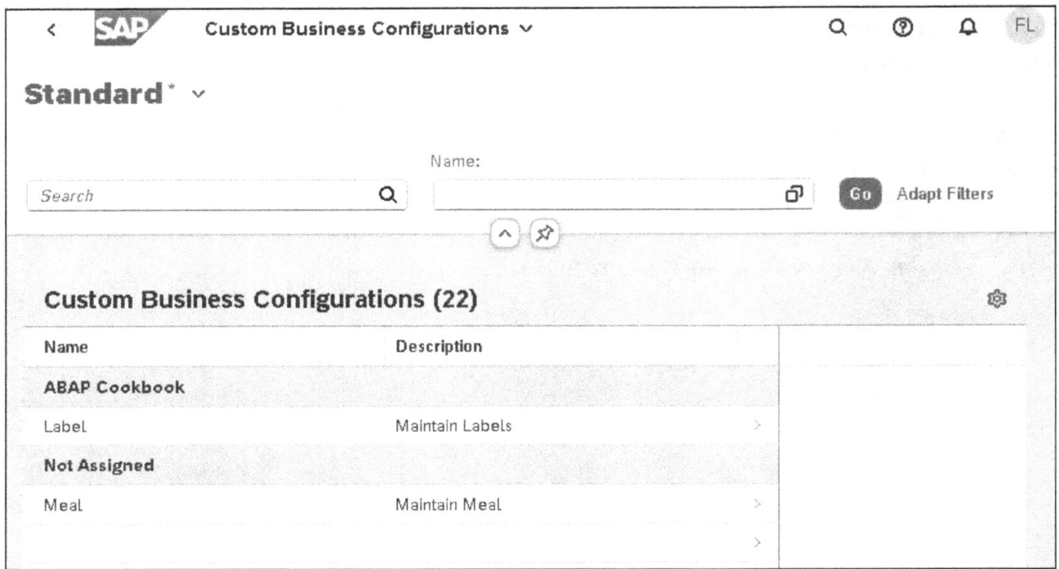

Figure 5.31 Grouping by Configuration Group in the Custom Business Configurations App

Adapting filters

You can add the configuration group to the filters using the **Customize filter** button. Note that there's currently no input help for the configuration groups. The **Group** tab can be opened via the settings ⚙ for the table. You can include the configuration group above this to display the table, as shown in Figure 5.31.

Assignment via data elements

To assign your business configuration maintenance object to a configuration group, you must enter a data element in the corresponding field. The data element object type is somewhat misused here to define an assignment with translatable text. The data element can then have the built-in CHAR dictionary type of length 1, for example, as it's technically never used for typing.

The four different texts aren't used either. Surprisingly, the text presented to users for the configuration group is the data element description from the **Properties** view.

You shouldn't reuse standard data elements, but create your own for the configuration groups. These should then be located in the same software component as the associated business configuration maintenance objects so that they can be used in ABAP Cloud without the need for a prior release.

Other settings options

In the other settings of the business configuration maintenance object, you can modify the page layout, the object pages, and the table structure as well as the buttons provided and the behavior when entries are added. These are largely options for the SAP Fiori interface that would be set up during the development of normal ABAP RESTful application programming model applications for application data in SAP Business Application Studio. If a desired function isn't available, you may therefore be able to activate it via SAP Fiori elements annotations in the metadata extensions or the projection views of the generated ABAP RESTful application programming model construct.

5.7 Lifecycle Management with Deprecation

Deprecation

The ABAP repository object generator for business configuration maintenance objects supports the option to generate actions for the *deprecation* of entries, that is, for marking them as obsolete. This involves the lifecycle management of Customizing settings. Obsolete settings can be marked as such.

The concept provides for a three-stage gradation of validity via the values listed in Table 5.5.

Value	Validity
(Blank character)	Valid
W	Obsolete (deprecated)
E	No longer valid (invalid)

Table 5.5 Values of the CONFIG_DEPRECATION_CODE Domain

A valid entry is available in all applications. Entries that are no longer valid generate a warning during foreign key checks. Obsolete entries generate an error message during foreign key checks and can then no longer be used.

If you use the deprecation function, the delete function will automatically be removed. From a technical point of view, this will then only be available as an internal action in the ABAP RESTful application programming model entity. Users aren't expected to delete entries, but to mark them as invalid or obsolete. You can easily memorize the technical values via the system behavior. `E` stands for error and `W` for warning.

Deprecation function in the app

In your business configuration app, the deprecation function is implemented in the new **Deprecate** and **Invalidate** buttons in the list report and on the object page. If you click on the **Deprecate** button, for example, you'll receive a message informing you of the effects (see Figure 5.32).

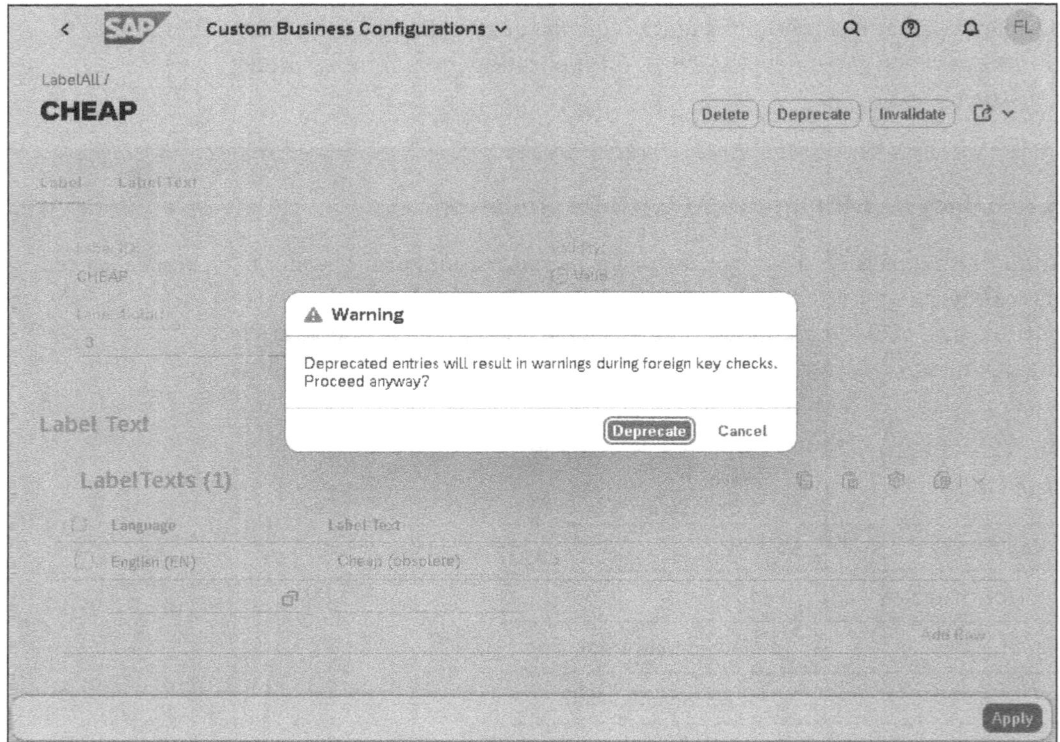

Figure 5.32 Warning Displayed When the Obsolete Status Is Set

Validity display

In the overview table, you can easily see the validity status, including a corresponding icon (see Figure 5.33).

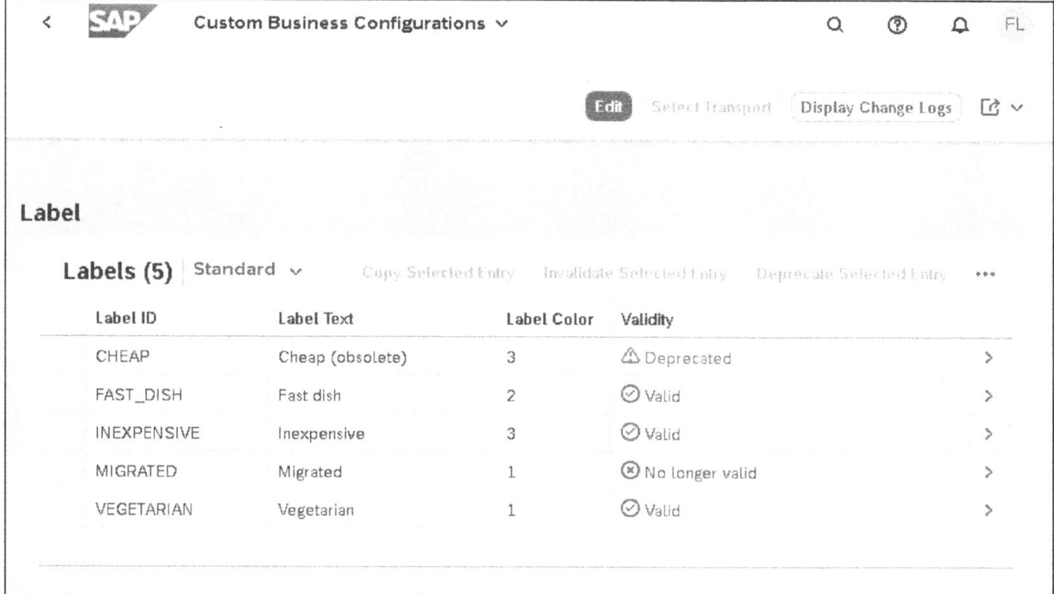

Figure 5.33 Validity Status in the Overview Table of a Business Configuration App

Activating the deprecation function

To activate the described function in the business configuration app, we added the CONFIGDEPRECATIONCODE field with the CONFIG_DEPRECATION_CODE data element as a data type in table ZACB_LABEL in Section 5.2. This field must have the exact name specified and use the described data element. In Section 5.3, we then selected the **Add Deprecate Actions** option. The generator took care of everything else.

Activating the Delete Function When Deprecation Is Enabled

The ABAP repository object generator for business configuration maintenance objects has deactivated the delete function on the user side because we've enabled the **Add Deprecate Actions** option. However, the generator is only used to create the objects and the underlying code as a template. If you want to use deprecation and deletion at the same time, you can modify the behavior manually. To do this, you need to remove the internal addition from the default delete operation in the behavior definition of the root entity and add use delete; in the projection behavior definition. These customizations are shown in Listing 5.7.

```
define behavior for ZACB_I_Label alias Label
...
{
  ...
```

```
  update( features : global );
// internal delete;
  delete;
  ...
}
define behavior for ZACB_C_Label alias Label
{
  ...
  use update( augment );
  use delete;
  ...
}
```

Listing 5.7 Customization of the Behavior Definition and Projection Behavior Definition to Use the Default Delete Operation When Deprecation Is Enabled

5.8 Documenting Business Configuration Maintenance Objects

Depending on the complexity of the setting options provided in your business configuration app, it may make sense to provide some documentation. If possible, app users should be able to find the documentation directly in the app instead of having to search for it on other platforms. You can provide such documentation in the app using *knowledge transfer documents* with Markdown syntax.

Creating the knowledge transfer document

Knowledge transfer documents are discussed in detail in Chapter 15. We'll therefore only briefly summarize here how these documents behave in a business configuration app. The first step is to create a new knowledge transfer document with reference to the business configuration object:

1. Right-click on the business configuration maintenance object in the project explorer, and select the **New Knowledge Transfer Document** option from the context menu (see Figure 5.34).

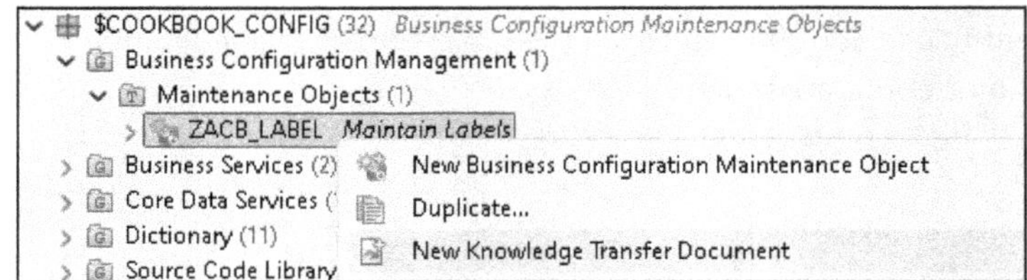

Figure 5.34 Creating a Knowledge Transfer Document with Reference to the Business Configuration Maintenance Object

2. Assign an identifier. This can correspond to the identifier of the object that is being documented. We've therefore decided to use the identifier ZACB_LABEL.
3. The package is determined automatically and can't differ from that of the documented object.
4. Finally, you want to enter a transport request and confirm the creation of the knowledge transfer document by clicking the **Finish** button.

After creating the documentation, you'll be directed to the Markdown-based text editor where you can maintain the text of your documentation. The editor's syntax and range of functions are also presented in Chapter 15.

Display in the business configuration app

As soon as an active knowledge transfer document exists for a business maintenance configuration object of a business configuration app, an additional **Show Documentation** button will be displayed in the app. Users can use this button to call the documentation they have created. The documentation is displayed visually by the web browser in a popup window, as you can see in Figure 5.35.

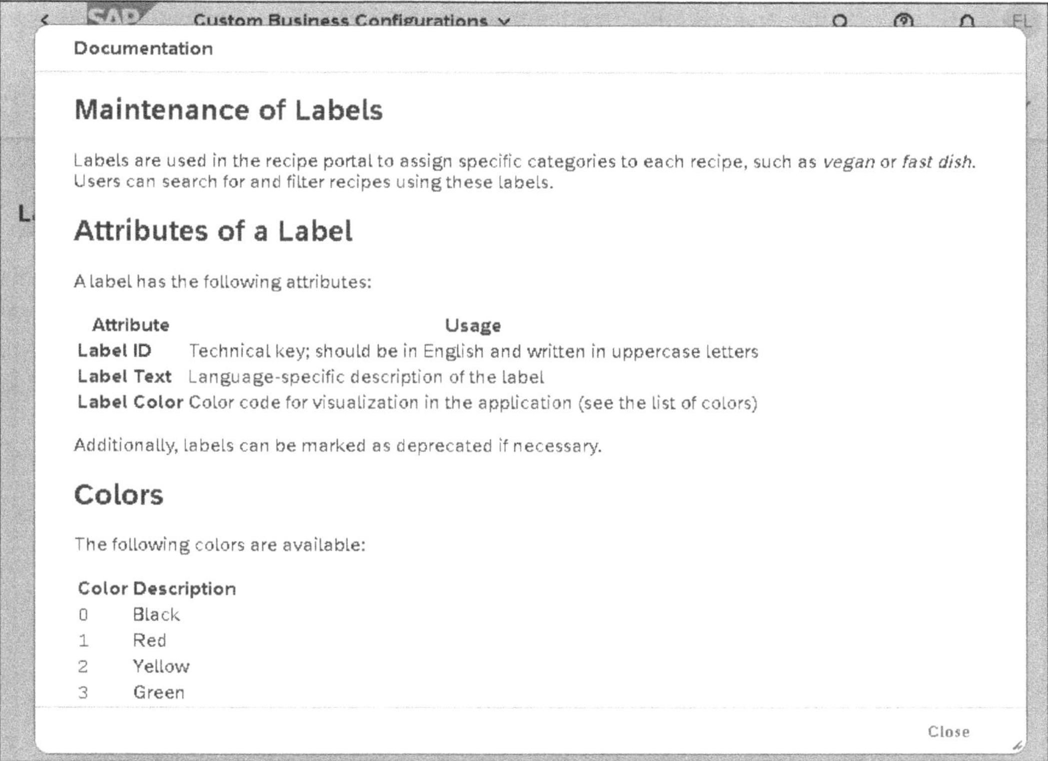

Figure 5.35 A Knowledge Transfer Document Displayed in a Business Configuration App

5.9 Summary

In this chapter, you've learned about the business configuration maintenance object as a way of providing maintenance dialogs for Customizing tables. For this purpose, we used an ABAP repository object generator and then modified the generated application according to our requirements. You've become familiar with some detailed functions such as deprecation and the provision of documentation.

Except for the automatic integration into the Custom Business Configurations app, the application created is largely a normal ABAP RESTful application programming model application. As such, it's very flexible and can be extended correspondingly. However, the generator already creates a more complex ABAP RESTful application programming model application with integration into the transportation system, validations, and additional actions. The generated implementation classes have several lines of boilerplate code, which is difficult for beginners to understand. In this context, it would be desirable to minimize the generated code and to be able to describe the majority of the functions declaratively via the behavior definition. In addition, at least one aspect of the generated application appears to be strongly geared toward public cloud environments: For SAP S/4HANA and SAP S/4HANA Cloud Private Edition, support for the `#RESTRICTED` value for the display and maintenance indicator by the generator would be very desirable. This is because the classic generic table maintenance tools could otherwise bypass the maintenance concept if you don't change the indicator again after generation.

There are also multiple other functions that couldn't be addressed in this chapter. For example, you can integrate the generated business configuration app in SAP S/4HANA and SAP S/4HANA Cloud Private Edition in the IMG via Transaction SPRO. Copying from or uploading Excel files is also available depending on the release.

Chapter 6
Application Logs

The application log for logging business process–relevant transactions has been given a new object-oriented API in ABAP Cloud. In this chapter, you'll learn how to use the relevant classes based on our sample application.

Application logs are a very popular solution for logging information about a specific program execution at runtime. The function provided for this in the standard SAP system is also named application log. The functional scope of the component is very extensive and has already been described frequently in other books on the subject.

Classic function modules

A disadvantage of application logs has always been the application programming interface (API), which is traditionally provided via function modules. These function modules with a BAL prefix, such as `BAL_LOG_CREATE` and `BAL_LOG_MSG_ADD`, can be used to create a log and add a log entry. This can then be persisted in the database using the `BAL_DB_SAVE` function module. For technical reasons, the calls of these function modules take up many lines of code during implementation and visually distract from the actual business logic due to their high line count alone. Many codebases therefore contain object-oriented wrappers around the old function modules to solve the problem using method calls and an object that represents the specific log. Such classes could even be found in various modules in the standard SAP system. Corresponding logging libraries are very popular in the open-source community.

The BALI API

ABAP Cloud provides a new object-oriented API in the standard SAP system. The objects of this API have the prefix `BALI`, which is why they are referred to collectively here as the *BALI API*. This chapter describes how you can use this new API.

Structure of this chapter

Section 6.1 provides a brief introduction to embedding the application log in our sample application. In Section 6.2, we establish the basis for log creation by creating application log objects. Section 6.3 provides an overview of the APIs provided. In Section 6.4, you'll get to know the techniques for adding log entries. Section 6.5 describes how you can also persist your generated logs. Finally, Section 6.6 shows you how the logs are displayed from a user's perspective.

6.1 Application Log for the Sample Application

Scenario: User data import

In this section, we describe step-by-step how you can implement logging when you import user data into the application in the context of our recipe portal. This involves the mass import of user master records, including first and last names, email addresses, and author and admin IDs using a text format. For this purpose, an HTTP service is provided in the SAP system, which receives the user data in CSV format in the payload of the incoming HTTP requests, converts the data into the internal format, and then persists it in the user database table. This is a typical scenario because logs are required in particular for long-running processes, in interface scenarios, or generally in background processing where no user is available for interaction and the actions performed must also be traceable at a later date. You can find a similar example, which also uses the application log but doesn't include it in terms of content, in Chapter 10, Section 10.2.

Architecture

The import component of the application already exists and is implemented very simply and directly in the HTTP handler to keep the example for this book as easy as possible. It would be cleaner to separate the HTTP level and the application logic via different objects.

Implementing the application

The starting point to the application is the ZCL_ACB_USER_IMPORT_HANDLER class, which implements the IF_HTTP_SERVICE_EXTENSION interface and is registered in the system via the ZACB_USER_IMPORT HTTP service. HTTP services are the ABAP Cloud–compliant successor concept to specially implemented SICF nodes.

Incoming requests are received in the handle_request method. Usually, a success text is returned in the body together with response code 201, Created. If an internal error occurs, only response code 500, Internal Server Error, will be returned instead. If an import error occurs due to missing or incorrect data, a response code 400, Bad Request, will be returned together with an error message.

handle_request method

Listing 6.1 shows the implementation of the handle_request method.

```
METHOD if_http_service_extension~handle_request.
  DATA(parameters) = request->get_form_fields( ).
  DATA(delimiter) = VALUE #(
    parameters[ name = 'delimiter' ]-value OPTIONAL ).
  DATA(drop_users) = VALUE #(
    parameters[ name = 'drop-users' ]-value OPTIONAL ).
  response->set_content_type(
    'text/plain; charset=utf-8' ).

  CASE request->get_content_type( ).
    WHEN 'application/csv'.
```

```
    IF delimiter IS INITIAL.
      response->set_status(
        if_web_http_status=>bad_request ).
      response->set_text(
        |Parameter DELIMITER (separator) was| &&
        | not supplied| ).
      RETURN.
    ENDIF.

    DATA(users) = convert_csv_to_users(
      csv       = request->get_text( )
      delimiter = delimiter ).

    IF users IS INITIAL.
      response->set_status(
        if_web_http_status=>bad_request ).
      response->set_text(
        |User data could not be converted| ).
      RETURN.
    ENDIF.

    IF to_lower( drop_users ) = 'true'.
      DELETE FROM zacb_user.
    ENDIF.

    MODIFY zacb_user FROM TABLE @users.
    IF sy-subrc <> 0.
      ROLLBACK WORK.
      response->set_status(
        if_web_http_status=>internal_server_error ).
      RETURN.
    ENDIF.

    DATA(modified_entries) = sy-dbcnt.
    response->set_status( if_web_http_status=>created ).
    response->set_text(
      |{ modified_entries NUMBER = USER } | &&
      |users were successfully adjusted| ).

    COMMIT WORK.

  WHEN OTHERS.
    response->set_status(
      if_web_http_status=>bad_request ).
    response->set_text(
```

```
        |Content Type not supported| ).
      RETURN.
  ENDCASE.
ENDMETHOD.
```

Listing 6.1 Implementation of the handle_request Method

[!]

Simplified Implementation

In this chapter, an HTTP service is used as an example because, unlike ABAP console applications, it allows simple parameterization without requiring a complete ABAP RESTful application program model application. We've simplified the implementation of this HTTP service here for presentation purposes, so it doesn't meet the requirements of productive web services. For example, there's no Cross-Site Request Forgery (CSRF) token validation to protect against *cross-site request forgery*. An example of the implementation of such a validation is shown in Listing 6.2.

```
METHOD if_http_service_extension~handle_request.
  CHECK cl_http_service_utility=>handle_csrf(
    request = request
    response = response ).
  ...
ENDMETHOD.
```

Listing 6.2 CSRF Token Validation in HTTP Handler Classes

User view of the process

In our case, the user data import interface is used by an API caller. However, you can also imagine that before the web API layer, an SAP Fiori frontend was implemented, for example, via an Excel upload function, which is actually used by an end user. You can trace the call and response in the SAP Gateway Client, Transaction /IWFND/GW_CLIENT (see Figure 6.1).

Problem

The current implementation fulfills the core requirement of importing user data from a file. However, there's a lack of traceability for application support, administration, and users. It's currently not possible to trace when and by whom the API was called. It's not possible to find out in detail why an import of user data didn't work. Debugging must be used directly to analyze errors. These problems are now supposed to be solved by means of a logging function.

[»]

Using the Sample Application

The user data import application, which has been fully extended to include logging, can be found in the LOG subpackage of the code examples accompanying this book. You can call the ZACB_USER_IMPORT HTTP service yourself or start the ZCL_ACB_USER_IMPORT_DEMO class in the ABAP development tools for Eclipse with predefined values using the [F9] function key or the context menu item **Run As • ABAP Application (Console)** as a console application. You can use a previously set breakpoint in the if_http_service_extension~handle_request method of the ZCL_ACB_USER_IMPORT_HANDLER class to trace the individual extensions that will be implemented in the debugger in the following sections.

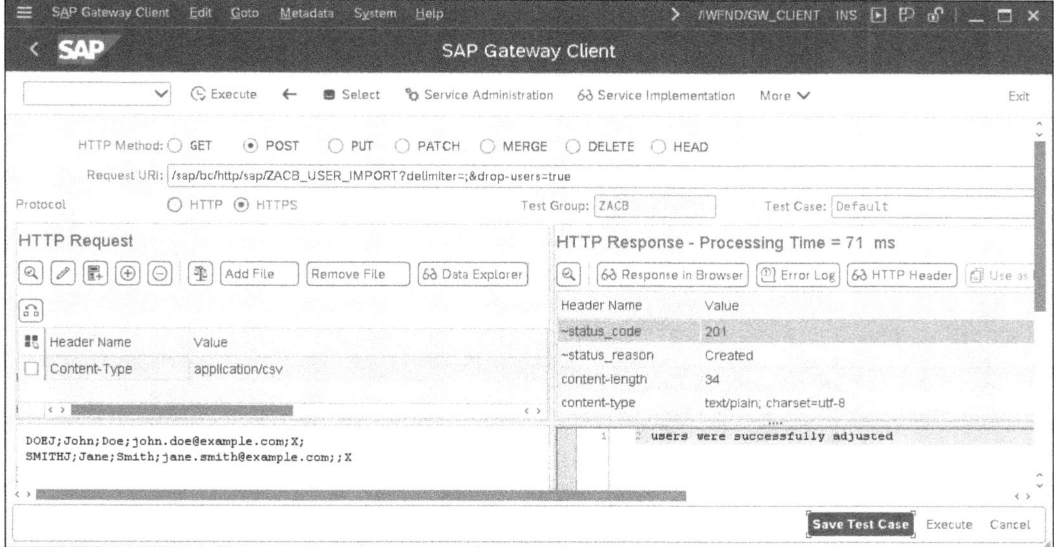

Figure 6.1 SAP Gateway Client with Request and Response of a User Data Import

6.2 Maintaining Application Log Objects and Subobjects

Repository object type APLO

To supplement logging in the application log, a storage option must first be created with which the logs to be generated are technically organized in the system. To do this, you can use application log objects. Optionally, you can also use application log subobjects to group the logs in a highly granular manner.

In ABAP Cloud, an option has been created to maintain these objects natively in the ABAP development tools. They are described using their own repository object type called APLO, which enables direct integration into the usual lifecycle of repository objects with object catalog entry, package assignment, versioning, and a workbench transport link. Application log objects and subobjects aren't new, but were maintained separately from the development environment in Transaction SLGO prior to ABAP Cloud.

The objects maintained in this transaction can't be used in ABAP Cloud. We therefore recommend that you use the new concept. Details of the differences can be found at the end of this section.

Creating an application log object

We want to create an application log object that logs the user data management of our recipe portal. The user data import subprocess is to be mapped via an application log subobject. Follow these steps to create the application log object:

1. In the context menu of our package, select the **New • Other • Repository Object** path.
2. In the popup window that appears, select **Others • Application Log Object** (see Figure 6.2). Confirm your selection by clicking the **Next** button.

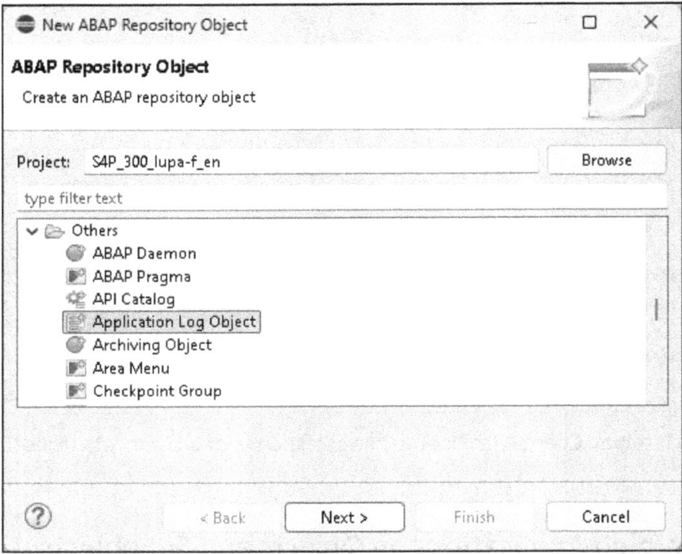

Figure 6.2 Object Type Selection "Application Log Object"

3. Assign a technical identifier and a description in the original language (see Figure 6.3). The selected package is already preassigned in the **Package** field.
4. Confirm this selection by clicking **Next**, and then select a transport request. Close the dialog by clicking the **Finish** button.

After completing the dialog, the maintenance screen for the new application log object opens (see Figure 6.4). You can find the description you've assigned in the **Properties** view in the ABAP development tools. Basically, the application log object could already be used in this way.

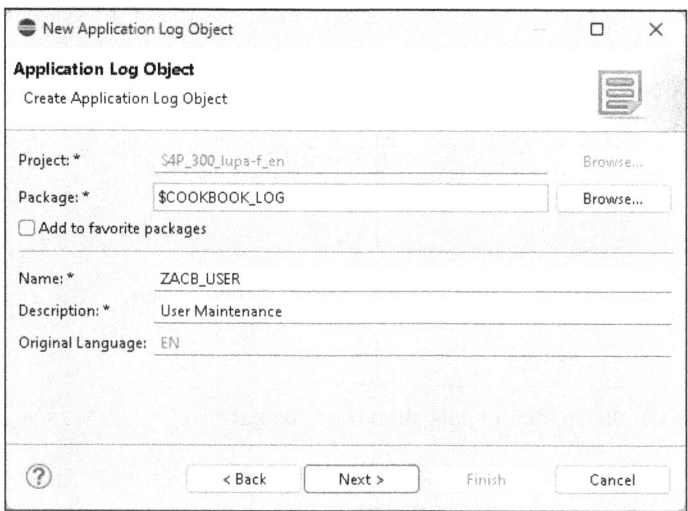

Figure 6.3 Creation Dialog for an Application Log Object

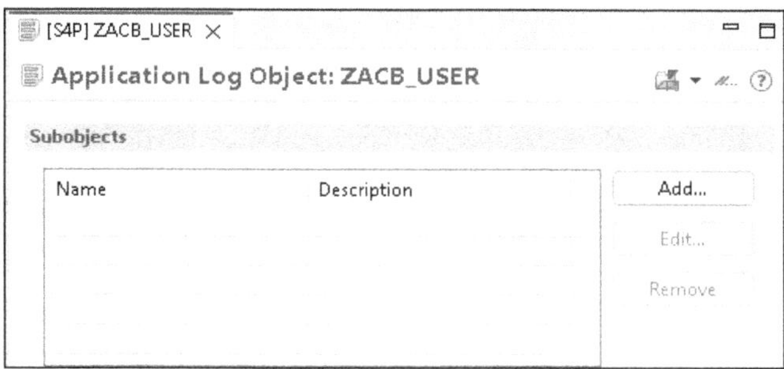

Figure 6.4 Maintenance Screen for an Application Log Object

Creating an application log subobject

We also want to create an application log subobject for the user data import to further refine the logs:

1. Click on the **Add** button in the maintenance dialog.
2. A dialog box opens for adding a subobject (see Figure 6.5).
3. Enter a name and a description in the original language, and confirm this by clicking the **Add** button. The name is specific to each application log object and doesn't require a namespace prefix.

   Now the subobject gets displayed in the **Subobjects** table (see Figure 6.6).
4. Save the change by clicking the **Save** icon at the top left, or use the Ctrl + S keyboard shortcut.

Application log objects don't need to be activated and can be used immediately after saving. You can use the **Edit** and **Remove** buttons to subsequently change or delete the subobject. Note that no relevant logs with reference to the object or subobject should have been created in the system or in subsequent systems if you subsequently change technical identifiers or delete (sub)objects.

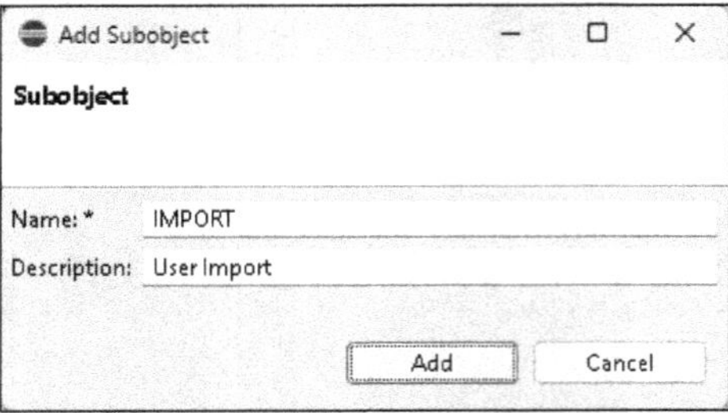

Figure 6.5 Creation Dialog for an Application Log Subobject

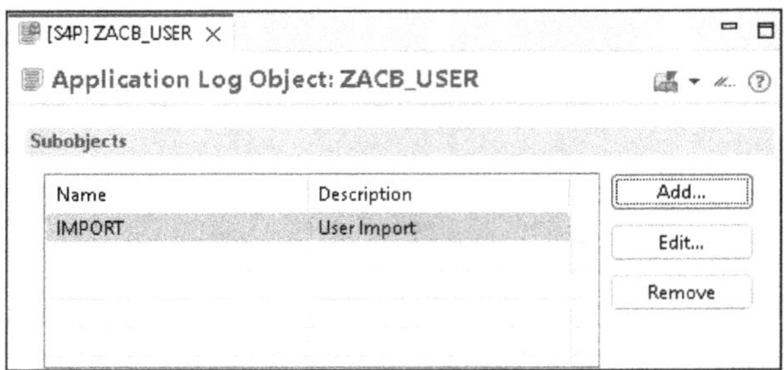

Figure 6.6 Maintenance Screen for an Application Log Object with an Application Log Subobject

Text-based maintenance

An alternative text-based editor is available in the ABAP development tools. You can access it via the **Open With • Source Editor** path in the context menu for the application log object in the project explorer. The object we've created in this editor is shown in Figure 6.7.

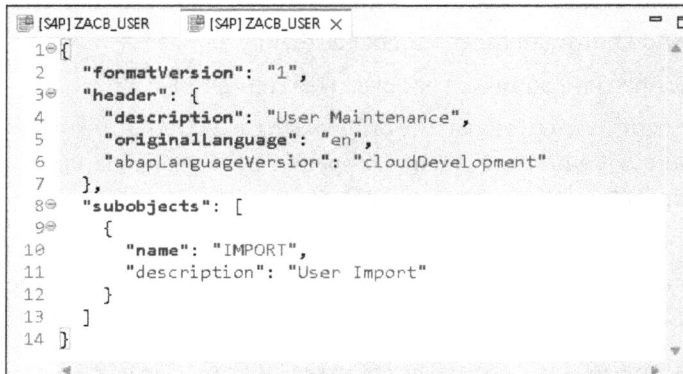

[S4P] ZACB_USER [S4P] ZACB_USER

```
{
  "formatVersion": "1",
  "header": {
    "description": "User Maintenance",
    "originalLanguage": "en",
    "abapLanguageVersion": "cloudDevelopment"
  },
  "subobjects": [
    {
      "name": "IMPORT",
      "description": "User Import"
    }
  ]
}
```

Figure 6.7 Maintenance of Application Log Objects and Subobjects Based on Source Code

You can only edit the subobjects in JavaScript Object Notation (JSON) format here. An auto-complete function is provided for the individual attributes such as `name` and `description` via the [Ctrl] + [space bar] keyboard shortcut.

Maintenance via Transaction SLGO

Traditionally, application log objects and subobjects were maintained via a maintenance view cluster that can be called in Transaction SLGO. However, this has always been subject to specific restrictions, and the following behavior had to be taken into account:

- The settings are treated as client-independent Customizing and are considered as such in the transport system.
- The objects don't have an original system and aren't a regular part of a software delivery.
- There's no technically tested namespace concept for the application log objects.

The new way of maintaining application log objects in the ABAP development tools using repository object type `APLO` shown in this chapter solves these problems. Objects created via Transaction SLGO can't be used in ABAP Cloud. Objects created via the ABAP development tools, on the other hand, can be used in all ABAP language versions. It's therefore highly recommended to switch to the new maintenance option if Transaction SLGO was previously used. Application log objects created in the ABAP development tools are displayed as read-only in Transaction SLGO, so that you have a complete overview there. You can migrate objects maintained there manually by deleting them in Transaction SLGO and creating them again with the same name in the ABAP development tools. Details can also be found in the long text of the initial message of the transaction.

6.3 The BALI API

Once we've created a new application log object, the first logs can be written. The new API for the application log is provided via objects whose identifiers contain the `BALI` abbreviation. In this section, we'll first provide an overview of these objects.

Design-time API

The API is divided into two areas, the design-time API and the runtime API. The application log objects and subobjects we created in Section 6.2 can use program-based maintenance via the components of the design-time API. Transportation is also taken into account. Here, design time means that the API supports the development process.

You can call the API by obtaining an instance using the `cl_bali_object_handler=>get_instance` method and then creating, changing, or deleting the objects using the methods of the `IF_BALI_OBJECT_HANDLER` interface. However, there are only a few use cases for program-based maintenance of these objects, which is why the design time API isn't discussed in detail here.

Runtime API The runtime API provides classes and interfaces that enable you to use application logs at runtime.

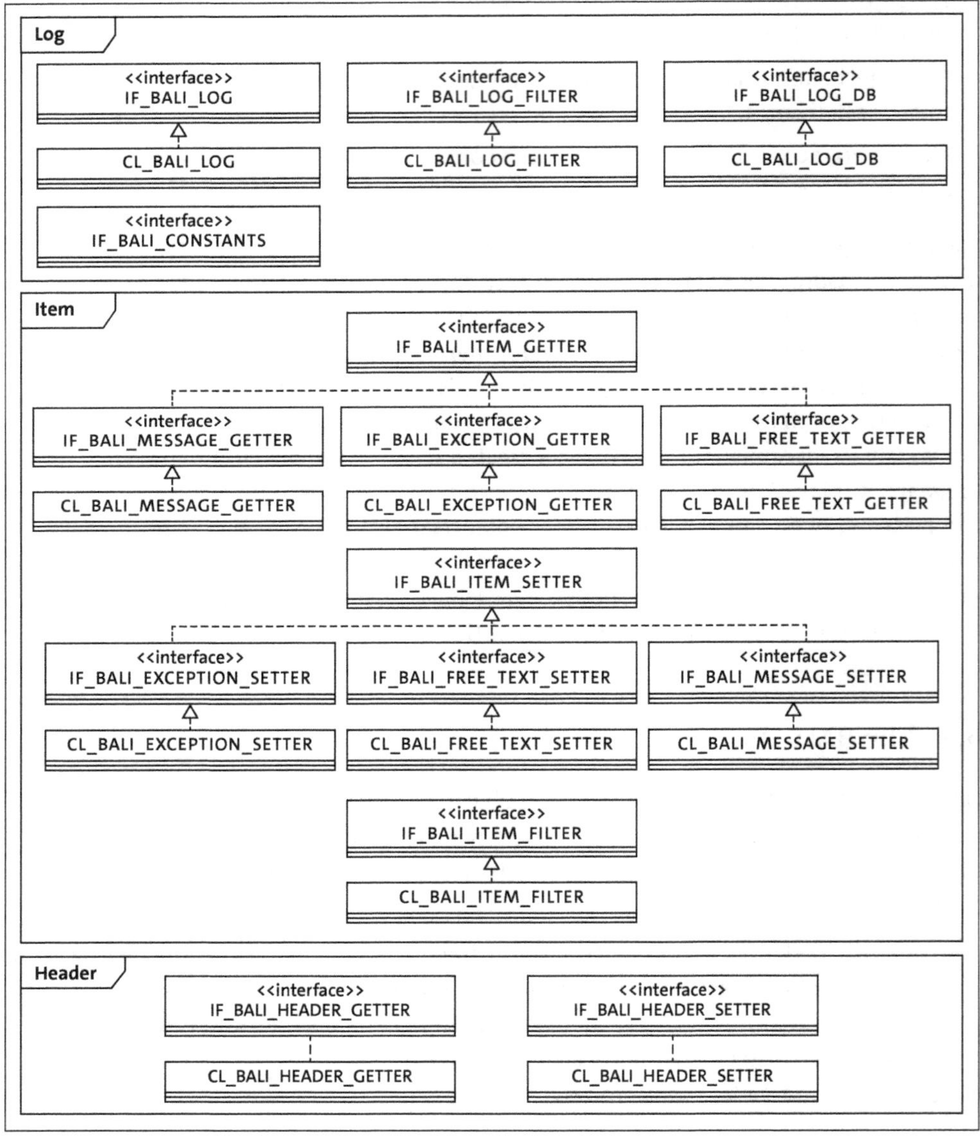

Figure 6.8 Overview of the Application Log Runtime API

The class diagram in Figure 6.8 provides a first overview of the relevant objects.

Roughly summarized, you can distinguish here between objects with the setter suffix, which sets values; objects with the getter suffix, which reads values; and filter options at the log entry and log header level. You can find out how to use the classes and interfaces in the following sections.

6.4 Creating a Log

A specific log is described via the IF_BALI_LOG interface. The instance of the class that implements the interface represents the log content at program runtime, which can then be persisted. You can obtain this instance via the static create and create_with_header methods of the CL_BALI_LOG class.

Integration into the user data import

In our user data import application, we add the call of the create_with_header factory method at an appropriate point. The responsibility for logging should lie high up in the call hierarchy to be able to decide at a central point what level of detail is appropriate and where and when logs are saved that have been enriched in the course of application execution. This aspect is comparable to transaction handling via COMMIT WORK and ROLLBACK WORK, which should also take place at a central point as high up in the call hierarchy as possible to avoid undesirable side effects.

Instantiating the application log

We therefore decide to create the instance of the application log directly at the beginning in the handle_request method. To do so, the method needs a little modifiying, as shown in Listing 6.3.

```
CLASS zcl_acb_user_import_handler DEFINITION
  PUBLIC FINAL
  CREATE PUBLIC.
  ...
  PRIVATE SECTION.
    DATA log TYPE REF TO if_bali_log. ❶
    ...
ENDCLASS.
METHOD if_http_service_extension~handle_request.
  TRY.
      log = cl_bali_log=>create_with_header(
        cl_bali_header_setter=>create(
          object      = 'ZACB_USER'
          subobject   = 'IMPORT' ) ). ❷
      DATA(parameters) = request->get_form_fields( ).
      ...
      COMMIT WORK.
```

```
    CATCH cx_bali_runtime INTO DATA(bali_error). ❸
      RAISE SHORTDUMP bali_error.
  ENDTRY.
ENDMETHOD.
```

Listing 6.3 Import Handler Class Extended by the Creation of a Log Instance

Changes made

In detail, the following adjustments were implemented in the import handler class ZCL_ACB_USER_IMPORT_HANDLER in Listing 6.3:

❶ **Addition of the log attribute with reference to the IF_BALI_LOG interface**
We've decided to include a log attribute for the log. This means that the log can also be used directly by other methods of the class without having to pass it on via parameters. Whether this option or the parameter transfer option is used is a design decision in each individual case.

❷ **Initialization of the log in the handle_request method**
The create_with_header method of the CL_BALI_LOG class is used to create a new instance and assign it to the log attribute. Because we also want to save the application log in the database within the application and not just keep it at runtime, the create_with_header method is used instead of the create method. The CL_BALI_HEADER_SETTER class provides the create method to supply the mandatory header parameter.

At this point, a reference is made to our application log object and subobject created in Section 6.2. For easier visibility, the ZACB_USER and IMPORT identifiers have been transferred as literals. In real life, constants would be appropriate at this point. Optionally, an external identification of the log can be specified via the external_id parameter, which can be used in the search, for example.

❸ **Handling the CX_BALI_RUNTIME exception**
The runtime APIs of the application log can trigger the CX_BALI_RUNTIME exception or its subclasses. Accordingly, this exception should be taken into account. The CX_BALI_RUNTIME class inherits from the CX_STATIC_CHECK class. Syntax warnings therefore appear if the exception hasn't been taken into account, and without a CATCH block or RAISING declaration to handle it, there's a risk of conversion to a CX_SY_NO_HANDLER exception with a subsequent runtime error. However, internal errors in logging aren't so easy to deal with. Where errors would normally be logged, there's now no log available in which to log them.

For our specific implementation in the application, we've made the decision that execution without a complete log isn't acceptable. Thus, in the event of an error, the RAISE SHORTDUMP command triggers a runtime error, including an implicit rollback of the logical unit of work (LUW).

Once you've made these three adjustments, the log is available via the log attribute in the class and can be used to generate log entries. Saving the log in the database hasn't yet been implemented here. This is described in Section 6.5.

[«]

Handling Internal Errors

Error handling is a complex topic that can only be touched on briefly here. In this chapter, the logging of errors is the primary approach of choice. In the case of internal errors, however, terminating the work process with a runtime error would also be a valid option.

Swallowing errors leads to subsequent problems and requires difficult subsequent root cause analyses. End users are generally unable to deal with internal error messages such as a database deadlock or technical issues when creating application logs. Deliberately triggering a short dump using RAISE SHORTDUMP can therefore also be a useful option in this context. It gives the system admin the chance to identify the error scenario as part of the monitoring process and take appropriate action.

6.4.1 Adding Free Text Messages

After creating the log object, you can now add log entries. The runtime API of the application log provides various methods and setter classes that you can use to write texts, messages, and exceptions to the log.

Adding log entries

The simplest type of log entries are free text messages. These are text-type variables or literals that you can add to the log. To create a new log entry, you need to call the add_item method of your created log object. This method is defined in the IF_BALI_LOG interface and accepts an *item*, that is, a log entry, in the form of an object that implements the IF_BALI_ITEM_SETTER interface. To create a log entry, you must use one of the setter classes provided. For free text messages, this is the CL_BALI_FREE_TEXT_SETTER class.

Adding free text messages

We're now going to expand the application to include the logging of free text messages. Start and stop messages are always useful to be able to track when the process was started and when it was properly ended. The following code is used to add such a start message:

```
log->add_item( cl_bali_free_text_setter=>create(
  'Starting user data import' ) ).
```

Note that the message text may be a maximum of 200 characters long. Anything beyond this will be truncated.

Severity of messages

In addition to the message text transferred as a literal, the `create` method of the `CL_BALI_FREE_TEXT_SETTER` class supports a further, optional parameter: `severity`. You can use this parameter to specify the *severity level* of the message. Without this information, the message will be interpreted as a *status message*.

From a technical perspective, the different severity levels are the classic message types of the `MESSAGE` statement. A status message corresponds to message type `S`. You can find suitable constants in the `IF_BALI_CONSTANTS` interface. They can each be recognized by the `c_severity` prefix. The constants available in the `IF_BALI_CONSTANTS` interface and the associated message types are listed in Table 6.1.

Constant	Message Type
C_SEVERITY_STATUS	S
C_SEVERITY_INFORMATION	I
C_SEVERITY_WARNING	W
C_SEVERITY_ERROR	E
C_SEVERITY_TERMINATION	A
C_SEVERITY_EXIT	X

Table 6.1 Constants for Severity Levels of Messages

Free text message in the event of an error

If we now add a free text message in case of an error, we override the default severity level `Status` with `Error`. We add a validation of the HTTP method and write an entry as an error in the log in the event of a violation (see Listing 6.4).

```
IF request->get_method( ) <> 'POST'.
  response->set_status(
    if_web_http_status=>bad_request ).
  log->add_item( cl_bali_free_text_setter=>create(
    severity = if_bali_constants=>c_severity_error
    text     = |Unsupported HTTP method | &&
               |{ request->get_method( ) } found| ) ).
ENDIF.
```

Listing 6.4 Free Text Message as Error

[!]

Avoiding Free Text Messages

Free text messages can be implemented with little effort. In the example of our error message, dynamic parts were also directly transferred to the message text via string templates, which is very convenient. Nevertheless, free text messages should only be used in exceptional cases. They can't be translated, can't be reused, and can't be found via the where-used list. The use of text elements doesn't help with translatability either because only the message is saved in the log in the logon language of the executing process. Anyone looking at the log then has to cope with this language and can't change it, even though translation work may have been done.

These are all problems that can be solved with little additional effort using message classes, which is the subject of the following section.

6.4.2 Adding Messages from Message Classes

Message classes are also the method of choice in ABAP Cloud for storing translatable, reusable texts for use in the application with the help of placeholders. In the development context, the native support in the class-based exception concept should be mentioned as well. The ABAP development tools provide a separate editor for maintaining message classes.

[«]

Restrictions for Message Classes in Cloud Systems

The maintenance of message long texts isn't yet supported in ABAP Cloud. In the SAP BTP ABAP environment and in SAP S/4HANA Cloud Public Edition, only existing long texts can be displayed. However, it's not possible to create new long texts or change existing ones. This is due to the SAP GUI dependency of long text maintenance.

Creating a message class

We create a new message class to store the messages of the user data management centrally and as translatable:

1. Right-click on the package in which you want to create the message classes, and select **New • Other Repository Object** from the context menu.
2. In the dialog box that opens, select the **Message Class** object type in the **Texts** folder (see Figure 6.9).
3. Confirm your selection by clicking the **Next** button and decide on a name and a description in the original language. The package is already preassigned. We enter "ZACB_USER" as the **Name** and "User Maintenance" as the **Description** (see Figure 6.10).
4. Confirm this again by clicking the **Next** button. Finally, select a transport request, and confirm this by clicking the **Finish** button.

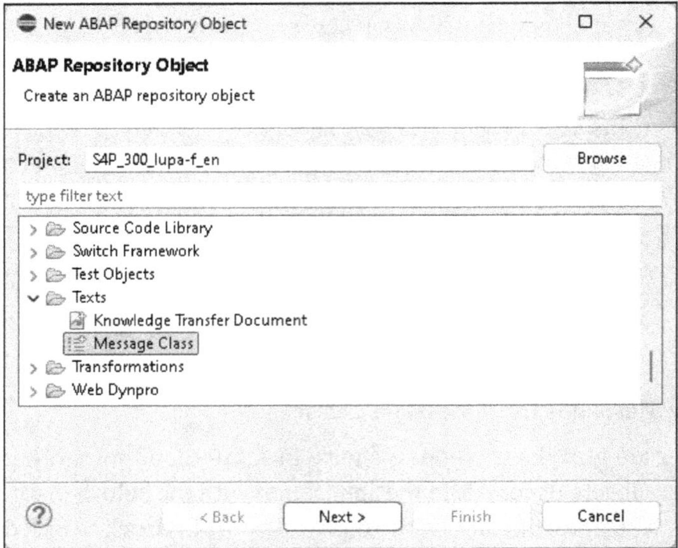

Figure 6.9 Selecting a Message Class Object Type

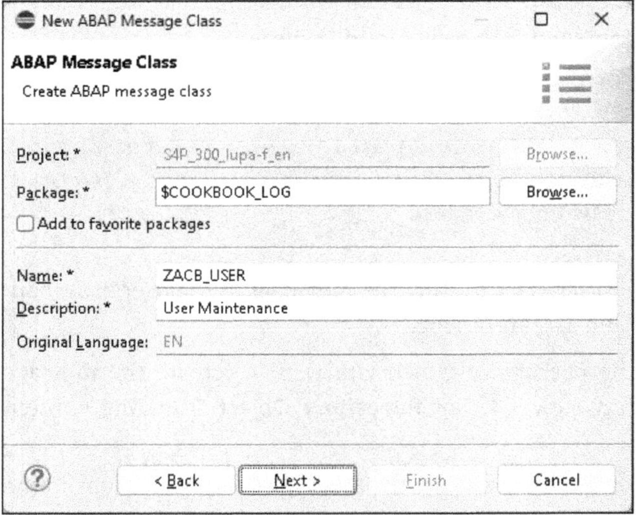

Figure 6.10 Creation Dialog for a Message Class

Maintaining messages

The maintenance screen for the message class opens automatically. There you can maintain messages with a three-digit numeric ID and a short text. We add a message to complete the user data import and a message containing a success message with the number of adjusted user master records (see Figure 6.11). The second message contains a variable message component: &1. Finally, you can save your changes by using the Ctrl + S keyboard shortcut.

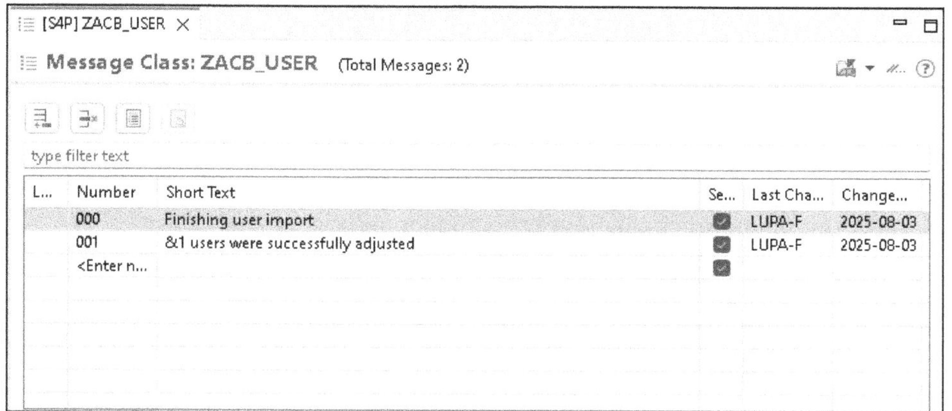

Figure 6.11 Maintaining Messages in a Message Class

Adding messages to the application log

We now use the created messages in the application by adding them to the log. Similar to the free text messages, you can use the add_item method of the log object and a setter class. The setter class for messages is called CL_BALI_MESSAGE_SETTER. In contrast to the CL_BALI_FREE_TEXT_SETTER class, it provides several static methods for creating a log entry:

- **create**
  The create method accepts the necessary parameters (message class and message number) and optionally message variables directly via formal parameters. It's very useful if this data is already in a variable.
- **create_from_sy**
  The create_from_sy method doesn't define any importing parameters. It reads the data for the message class from the system fields (sy-msgid, sy-msgno, sy-msgty, and sy-msgv1 through sy-msgv4). This method is therefore always suitable if the system fields in the statement were set via an ABAP statement or the call of a procedure with a classic exception concept before logging.
- **create_from_bapiret2**
  The create_from_bapiret2 method accepts a structured data object with the BAPIRET2 dictionary structure type. Such an object is very often used in Business Application Programming Interfaces (BAPIs) or other function modules.

[+]

Logging BAPI Logs

To add the rows of an internal table of type BAPIRET2 to a log, you can use the add_messages_from_bapirettab method, which isn't provided in the setter class, but directly in the log object (IF_BALI_LOG).

Where-used list for messages

As a rule, you'll have to choose between the `create` and `create_from_sy` methods. We recommend that you use the `create_from_sy` method wherever possible, especially if you define the message yourself and don't just receive and log it. If the message class and message number are transferred as a literal, the system can't establish a reference between the code line and the message. You're therefore unnecessarily losing the where-used list at this point, which can be very helpful for messages in particular.

Message via MESSAGE statement

To avoid losing the where-used list, we use the `MESSAGE` statement in combination with the `create_from_sy` method. You can use the `INTO` addition to the `MESSAGE` statement to specify a dummy variable and thus disable the flow control by message types, but still set the system variables:

```
MESSAGE s000(zacb_user) INTO DATA(dummy).
log->add_item(
  cl_bali_message_setter=>create_from_sy( ) ).
```

A slight disadvantage of this technique is that a variable must be specified that isn't otherwise used by the program. Accordingly, this can then be flagged as an unused variable in ABAP Test Cockpit checks. You can use the `##NEEDED` pragma to prevent this type of false positives.

The inline declaration, here of the `dummy` variable, is inevitably only possible once with the identifier in the current scope. If this technique is used multiple times, the quick fix in the ABAP development tools can be used to classically declare the variable as reusable via an explicit declaration (see Figure 6.12).

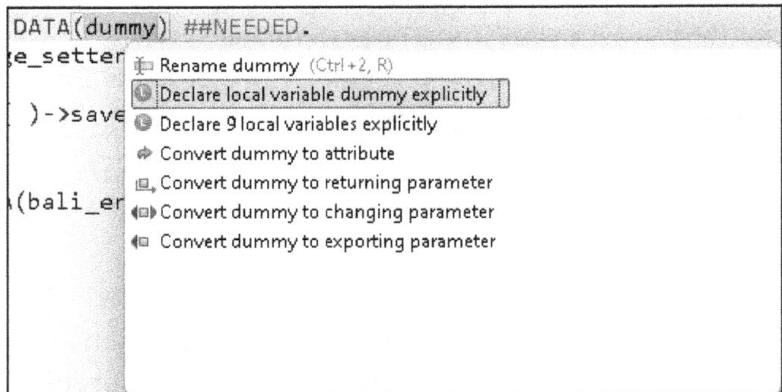

Figure 6.12 Quick Fix in the ABAP Development Tools for Refactoring an Inline Declaration to an Explicit Declaration

Messages with variables

The logging of messages with variables works just as well. In our example, the success message contains the number of processed user master records. Previously, the message was generated using a string template and was neither reusable nor translatable. We now change this to the second message we've created in the message class (see Listing 6.5).

```
DATA(modified_entries) = sy-dbcnt.
response->set_status( if_web_http_status=>created ).
MESSAGE s001(zacb_user)
  WITH modified_entries
  INTO DATA(success_message).
log->add_item(
  cl_bali_message_setter=>create_from_sy( ) ).
response->set_text( success_message ).
```

Listing 6.5 Adding a Message with Variable to the Application Log

In this specific case, the target variable is even used in the MESSAGE statement because the formatted text is returned to the API caller in the HTTP body.

6.4.3 Adding Messages from Exception Classes

In modern ABAP applications, errors are communicated via *class-based exceptions*. Using this technique, all information about an error can be collected in the instance of an exception class and passed through the call hierarchy with its own control flow for the error case. In contrast to the classic exception concept, which you could already take into account in the application log using the technique from Section 6.4.2, with class-based exceptions, you receive object instances that you want to log. A setter class with the add_item method is also provided for this purpose. The name of the class is CL_BALI_EXCEPTION_SETTER.

Triggering class-based exceptions

In our example, we want to map errors in the convert_csv_to_users method using an exception class. We've already created a suitable exception class named ZCX_ACB_USER_CONVERSION_ERROR in advance. The implementation already checks for error scenarios, but in the event of an error the method is only terminated prematurely using the RETURN statement with an initial return value. We now change this and trigger the exception using RAISE EXCEPTION. We also include it in the signature of the method because it's of type CX_STATIC_CHECK. You can find these adjustments in Listing 6.6.

```
METHODS convert_csv_to_users
  IMPORTING csv           TYPE string
            delimiter     TYPE string
  RETURNING VALUE(result) TYPE users
  RAISING   zcx_acb_user_conversion_error.
```

```
METHOD convert_csv_to_users.
  IF delimiter IS INITIAL.
    RAISE EXCEPTION NEW zcx_acb_user_conversion_error(
      textid =
zcx_acb_user_conversion_error=>delimiter_not_specified ).
  ENDIF.
  DATA(text) = csv.

  REPLACE ALL OCCURRENCES OF
    cl_abap_char_utilities=>cr_lf(1)
          IN text WITH ''.
  SPLIT text AT |\n| INTO TABLE DATA(lines).

  LOOP AT lines ASSIGNING FIELD-SYMBOL(<line>).
    TRY.
        SPLIT <line> AT delimiter INTO TABLE DATA(fields).

        INSERT VALUE #(
          username   = EXACT #( fields[ 1 ] )
          first_name =
            EXACT #( VALUE #( fields[ 2 ] OPTIONAL ) )
          last_name  =
            EXACT #( VALUE #( fields[ 3 ] OPTIONAL ) )
          ...
        ) INTO TABLE result.
      CATCH cx_sy_itab_line_not_found
            cx_sy_conversion_error
            INTODATA(format_error).
        RAISE EXCEPTION TYPE zcx_acb_user_conversion_error
              MESSAGE e004(zacb_user)
              EXPORTING previous = format_error.
    ENDTRY.
  ENDLOOP.
ENDMETHOD.
```

Listing 6.6 Modification of the convert_csv_to_users Method

Handling class-based exceptions

The ZCX_ACB_USER_CONVERSION_ERROR exception of the convert_csv_to_users method must now be handled in our handle_request method. We supplement a TRY control structure with a CATCH block for the exception. The TRY block deliberately contains not only the method call but also the remaining processing logic that is to be run through in the event of success, that is, if the exception isn't triggered. In the event of an error, however, it should be skipped, which can be mapped as shown in Listing 6.7.

```
TRY.
    CASE request->get_content_type( ).
      WHEN 'application/csv'.
        ...
        DATA(users) = convert_csv_to_users(
          csv       = request->get_text( )
          delimiter = delimiter ).
        ...
        MESSAGE s001(zacb_user)
          WITH modified_entries
          INTO DATA(success_message).
        log->add_item(
          cl_bali_message_setter=>create_from_sy( ) ).
        response->set_text( success_message ).
        COMMIT WORK.
        ...
    ENDCASE.
  CATCH zcx_acb_user_conversion_error
    INTO DATA(conversion_error).

    ROLLBACK WORK.

    log->add_item( cl_bali_exception_setter=>create(
      severity  = if_bali_constants=>c_severity_error
      exception = conversion_error ) ).

    response->set_status(
      if_web_http_status=>bad_request ).
    response->set_text( conversion_error->get_text( ) ).
ENDTRY.
MESSAGE s000(zacb_user) INTO DATA(dummy) ##NEEDED.
log->add_item(
  cl_bali_message_setter=>create_from_sy( ) ).
```

Listing 6.7 Adding an Exception Object to the Application Log

In Listing 6.7, you can see the added exception handling. The instance of the exception class is written to the conversion_error variable and then added to the log. To log class-based exceptions, you can use the static create method of the CL_BALI_EXCEPTION_SETTER setter class. The object created in this way can be added to the log like other log entries using the add_item method. We also supply the severity parameter so that the error is also taken into account as such in the log. Surprisingly, exceptions are also logged by default as status/success messages (type S).

[»]

Nesting TRY Control Structures

In Listing 6.7, it was conveniently omitted that the existing TRY control structure for handling CX_BALI_RUNTIME_ERROR already exists around the supplementary TRY control structure. Usually, you would extend this outer structure with another CATCH block, save the inner one and thus be able to handle all exceptions in the same way. Unfortunately, however, this doesn't work in our case because it's our intention to log the exceptions in the CATCH blocks. However, the logging itself can trigger the CX_BALI_RUNTIME_ERROR exception and then no longer has a surrounding TRY block to handle it. You would receive a syntax warning.

This example is therefore not entirely satisfactory with the nested control structures. In real life, you could swap out subfunctions to methods and only deal with the internal logging errors at the top level to avoid this.

6.5 Saving a Log

Now that you've created log entries using various techniques and added them to the log instance, you probably also want to persist these entries. This doesn't happen automatically. There are also use cases that you don't want to persist at all, for instance, because you want to display the log directly to the users. As a rule, however, persistence does make sense.

Reference to the application log object

We've already prepared the subsequent saving when we generated the log. An optional header has been added there, which establishes the reference to the application log object and optionally the subobject (see Listing 6.8). Note that when using the ABAP for cloud development language version, the application log object transferred in the object parameter must be in the same software component as the logging logic or it must have been released for external use via a release contract.

```
log = cl_bali_log=>create_with_header(
  cl_bali_header_setter=>create(
    object     = 'ZACB_USER'
    subobject  = 'IMPORT' ) ).
```

Listing 6.8 Creating the Log Instance with Reference to the Application Log

Saving via "save"

To persist a log prepared in this way in the database, you can use the CL_BALI_LOG_DB class. It contains the static get_instance method which returns an object that implements the IF_BALI_LOG_DB interface. The interface defines various methods for reading logs from the database, saving them in the database, deleting them, or setting and releasing locks (see Figure 6.13).

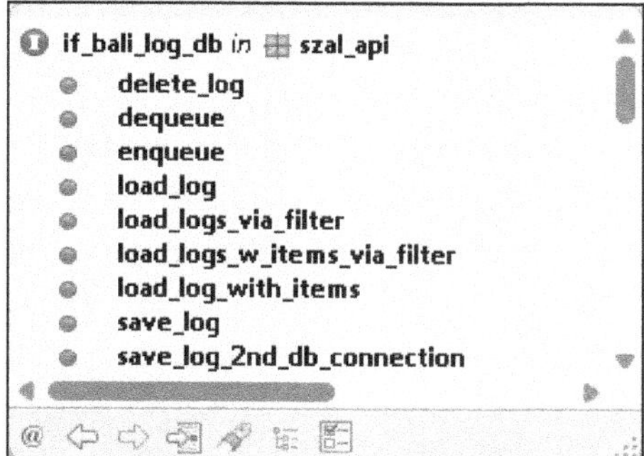

Figure 6.13 Methods in the IF_BALI_LOG_DB Interface

We only want to save the log and therefore use the `save_log` method. For this purpose, you must transfer your object of type `IF_BALI_LOG` to it. Depending on the context, you should then trigger a `COMMIT WORK` if this isn't taken over by the surrounding framework or the extended standard SAP system (see Listing 6.9).

```
TRY.
    ...
    MESSAGE s000(zacb_user) INTO DATA(dummy) ##NEEDED.
    log->add_item(
      cl_bali_message_setter=>create_from_sy( ) ).

    cl_bali_log_db=>get_instance( )->save_log( log ).
    COMMIT WORK.
  CATCH cx_bali_runtime INTO DATA(bali_error).
    RAISE SHORTDUMP bali_error.
ENDTRY.
```

Listing 6.9 Saving an Application Log

In this way, you can also decide whether the logs shouldn't be saved in the event of a reversal via `ROLLBACK WORK` or whether they should be available even in this case. The alternative method exists precisely for this case. `save_log_2nd_db_connection` uses a service connection to the database and is therefore independent of the regular LUW. The method corresponds to the supply of the `i_2th_connection` and `i_2th_connect_commit` parameters of the classic `BAL_DB_SAVE` function module. This is particularly useful in the context of ABAP Cloud because service connections to the database aren't permitted in the ABAP for cloud development language version. For example, details of the cause of errors in ABAP RESTful application program model applications can be logged even though the framework triggers a rollback.

6.6 Displaying Logs

SAP Fiori apps for log display

You can use the Application Logs app (app ID F1487) to display the logs from an administration perspective. It provides a range of functions comparable to the classic Transaction SLG1 with filter criteria and detailed view (see Figure 6.14 and Figure 6.15), and it allows you to view all application log objects that you're authorized to display.

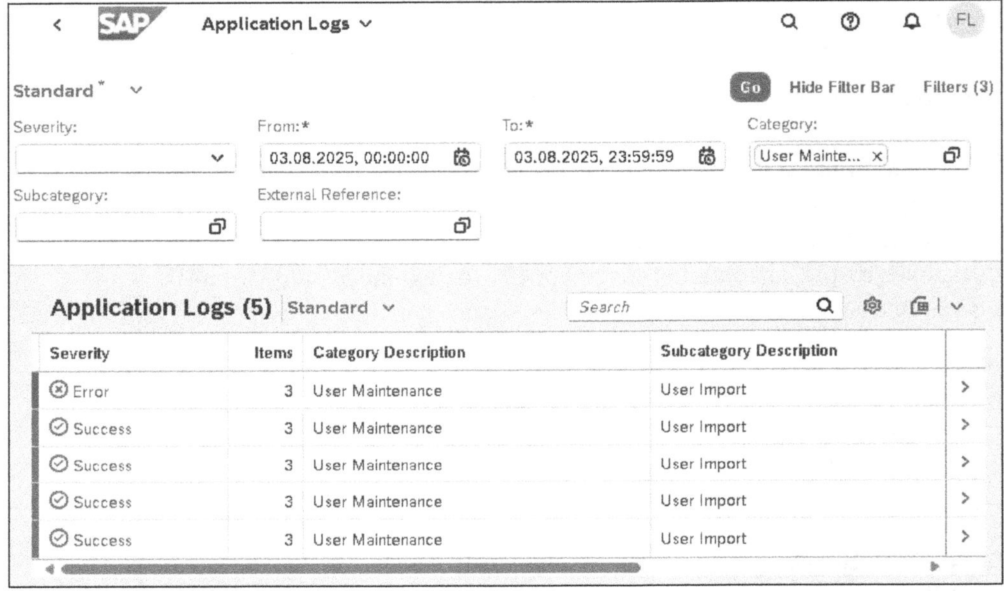

Figure 6.14 Log Overview in the Application Logs App

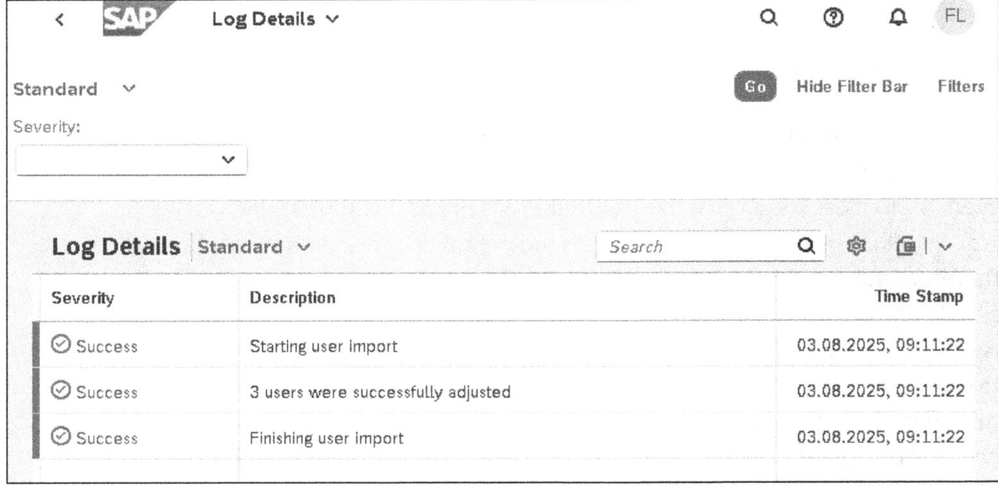

Figure 6.15 Log Details in the Application Logs App

In our sample application, this solved the problem of the lack of traceability. All calls of the API, including possible error messages, can be viewed centrally in the app.

Integration into custom SAP Fiori apps

In addition to this more administrative display, the *Reuse Library* also provides reusable components for integrating the application log display into your own SAP Fiori apps. Both freestyle apps and apps based on SAP Fiori elements with OData V2 or OData V4 are supported. Details can be found via the library name `sap.nw.core.applogs.lib.reuse`, the app ID F1488, or in the SAP Help Portal at *http://s-prs.de/v1064805*.

Log display in Transaction SLG1

The generated application logs are also persisted in the classic database tables via the new API. The classes presented in this chapter only serve as ABAP cloud–enabled wrappers. This means that the data can still be found in database tables BALHDR and BALDAT, and in SAP S/4HANA and SAP S/4HANA Cloud Private Edition. The classic Transaction SLG1 can also be used to display the logs (see Figure 6.16).

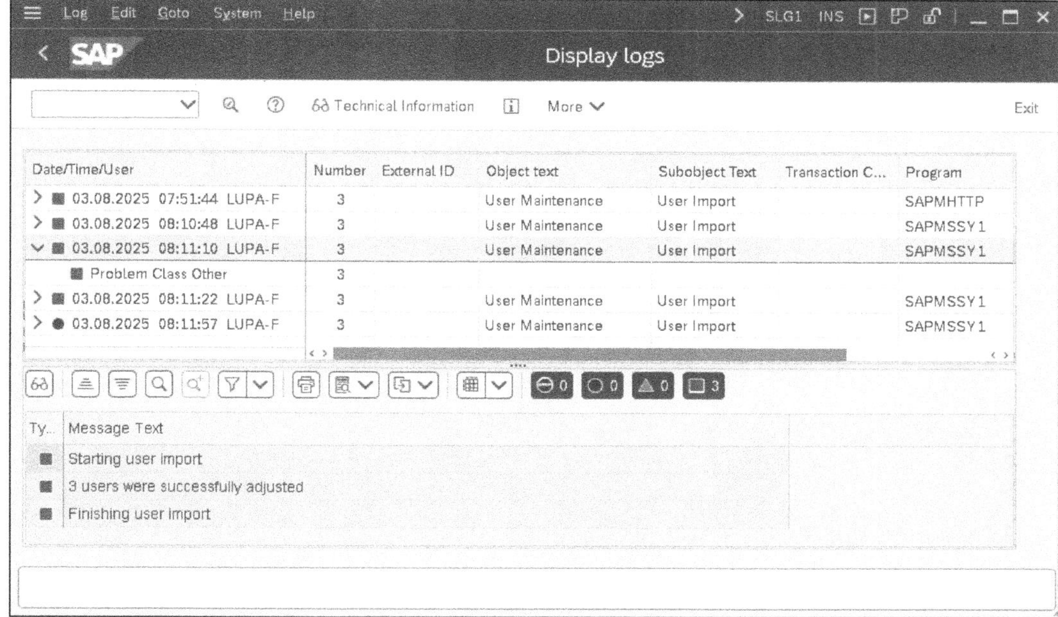

Figure 6.16 Log Display in Transaction SLG1

6.7 Summary

Functional scope of the BALI API

In this chapter, you've learned how to use the BALI API to write application logs. The API has been released and can be used in ABAP Cloud. The range of functions is already quite large and covers most use cases. Even some niche topics that we haven't covered here, such as the context or the level of detail, have been implemented by SAP. Extensive options for program-based reading of logs, including filters, are provided as well. You can even create application log objects and subobjects using a program-based approach, should this ever be necessary.

Handling the log as an object

The handling of the log object, which is typified via the IF_BALI_LOG interface, can be somewhat unfamiliar and may require workarounds in some places. Our small example has already shown that the CX_BALI_RUNTIME exception can be cumbersome to use. The fact that it's of type CX_STATIC_CHECK and thus forces every user of the logging API to handle it in a local CATCH block is surprising and awkward. Logging is an implementation detail that shouldn't normally be visible to the outside world. When using the API, you must therefore decide at each point whether you want to continue with the execution or terminate the application entirely. This freedom is generally a good thing, but a one-time decision when creating the instance would save users a lot of work, especially because the exception seems to have been added to every method of the API across the board and isn't always triggered. If, for example, you log a message of a message class via the CL_BALI_MESSAGE_SETTER class, this class doesn't check the existence of the transferred message.

Chapter 7
Change Documents

Changes to application data can also be easily logged in ABAP Cloud via change document objects. This chapter introduces you to the new class-based call.

Traceability of changes

In almost every transactional business application, there's a need to be able to trace changes, such as who has stored a special indicator in the customer master record that is now causing issues? When was this change made? What value was previously stored? Tracing these kinds of changes may even be a regulatory requirement.

To answer these and similar questions, the concept of *change documents* has long existed in ABAP development. You can use a change document object to define what the data to be logged looks like with reference to the relevant database tables. You then generate a function module and call it at a suitable point in your application. There are even reusable function modules for displaying change documents in SAP GUI.

Class-based API

In modern ABAP-based applications developed using the ABAP RESTful application programming model and SAP Fiori elements, the same questions continue to arise. Function modules and SAP GUI, on the other hand, are no longer up to date. In this chapter, you'll learn how to use the new class-based API for change documents and how to integrate it into your ABAP RESTful application programming model business objects.

Structure of this chapter

A change document object is created as a technical basis in Section 7.1. We then implement the call of the change document update in a small application within our application scenario in Section 7.2. Section 7.3 presents a more complex scenario with the integration of change documents in ABAP RESTful application programming model-based applications. In Section 7.4, we take a brief look at the display of change documents.

[«]

Difference Between Change Documents, Change Logging, and the Application Log

At this point, we'll explicitly discuss the technology used to log changes to application data. This master and transaction data changes frequently and directly in the respective system. The solution provided in the ABAP platform for tracing changes comprises the change document objects and change documents generated through them.

A technically differently implemented solution exists for Customizing changes. For Customizing tables, you can activate change logging in the technical settings. In this case, changes will be automatically logged in database table DBTABLOG. This technique is intended for Customizing and not for application data because it has a performance drawback, and it's more difficult to implement a program-based data analysis. Details on change logging in Customizing can also be found in Chapter 5.

The application log from Chapter 6 represents another logging technique. However, this isn't about changes to data, but about logging process steps, decisions in the program sequence, and messages.

Change documents for our sample application

With regard to our recipe portal, this section implements the change document update for recipes and their ingredients. Changes haven't yet been logged. Only administrative data such as the creation time, the creating user, the last modification time, and the last modifying user are currently saved. To close this gap, we extend the implementation of our ABAP RESTful application programming model business object in Section 7.3. Prior to that, in Section 7.2, we implement a program in a simpler example that updates or technically migrates the units of measure for ingredients.

Using the Sample Application

The development objects for this chapter can be found in the CHANGEDOCUMENTS subpackage. The integration into the ABAP RESTful application programming model Recipe business object can be found in the corresponding behavior definition and implementation in the DATAMODEL subpackage. You can also run the ZCL_ACB_UNIT_UPDATER demo application in the ABAP development tools for Eclipse as an ABAP console application via the F9 key and view it in the debugger. It performs a unit of measure change and then removes it again to remain reproducible.

7.1 Maintaining Change Document Objects in the ABAP Development Tools

Creating the change document object

First, you need to create a new change document object, which defines the format of the changes to be logged on the basis of database table definitions or structure types:

1. Right-click on the desired package in the ABAP development tools, and select the **New • Other Repository Object** path from the context menu.

2. In the dialog box that opens, select the **Change Document Object** item in the **Change Document Management** folder.
3. In the steps that follow, you need to assign a technical identifier and the description. We entered the "ZACB_RECIPE" as **Name** and the "Recipe" as **Description** (see Figure 7.1).
4. Finally, select a transport request, and end the creation wizard by clicking the **Finish** button.

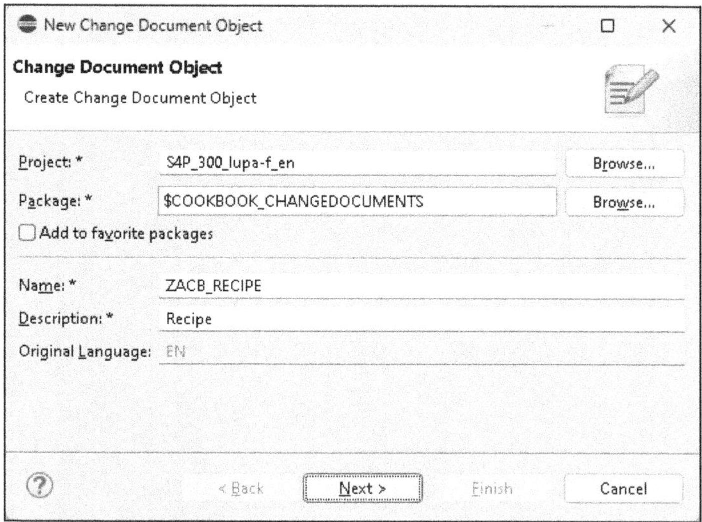

Figure 7.1 Creating a Change Document Object

Maintaining the change document object

The maintenance view for the new change document object opens automatically (see Figure 7.2).

Figure 7.2 Maintenance UI for a Change Document Object

If you're familiar with the SAP GUI–based maintenance of change document objects in Transaction SCDO, the user interface (UI) may seem unfamiliar at first. However, it has been greatly simplified and made more intuitive in comparison.

Adding database tables or structure types

You can use the **Add** button to add database tables or structure types. You can define the structure of the changes to be logged and also the parameters in the subsequent method call. After clicking on the **Add** button, a form for specifying the necessary parameters for the table or structure will get displayed on the right (see Figure 7.3). We opt for table ZACB_RECIPE here and don't fill the **Reference Table** field. If your table or structure contains fields with unit of measure or currency references to another table or structure, this should be entered here.

Table or Structure Details

Name: * ZACB_RECIPE Browse...

Reference Table: Browse...

☐ Log Multiple Changes

Database Insertions

☐ Log Field Values

☐ Log Initial Values

Database Deletions

☐ Log Field Values

☐ Log Initial Values

Figure 7.3 Adding a Table to a Change Document Object

For each table or structure, you can specify whether multiple entries are expected when writing change documents or just one. The parameters in the generated API will be typed depending on this. For header tables, a single entry makes more sense, so we decide against the **Log Multiple Changes** option here.

We don't activate the additional options for **Database Insertions** and **Database Deletions**. They would make it possible to log the individual fields of a data record when it's created or deleted. Usually, this isn't necessary because the data has already been logged elsewhere. The values before deletion and the values for new creation can be traced via the individual field logging when changes are made, although this is somewhat more complicated.

We repeat this process and add table ZACB_INGREDIENT. It's our item table with the ingredients for a recipe header record. It therefore makes sense to log it in the same change document object. In contrast to table ZACB_RECIPE, we activate the **Log Multiple Changes** checkbox for this table because multiple entries are expected for this table when changes are updated.

Activating the change document object

Finally, you must activate the change document object using the keyboard shortcut Ctrl + F3 or by clicking the **Activate** button. The maintenance screen should then look like the one shown in Figure 7.4.

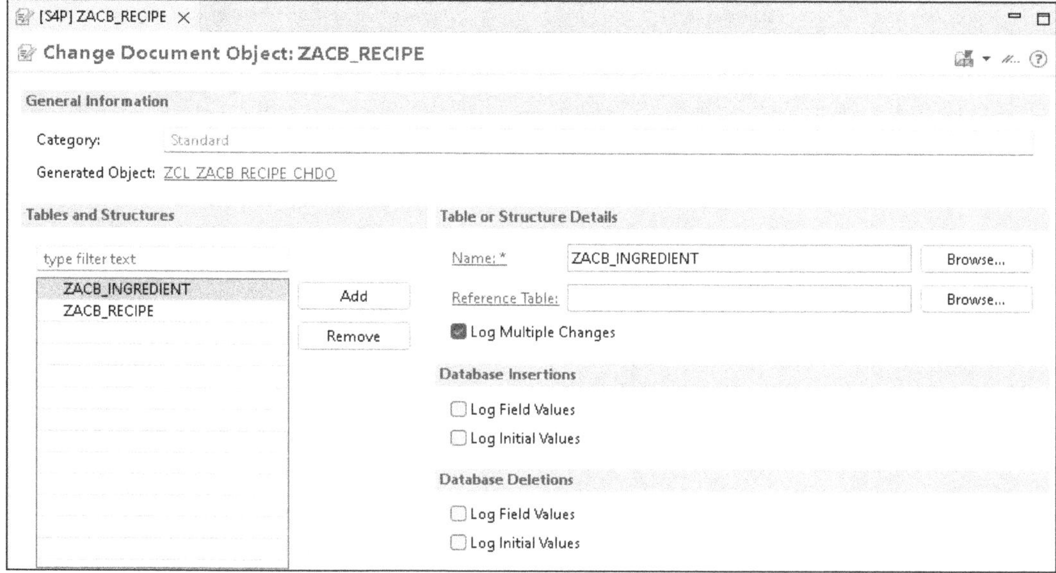

Figure 7.4 Maintenance Screen of a Change Document Object After Activation

Under **Generated Object**, you'll find the name of the generated class that serves as an API for updating changes based on this change document object.

[!]

Different Scope of Functions in the ABAP Development Tools and Transaction SCDO

The options for change document object maintenance in the ABAP development tools and in Transaction SCDO in SAP GUI differ slightly. While a maintenance in the ABAP development tools always generates a class as an API, in the SAP GUI transaction, you're given the choice of whether you want to generate a class or a function module. Depending on the application scenario, the function module may also be useful because it can be used to bundle changes in a single database logical unit of work (LUW).

Activating change document relevance

As a prerequisite for updating change documents, the tables or structures specified in the change document object must use data elements that have been marked as relevant for change logging. In the ABAP development tools, you'll find the corresponding indicator named **Change Document Logging** in the data element maintenance in the expandable **Additional Properties** area (see Figure 7.5).

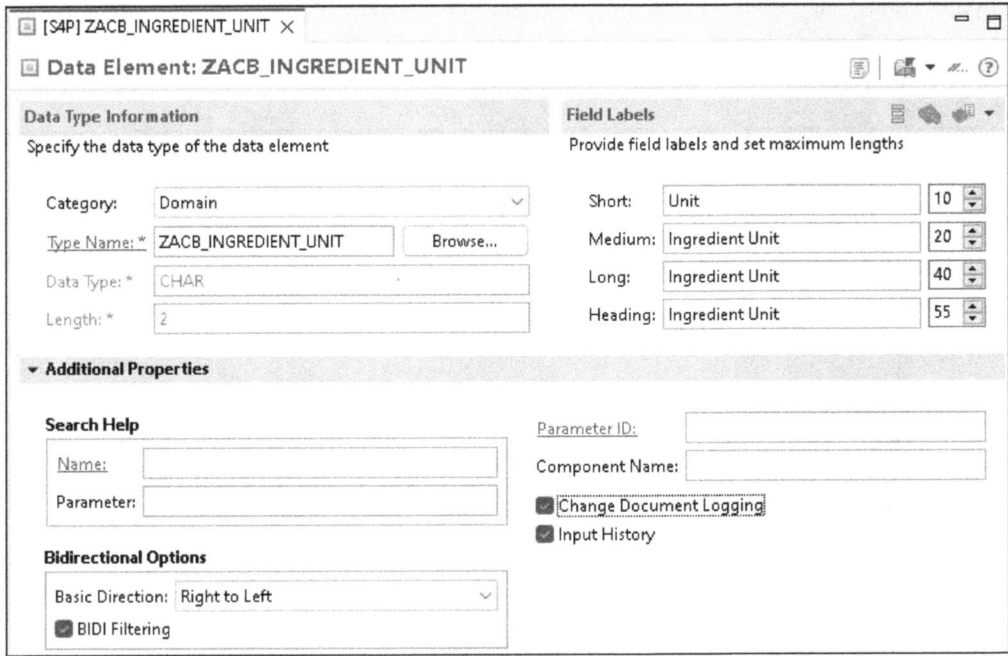

Figure 7.5 Indicator for Change Document Update in the Data Element

We recommend that you activate this indicator as a rule because most data elements define business-relevant data whose changes should be logged. The only exception is technical data, such as hash values. Note that follow-up processes can also be triggered on the basis of change documents using change pointers.

7.2 Calling the Logging Function via the Generated Class

Generated class

When the change document object is activated, the system generates a global class. It follows the naming convention `<namespace>CL_<change document object>_CHDO` and is automatically created in the same package as the change document object. In object maintenance, you can also navigate directly to the global class by clicking on the object name.

[!]

Customizing Generated Objects

The global class only contains generated code and gets overwritten when the change document object is changed. This isn't a template like the objects created by an ABAP repository object generator. You should therefore not change this class manually, as your changes will otherwise be lost.

The write method

The generated class contains a static method named `write`. You can use this method to generate change documents. It also contains some relevant type definitions and the name of the change document object as a read-only static variable. The structure of the generated class is shown in Figure 7.6.

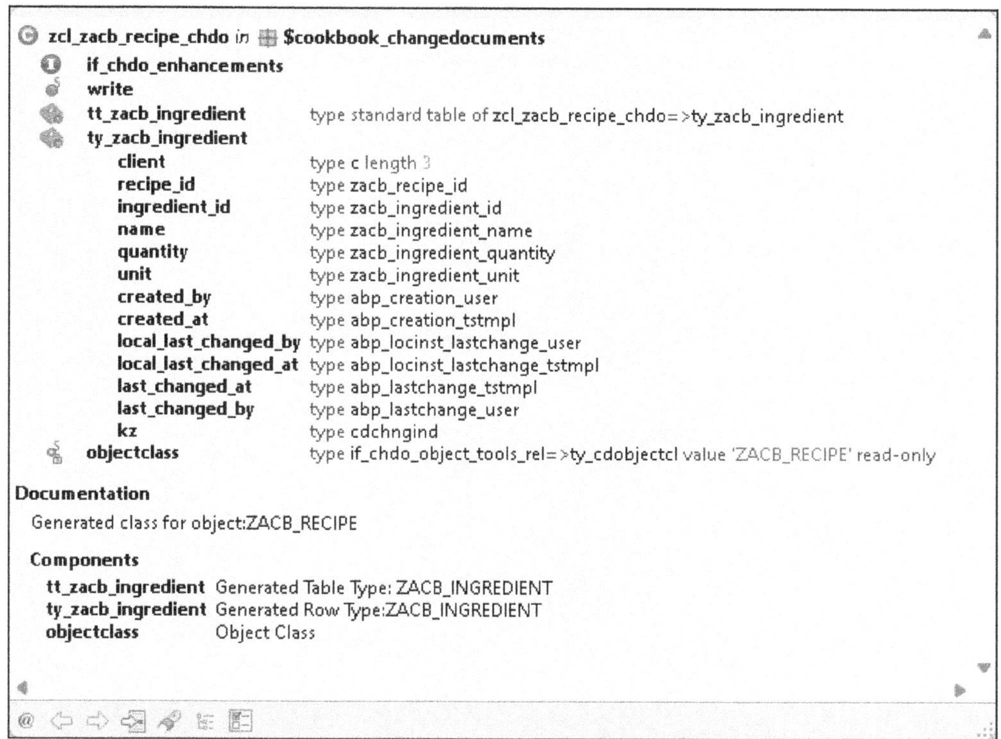
zcl_zacb_recipe_chdo in $cookbook_changedocuments
if_chdo_enhancements
write
tt_zacb_ingredient type standard table of zcl_zacb_recipe_chdo=>ty_zacb_ingredient
ty_zacb_ingredient
client type c length 3
recipe_id type zacb_recipe_id
ingredient_id type zacb_ingredient_id
name type zacb_ingredient_name
quantity type zacb_ingredient_quantity
unit type zacb_ingredient_unit
created_by type abp_creation_user
created_at type abp_creation_tstmpl
local_last_changed_by type abp_locinst_lastchange_user
local_last_changed_at type abp_locinst_lastchange_tstmpl
last_changed_at type abp_lastchange_tstmpl
last_changed_by type abp_lastchange_user
kz type cdchngind
objectclass type if_chdo_object_tools_rel=>ty_cdobjectcl value 'ZACB_RECIPE' read-only
Documentation
Generated class for object:ZACB_RECIPE
Components
tt_zacb_ingredient Generated Table Type: ZACB_INGREDIENT
ty_zacb_ingredient Generated Row Type:ZACB_INGREDIENT
objectclass Object Class

Figure 7.6 ABAP Element Information on the Generated Class of the Change Document Object

Application for updating units of measure

To update changes via the change document object, you must call the `write` method at the appropriate point. For a simple call, we'll first implement a sample program that adjusts the unit of measure for some ingredients. The background to this is a technical change in the key for the units of measure. The developed program initially takes care of converting the old unit values into the new ones close to the database.

Determining the data records

The first step is to determine the ingredient data records to be changed (see Listing 7.1). This is done using a range table, which serves as a filter based on a mapping table.

```
DATA(mapping) = VALUE unit_mapping_tab(
  ( old_unit = 'G' new_unit = 'GR' ) ).
DATA(unit_filter) = get_old_unit_rnge_from_mapping(
  mapping ).
```

```
SELECT FROM zacb_ingredient
  FIELDS DISTINCT recipe_id
  WHERE unit IN @unit_filter
  ORDER BY recipe_id
  INTO TABLE @DATA(recipes_to_update).
out->write(
  |{ lines( recipes_to_update ) NUMBER = USER }| &&
  | recipes to update were found| ).
```

Listing 7.1 Data Selection for Ingredient Updating

Single record processing

The individual data records are then processed via a loop. We've opted for single set record processing with the structure from Listing 7.2.

```
LOOP AT recipes_to_update
     ASSIGNING FIELD-SYMBOL(<recipe>).
  TRY.
      out->write(
        |Updating recipe { <recipe>-recipe_id }| ).
      " Data selection
      ...
      " Data processing
      ...
      " Update in the database
      ...
      COMMIT WORK.
    CATCH lcx_error INTO DATA(error).
      ROLLBACK WORK.
      out->write( error->get_text( ) ).
  ENDTRY.
ENDLOOP.
```

Listing 7.2 Single Record Processing of the Ingredient Update

As you can see, class-based exception handling has been implemented within the loop and a local exception class named LCX_ERROR has been specified in the CATCH block. This control structure is useful because it allows the LUW to be delimited for each individual record and completed with a COMMIT WORK or ROLLBACK WORK regardless of the specific error case. Possible errors include locking conflicts, errors when saving to the database or errors in data validation. Using the local exception class, all exceptions can be handled equally via the control structure instead of having to mix class-based exceptions and return codes with IF blocks via the sy-subrc system field.

Data selection

The ingredients to be migrated are first read for each recipe determined. The recipe header record is also read from the database. Both are required because the header record contains admin data on the last change, which should be changed together with the units of measure in the ingredient item. It's essential to select the data in the status prior to the change because the change document update can't determine the old status itself. This must be provided by you. Correspondingly, the identifiers of the variables are therefore given an `old` suffix. If the old status can't be found in the database, parallel changes have been made during program execution. In this case, the recipe will be skipped. The complete selection logic is shown in Listing 7.3.


```
SELECT SINGLE FROM zacb_recipe
  FIELDS *
  WHERE recipe_id  = @<recipe>-recipe_id
  INTO @DATA(recipe_old).
SELECT FROM zacb_ingredient
  FIELDS *
  WHERE recipe_id  = @<recipe>-recipe_id
    AND unit       IN @unit_filter
  ORDER BY PRIMARY KEY
  INTO TABLE @DATA(ingredients_old).
IF recipe_old IS INITIAL OR ingredients_old IS INITIAL.
  cl_message_helper=>set_msg_vars_for_clike(
    |Skipping recipe, parallel changes| ).
  RAISE EXCEPTION TYPE lcx_error USING MESSAGE.
ENDIF.
```

Listing 7.3 Data Selection in Single Record Processing

Data processing

The selected data is then copied to new variables with a `new` suffix and then updated. For this purpose, the admin fields for changes are overwritten in the header record and the actual change to the units of measure is made in the relevant items (see Listing 7.4).

```
DATA(recipe_new) = VALUE #( BASE recipe_old
  last_changed_by =
    cl_abap_context_info=>get_user_technical_name( )
  last_changed_at =
    cl_abap_tstmp=>utclong2tstmp( utclong_current( ) ) ).
DATA(ingredients_new) = ingredients_old.
LOOP AT ingredients_new ASSIGNING FIELD-SYMBOL(<ingredient>).
  <ingredient>-unit                =
    mapping[ old_unit = <ingredient>-unit ]-new_unit.
  <ingredient>-last_changed_by     =
```

```
    cl_abap_context_info=>get_user_technical_name( ).
  <ingredient>-last_changed_at       =
    cl_abap_tstmp=>utclong2tstmp( utclong_current( ) ).
  <ingredient>-local_last_changed_by =
    <ingredient>-last_changed_by.
  <ingredient>-local_last_changed_at =
    <ingredient>-last_changed_at.
ENDLOOP.
```

Listing 7.4 Updating Recipe and Ingredients Data

Update in the database

The changed data can then be transferred to the database using the UPDATE command (see Listing 7.5). The application performs a technical migration and therefore deliberately bypasses the ABAP RESTful application programming model facade of the Recipe business object at this point. This simplifies the presentation here. In real life, this approach should be an exception and the integration of the change document update should take place centrally in the ABAP RESTful application programming model business object, as described in Section 7.3.

```
UPDATE zacb_recipe FROM @recipe_new.
IF sy-subrc <> 0.
  cl_message_helper=>set_msg_vars_for_clike(
    |Error updating ZACB_RECIPE: { sy-subrc }| ).
  RAISE EXCEPTION TYPE lcx_error USING MESSAGE.
ENDIF.

UPDATE zacb_ingredient FROM TABLE @ingredients_new.
IF sy-subrc <> 0.
  cl_message_helper=>set_msg_vars_for_clike(
    |Error updating ZACB_INGREDIENT: { sy-subrc }| ).
  RAISE EXCEPTION TYPE lcx_error USING MESSAGE.
ENDIF.

COMMIT WORK.
```

Listing 7.5 Update of the Database Tables

Calling the "write" method

To implement the change document update, we add the call of the write method from the generated class at a suitable point—in our case, its name is ZCL_ZACB_RECIPE_CHDO. The appropriate position is exactly between the last write database interaction with UPDATE and the completion of the transaction with COMMIT WORK. This means that the change documents are included in the LUW and only saved together with the associated changes. Conversely, the changes to the application data aren't made if the change document update fails. The complete call can be found in Listing 7.6, which we explain in more detail next.

```
DATA(object_id) = EXACT
  if_chdo_object_tools_rel=>ty_cdobjectv(
    sy-mandt && <recipe>-recipe_id ).

zcl_zacb_recipe_chdo=>write(
  objectid             = object_id
  utime                =
    cl_abap_context_info=>get_system_time( )
  udate                =
    cl_abap_context_info=>get_system_date( )
  username             = EXACT #(
    cl_abap_context_info=>get_user_technical_name( ) )
  o_zacb_recipe        = recipe_old
  n_zacb_recipe        = recipe_new
  upd_zacb_recipe      = 'U'
  xzacb_ingredient     = CORRESPONDING #( ingredients_new )
  yzacb_ingredient     = CORRESPONDING #( ingredients_old )
  upd_zacb_ingredient = 'U' ).

COMMIT WORK.
```

Listing 7.6 Calling the write Method

Object ID

To identify the object affected by the change, you must first define an object ID. This is usually the concatenation of the primary key from the primary table of the respective business object. In our case, it's the client and the recipe ID. This object ID is first assigned in Listing 7.6. The CDOBJECTTV data element isn't released in ABAP Cloud, so typing is performed indirectly via the IF_CHDO_OBJECT_TOOLS_REL=>TY_CDOBJECTTV type.

[!]

Object ID for Change Documents

If you're implementing an integration with an existing business object in the standard SAP system, you can use the object ID definition selected there as a guide. Some objects omit the client in the object ID (e.g., BUPA_BUP for the business partner), while others use it (e.g., BELEG for accounting documents).

Parameters of the "write" method

The static write method is then called in the ZCL_ZACB_RECIPE_CHDO class generated by the change document object. The method signature contains many parameters, some of which are optional or have default values. The meaning of the individual parameters is described in Table 7.1.

Parameter	Meaning
objectid	Key of the object affected by the change (primary table)
utime, udate	Change date and time in system time (can, e.g., be queried via the CL_ABAP_CONTEXT_INFO class)
username	Changing user (e.g., can be queried via the CL_ABAP_CONTEXT_INFO class)
object_change_indicator	Change indicator for the type of change (see Table 7.2)
o_<table>	Old data of a table with single record transfer
n_<table>	New data of a table with single record transfer
x<table>	New data of a table with transfer of multiple data records
y<table>	Old data of a table with transfer of multiple data records
upd_<table>	Change indicator of a table

Table 7.1 Parameters of the "write" Method

Other optional parameters are intended for planned changes and the deactivation of change pointers. Additional parameters are provided for text tables or long texts to handle these separately. The signature of the write method can be a little daunting at first due to this abundance of parameters. In most cases, however, you can simply apply the following rules:

- Transfer the object ID, the date and time of the change, and the user name as described.
- Transfer all tables to which you've made changes, including old and new data via the o_<table>, n_<table>, y<table>, and x<table> parameters.
- Use the U value for the change indicator of tables upd_<table> for the transferred tables. The change document update is smart enough to compare o_<table> and n_<table> or y<table> and x<table> with each other to distinguish changes from new entries.
- The y<table> and x<table> tables also contain a kz component to identify changes at row level. As a rule, you don't need to fill in this field, and the type of change is automatically derived. It would only be relevant in special cases, for example, if you delete a data record within the same LUW and create a new one with an identical key. In the code example from Listing 7.6, the CORRESPONDING constructor expression is used to supply the parameter appropriately.

Change indicator

The possible change indicators are listed in Table 7.2.

Value	Meaning
I	Insert
J	Insert (single field documentation)
U	Change
D	Delete
E	Delete (single field documentation)

Table 7.2 Meaning of the Change Indicators

Error handling

The method can trigger a class-based exception of type CX_CHDO_WRITE_ERROR. You should handle this exception in such a way that changes aren't inadvertently made without logging. We therefore want to adjust the CATCH block slightly (see Listing 7.7).

```
LOOP AT recipes_to_update ASSIGNING FIELD-SYMBOL(<recipe>).
  TRY.
      ...
      zcl_zacb_recipe_chdo=>write( ... ).
      COMMIT WORK.
    CATCH cx_chdo_write_error
          lcx_error INTO DATA(error).
      ROLLBACK WORK.
      out->write( error->get_text( ) ).
  ENDTRY.
ENDLOOP.
```

Listing 7.7 Error Handling When Updating Change Documents

Your change documents should then be updated. Don't forget to test this too. The indicators on the data element or the transfer of change indicators are often forgotten. The display of change documents is described in Section 7.4.

7.3 Change Document Update Using the ABAP RESTful Application Programming Model

Call points for the change document update

In the previous section, we implemented an application that performs a technical migration of the units of measure close to the database. In this simple example, the change documents were written directly after the database operations. In real life, this would be an exception for such technical migrations or mass data processing. Technically, change document creation should be centralized in the implementation of your business object. Depending on the programming model used, it's located in different places. In this section, we want to look at the integration into the ABAP RESTful application programming model, the current programming model in ABAP Cloud.

Unit of measure update

In applications based on this programming model, the ABAP RESTful application programming model business object should only provide one API for consumption inside and outside the system. It's therefore often referred to as an *ABAP RESTful application programming model facade*, which hides details of the persistence layer or logging from the API user.

With regard to our sample application, we'll make the same change to the units of measure as in the previous section, but in a different direction so that the application remains executable for test purposes and the original state can always be restored. This time, we use core data services (CDS) views and Entity Manipulation Language (EML) statements.

Determining data records via CDS views

The data selection is initially very similar to the previous example. However, instead of selecting directly from the database tables, this time we use a view of the ABAP RESTful application programming model application to determine the recipes to be changed (see Listing 7.8).

```
DATA unit_updates TYPE TABLE FOR UPDATE ZACB_R_Ingredient.

DATA(mapping) = VALUE unit_mapping_tab(
  ( old_unit = 'GR' new_unit = 'G' ) ).
DATA(unit_filter) = get_old_unit_rnge_from_mapping(
  mapping ).

SELECT FROM ZACB_R_Ingredient
  FIELDS DISTINCT RecipeId
  WHERE Unit IN @unit_filter
  ORDER BY RecipeId
  INTO TABLE @DATA(recipes_to_update).
```

Listing 7.8 Determining the Recipes to Be Changed via a CDS View

[»]

Selection and EML Statements Without an Interface View

To simplify the sample application, we work directly with the root entity—ZACB_R_Ingredient—of our Recipe business object, which isn't actually intended for selections and EML statements outside the implementation of the behavior definition. In your use cases, you should consider creating interface views as a projection that can be shared and define a stable point of entry into your ABAP RESTful application programming model application from the outside.

Read operation via "READ ENTITIES"

After determining the data records to be changed, we read them in their current state using their keys via the READ ENTITIES statement, which is part of the EML scope (see Listing 7.9). This statement enables access to ABAP RESTful application programming model business objects within the system.


```
READ ENTITIES OF ZACB_R_Recipe
  ENTITY Recipe BY \_Ingredient
  FIELDS ( Unit )
  WITH CORRESPONDING #( recipes_to_update )
  RESULT DATA(ingredients).
```

Listing 7.9 Read Operation via "READ ENTITIES"

ABAP RESTful application programming model business objects have been designed to be multi-instance capable. For this reason, unlike the implementation from the previous section, we can pass our determined recipe IDs directly to the statement and don't need to call them in a loop. Using a read-by-association, we can obtain the values of the Ingredient child entity, although we specify key values for Recipe. The key values and the Unit field requested via the FIELDS statement are filled in the ingredients return value (see Figure 7.7).

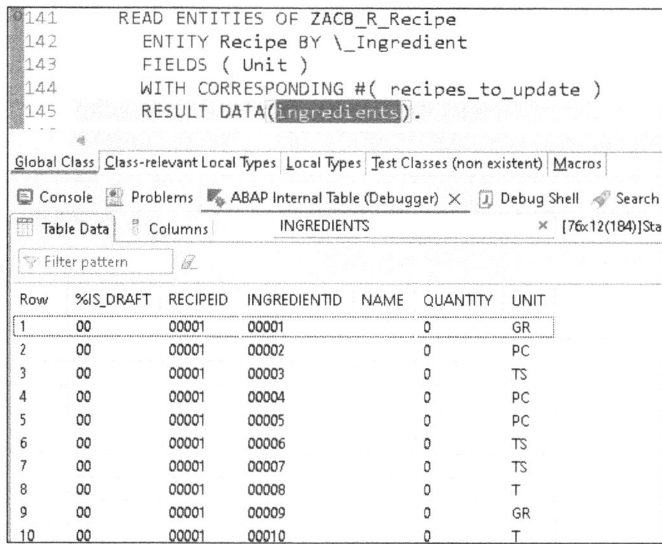

Row	%IS_DRAFT	RECIPEID	INGREDIENTID	NAME	QUANTITY	UNIT
1	00	00001	00001		0	GR
2	00	00001	00002		0	PC
3	00	00001	00003		0	TS
4	00	00001	00004		0	PC
5	00	00001	00005		0	PC
6	00	00001	00006		0	TS
7	00	00001	00007		0	TS
8	00	00001	00008		0	T
9	00	00001	00009		0	GR
10	00	00001	00010		0	T

Figure 7.7 Return of the READ ENTITIES Statement in the Debugger

Write operation via "MODIFY ENTITIES"

We now want to update some of the returned ingredients and change their units of measure. The update via EML is carried out using the MODIFY ENTITIES statement. This statement requires a special internal table that contains additional fields, such as a structure for marking changed components. Such an internal table is declared using the special TABLE FOR UPDATE addition. We can use the CORRESPONDING constructor expression to transfer the imported data to the new table and then make our changes. The changes are completed using the COMMIT ENTITIES statement (see Listing 7.10).

```
DATA unit_updates TYPE TABLE FOR UPDATE ZACB_R_Ingredient.
unit_updates = CORRESPONDING #( ingredients ).

LOOP AT unit_updates ASSIGNING FIELD-SYMBOL(<update>)
  WHERE Unit IN unit_filter.
  <update>-Unit =
    mapping[ old_unit = <update>-Unit ]-new_unit.
ENDLOOP.

MODIFY ENTITIES OF ZACB_R_Recipe
  ENTITY Ingredient
  UPDATE FIELDS ( Unit )
  WITH unit_updates
  REPORTED DATA(reported)
  FAILED DATA(failed).

IF failed IS NOT INITIAL.
  ROLLBACK ENTITIES.
ELSE.
  COMMIT ENTITIES RESPONSE OF ZACB_R_Recipe
    REPORTED DATA(reported_late)
    FAILED DATA(failed_late).
  ...
ENDIF.
```

Listing 7.10 Write Operation via MODIFY ENTITIES

The managed implementation of the ABAP RESTful application programming model business object now changes the data in the underlying database tables after execution. In addition, implemented validations and lock handling prevent possible inconsistencies. However, no change documents have yet been written.

[«]

Release-Dependent Integration of Change Documents

The following steps described next for a manual integration of the change document update in ABAP RESTful application programming model applications are very time-consuming and prone to errors. Late in the development of the ABAP RESTful application programming model, SAP planned to provide a native integration that would generate change documents without any implementation effort. However, this isn't yet available in SAP S/4HANA 2023. At the time this book was written, the corresponding road map item was scheduled for SAP S/4HANA 2025 or SAP S/4HANA Cloud Private Edition 2025 and SAP BTP ABAP environment 2602 or SAP S/4HANA Cloud Public Edition 2602. The planned new features include a statement in the behavior definition with reference to the change document object, a quick fix for creating a suitable change document object, and an SAP Fiori elements display of the change documents.

Additional save

To update additional data alongside the actual application data, the ABAP RESTful application programming model provides the **Additional Save** option. To use it, you must first add the `with additional save` statement in the behavior definition of your business object for the entities for which you want to update change documents. We add this statement accordingly in the `ZACB_R_Recipe` behavior definition for our business objects (see Listing 7.11).

```
define behavior for ZACB_R_Recipe alias Recipe
persistent table zacb_recipe
draft table zacb_recipe_d
with additional save
...
define behavior for ZACB_R_Ingredient alias Ingredient
persistent table zacb_ingredient
draft table zacb_ingredien_d
with additional save
...
```

Listing 7.11 Activation of Additional Save in the Behavior Definition

Implementing the SAVE_MODIFIED method

The supplemented statements are automatically provided with a warning by the syntax check that no implementation of the `save_modified` method exists yet. You can use the keyboard shortcut [Ctrl]+[1] to select a quick fix to create this method automatically (see Figure 7.8).

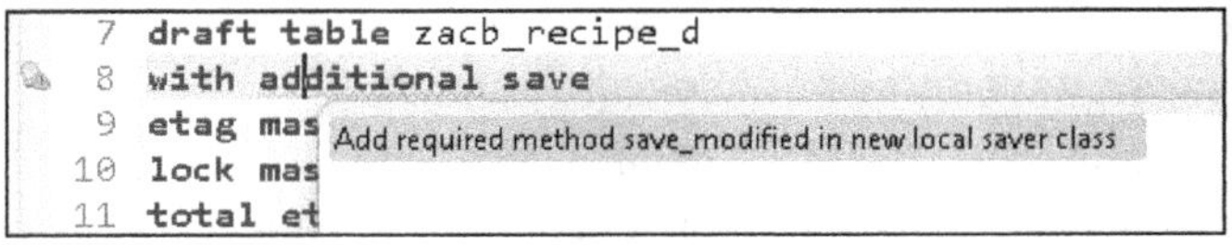

```
draft table zacb_recipe_d
with additional save
etag mas
lock mas
total et
```

Add required method save_modified in new local saver class

Figure 7.8 Quick Fix for Creating the save_modified Method

You're then automatically forwarded to the local types include, where a local saver class has been created, which inherits from the CL_ABAP_BEHAVIOR_SAVER class and provides an empty method implementation of the redefined save_modified method. We adapt this method right away. To link the change document update and the persistence of the application data in a transaction-safe manner, the entire LUW must be rolled back in the event of an error in the write method. To do this in the save_modified method, we change the base class to CL_ABAP_BEHAVIOR_SAVER_FAILED. The adapted class structure is shown in Listing 7.12.

```
CLASS lsc_zacb_r_recipe DEFINITION
  INHERITING FROM cl_abap_behavior_saver_failed.
  PROTECTED SECTION.
    METHODS save_modified REDEFINITION.
ENDCLASS.
CLASS lsc_zacb_r_recipe IMPLEMENTATION.
  METHOD save_modified.
  ENDMETHOD.
ENDCLASS.
```

Listing 7.12 Structure of the Local Saver Class

You can use the create, update, and delete parameters to obtain internal tables with the changes to the entities marked with additional save. Messages can be returned via the reported parameter. The failed parameter added by changing the base class makes it possible to force an abort.

Implementing the "write" method

It's now your task to convert the changes provided via these parameters into the database-oriented format of the change document objects to enrich them with additional data and to determine the previous data to then correctly call the write method of your change document object for the creation, modification, and deletion of instances of your business object, including subordinate item tables. This implementation is so extensive that printing it in its entirety would go beyond the scope of this chapter. It's therefore included in full with the sample objects in the download material. You can find it in the local types include of the ZBP_ACB_R_RECIPE class in the local LSC_ZACB_R_RECIPE class.

In the following, we only provide a few tips for implementation:

- You can use the MAPPING FROM ENTITY addition of the CORRESPONDING constructor expression to automatically compare the identifiers and convert between Pascal case and Snake case:

```
recipe_new = CORRESPONDING #(
  <create_recipe> MAPPING FROM ENTITY  )
```

- You can use the MAPPING FROM ENTITY USING CONTROL addition to take the control structures into account so that only changed fields are transferred to the target data object:

```
<change>-recipe_new = CORRESPONDING #(
  BASE ( <change>-recipe_old )
  <update_recipe> MAPPING FROM ENTITY USING CONTROL )
```

- The with additional save statement offers a with full data addition. When you use it, all fields will be transferred to the save_modified method, not just the changed fields. This can help you with the implementation.
- By changing the base class to CL_ABAP_BEHAVIOR_SAVER_FAILED, you can abort in the event of an error by supplying the failed parameter.

7

7.4 Displaying Change Documents

You can display the written change documents in SAP S/4HANA and SAP S/4HANA Cloud Private Edition in the classic way via Transaction RSSCD100 (see Figure 7.9). From a technical perspective, only a wrapper around the old function modules has been implemented here, so that the data is still stored in database tables CDHDR and CDPOS.

Reusable component

A comparable SAP Fiori app doesn't exist. A reusable component named sap.nw.core.changedocs.lib.reuse.changedocscomponent is offered in the Reuse Library, which you can integrate into your custom SAP Fiori app.

Release-Dependent Integration of Change Documents

With the announced native integration of change documents in ABAP RESTful application programming model business objects, it should be possible in the future to integrate change documents directly into your apps via SAP Fiori elements and annotations in the metadata extension.

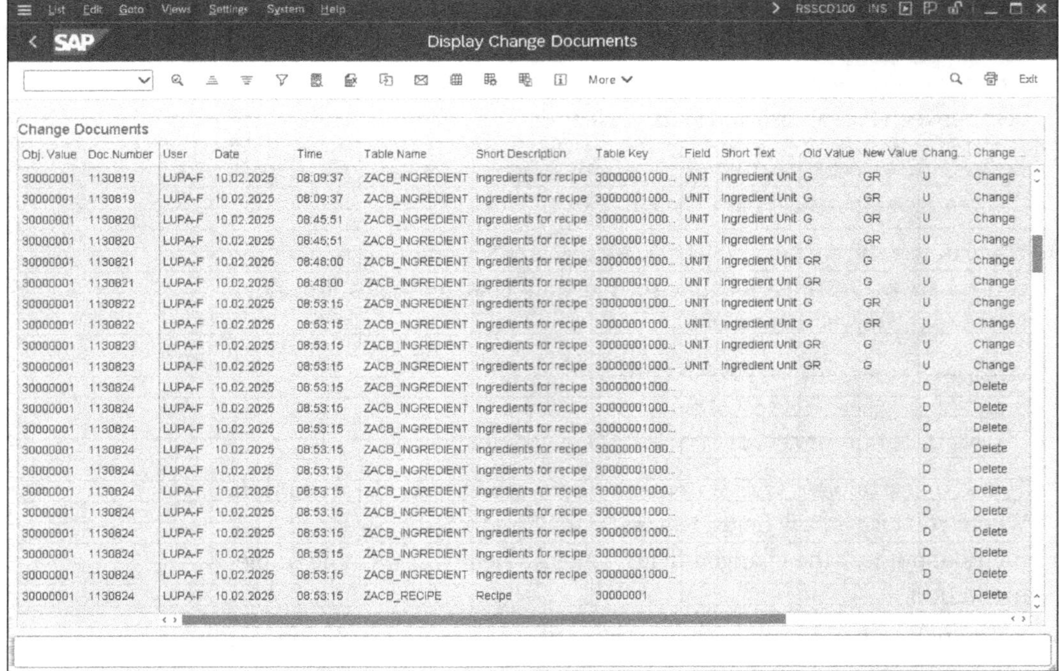

Figure 7.9 Display of Change Documents in Transaction RSSCD100

Program-based analysis of change documents

An alternative approach is the program-based analysis of change documents. The CL_CHDO_READ_TOOLS class is provided for this purpose. In Listing 7.13, the changes of the current day are selected and displayed.

```
TRY.
    cl_chdo_read_tools=>changedocument_read(
      EXPORTING
        i_objectclass    =
          zcl_zacb_recipe_chdo=>objectclass
        i_date_of_change =
          cl_abap_context_info=>get_system_date( )
      IMPORTING
        et_cdredadd_tab  = DATA(change_document_lines) ).
    out->write( change_document_lines ).
  CATCH cx_chdo_read_error INTO DATA(error).
    out->write( error->get_text( ) ).
ENDTRY.
```

Listing 7.13 Read Access to Change Documents

7.5 Summary

In this chapter, you've learned about the options for creating change documents using change document objects in the system. After some familiarization, the generated class provides a simple method that takes care of the manual comparison of old and new data and generates the documents. In the more complex scenario of implementing a ABAP RESTful application programming model object, this simple method is much more difficult to provide, as generically all types and combinations of changes to the underlying tables must be taken into account. The native integration announced by SAP should make this much easier in future releases.

Chapter 8
Lock Objects

An SAP database is used by many users and programs at the same time. Conflicts often arise when transactions are carried out on the same database objects.

Lock mechanisms are used to prevent conflicts when database objects are accessed simultaneously by multiple users. Lock mechanisms follow the *ACID principle* according to which a database transaction is indivisible (atomicity) and isolated from other transactions (isolation). Once the transaction has been completed, the data is in a consistent state (consistency), and the changes are permanent (durability).

Structure of this chapter

In this chapter, we first present the general concept of lock mechanisms in Section 8.1. In Section 8.2, we then describe how the lock mechanisms are implemented in ABAP Cloud. Section 8.3 describes how you can create lock objects in ABAP Cloud. Before the integration of the lock mechanisms in our recipe portal sample application is explained in Section 8.5, we present the new API for locks in ABAP Cloud in Section 8.4.

Simultaneous editing of recipes

For this chapter, we're expanding our application scenario so that it's no longer possible for multiple users to edit entries in the recipe portal at the same time. To do this, we create a lock object based on the ZACB_RECIPE database table. In addition to the standard integration of this lock object by the ABAP RESTful application programming model framework, we describe a possible lock mechanism in the unmanaged scenario on this basis.

[«]

Using the Sample Application

The objects created in this chapter are located in the LOCK subpackage.

8.1 Lock Mechanisms in the Database Environment

Lost update anomaly

Figure 8.1 shows which problems would occur without locks in our database table for the ingredients.

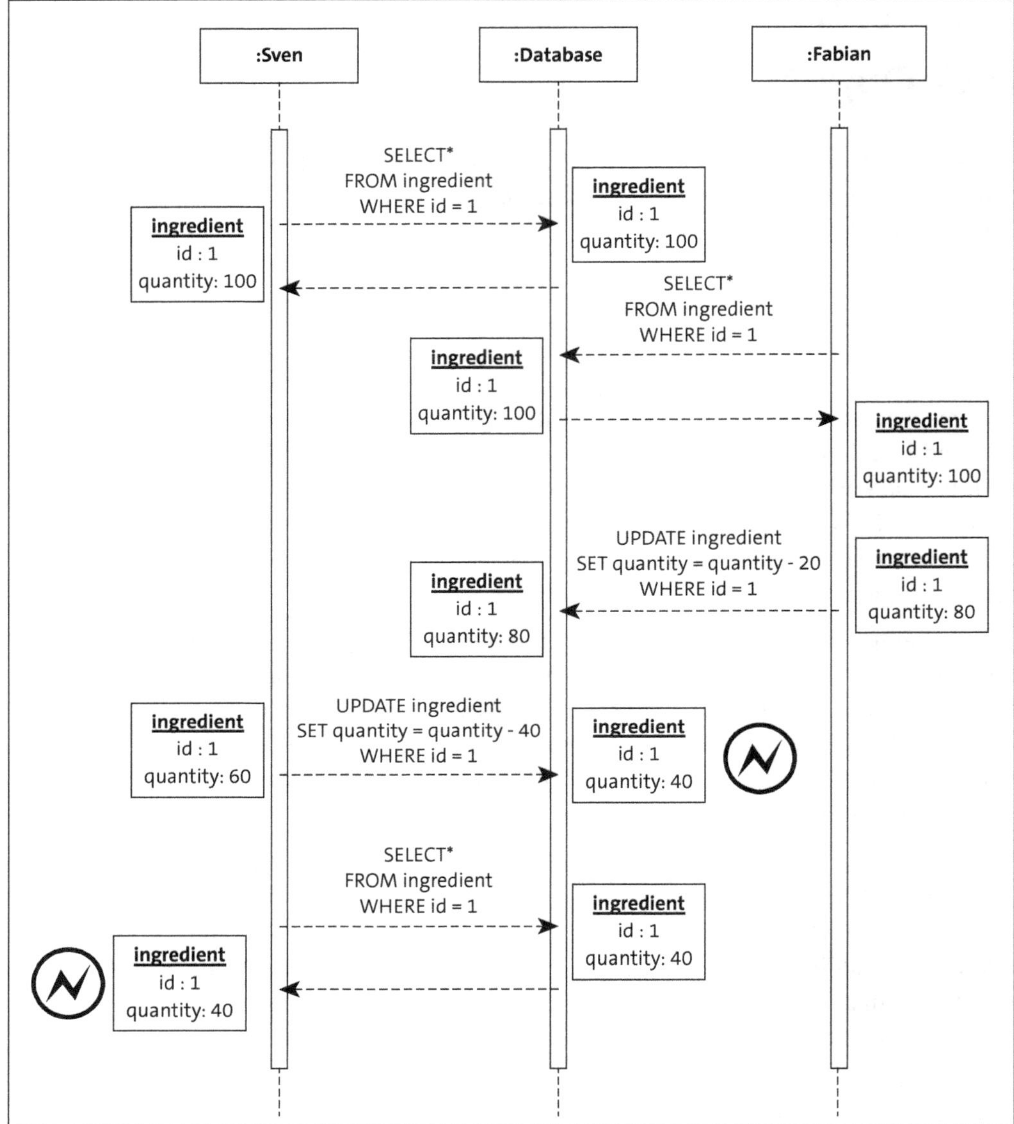

Figure 8.1 The Lost Update Anomaly in Our Sample Application

The diagram shows the following situation:

1. Sven reads the quantity of ingredients with the value 100.
2. Fabian reads the quantity of ingredients, changes the quantity from 100 to 80 and saves his changes.
3. Meanwhile, Sven's transaction continues. He thinks the quantity of ingredients is still 100. He subtracts a quantity of 40 and therefore assumes that the new quantity of ingredients is 60.

4. As Fabian's change has been saved, a quantity of 40 ingredients is deducted from the saved quantity of 80. This reduces the quantity stored in the database to 40.

To prevent such a *lost update anomaly* from occurring, you can use database locks. This way, you can reserve parts of a database for a transaction. There are two different lock mechanisms: *pessimistic* and *optimistic*.

Pessimistic lock mechanism

The pessimistic lock mechanism ensures that no conflict can arise before a transaction starts. The lock isn't removed until the transaction has been completed. This lock mechanism is shown in the context of our example in Figure 8.2. The aim of the pessimistic lock is to avoid conflicts through the use of locks.

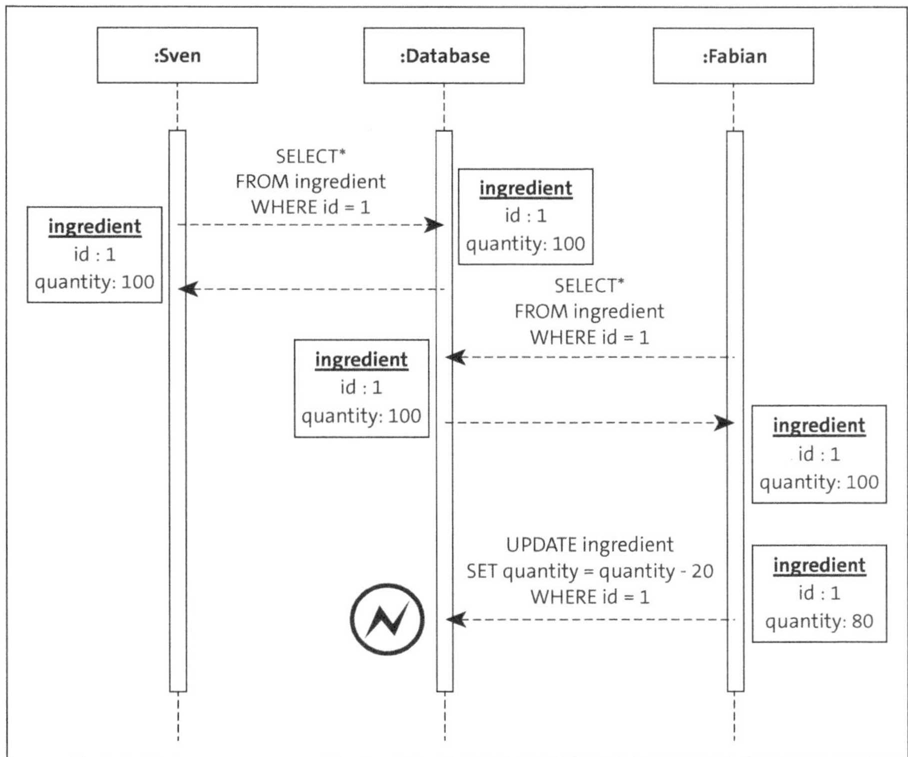

Figure 8.2 Pessimistic Lock in Our Recipe Portal

Both Sven and Fabian receive a *read lock* (shared lock) for the relevant row of the ingredient table when reading. As both Sven and Fabian have the read lock for the ingredient data record with ID 1, neither of them can change the ingredient quantity until one of them releases the read lock by completing the editing of the data record. This is because a write operation requires the acquisition of a write lock (exclusive) and read locks prevent write locks. For this reason, Fabian's update is locked until Sven releases the shared lock.

Optimistic lock mechanism

With the optimistic lock mechanism, on the other hand, it's assumed that there's no conflict during execution and all changes get implemented. The validity of the transaction isn't checked until it's committed, and it will be canceled if there are any conflicts.

The optimistic lock mechanism (see Figure 8.3) enables conflicts, but these must be recognized at the time of writing.

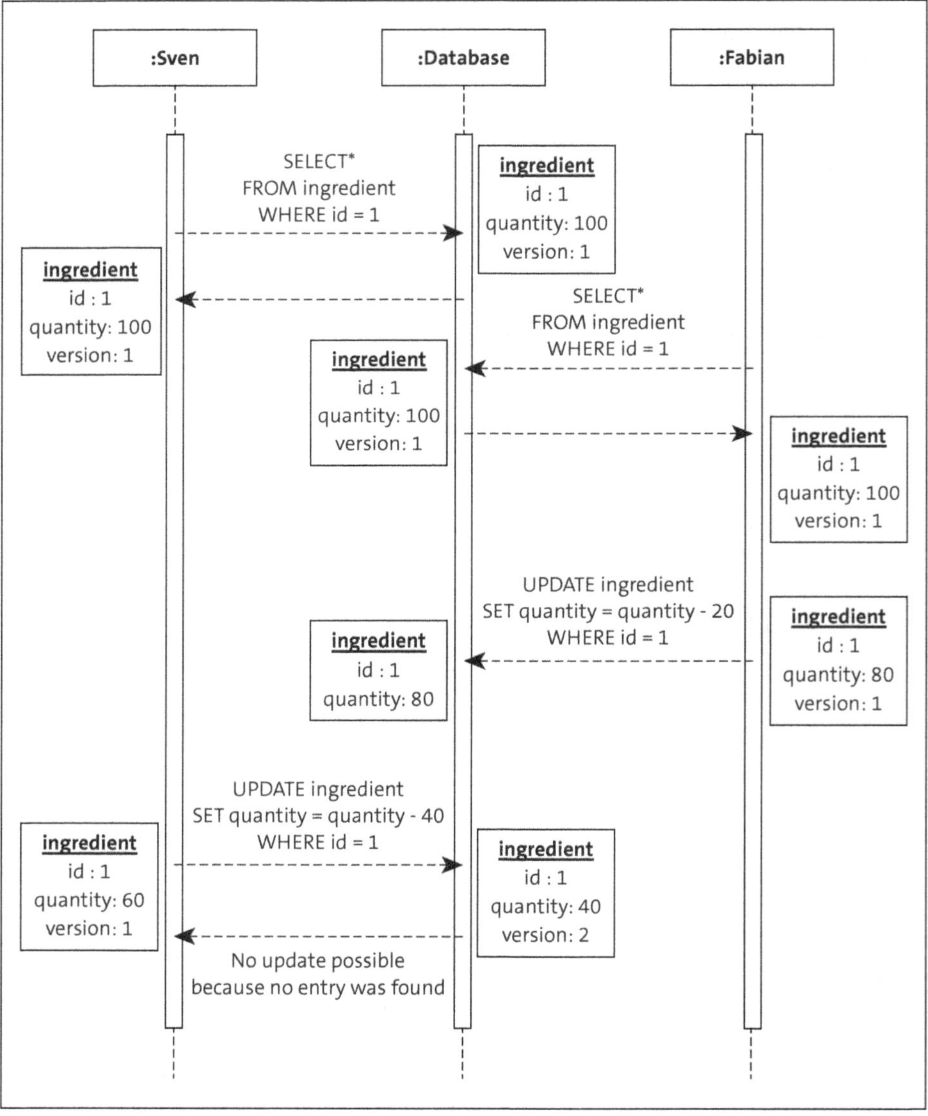

Figure 8.3 Optimistic Lock in Our Recipe Portal

For this purpose, the database uses a lock mechanism. In our example, we recreate this lock mechanism with a version column. This lock mechanism is executed each time an UPDATE or DELETE statement is run. It's also used for simultaneous adjustment during a parallel modification operation.

When reading the ingredient data record, both users read the current version of this data record. However, if Fabian changes the quantity of ingredients, he also changes the version of the data record from 1 to 2. If Sven then wants to change the ingredient quantity, his UPDATE statement doesn't match any data record because the value in the version column is no longer 1 but 2. The mechanism communicates this, and the UPDATE operation gets reset. Thus, the subsequent update is prevented by resetting the following transactions that use status data.

8.2 Locks on the ABAP Platform

Locks are also used in SAP applications. The system sets a database lock during the execution of an application, whereby the database locks are kept as short as possible. An SAP lock can exist across multiple database LUWs. In contrast, a database lock only exists during a database LUW in which the changes made in the SAP system are actually posted.

[«]

Logical Unit of Work (LUW)

The smallest unit in which something is updated in the database is referred to as a *database logical unit of work (database LUW)*. A logically coherent unit whose changes are made within a single database LUW is referred to as an *SAP LUW*.

Lock server

The *lock server* (ENQUEUE server) manages the locks on the ABAP platform. The lock server receives the locks and uses the *lock table*, which stores all active locks, to check whether the lock request collides with a lock that has already been set. If it already appears in the lock table, the request will be rejected. Otherwise, the lock will be set and make the corresponding entry in the lock table.

Lock objects in the SAP system

To enable you to use these mechanisms for your customer-specific applications, the ABAP platform provides *lock objects*. The tables in which data records are to be locked with a lock request are specified with their key fields in a lock object.

In the simplest case, a lock object consists of exactly one table, and the *lock argument* of this table is the primary key of this table. The lock argument consists of the key fields of the database table to be locked. It's possible that multiple tables are included in a lock object. A lock request can then be used to not only lock a record from a table, but an entire logical object. In our example, this could be a recipe with its ingredients.

8.3 Using Lock Objects

Creating a lock object

To create a lock object in the system, the following steps must be carried out:

1. Right-click on the package, and select **New • Other ABAP Repository Menu** from the context menu.
2. In the dialog box that opens (see Figure 8.4), you can enter the keyword search to search for the ABAP Dictionary object "Lock Object". Select **Lock Object**, and click on the **Next** button.

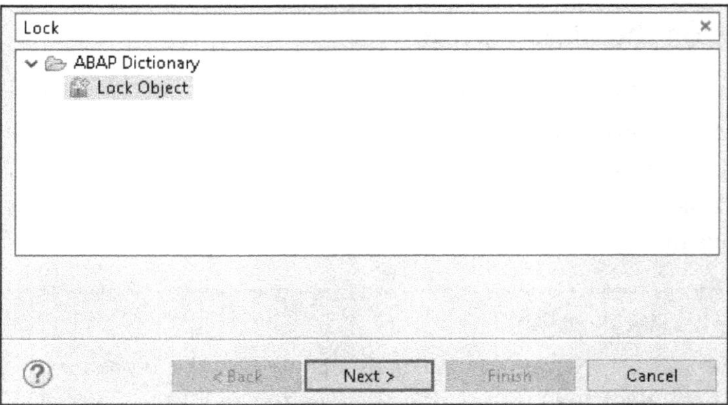

Figure 8.4 Searching for the Lock Object

3. Enter a suitable name (in our example, "EZACB") and a description. In the **Primary Table** field, enter the name of the database table for which you want to create a lock object (see Figure 8.5).

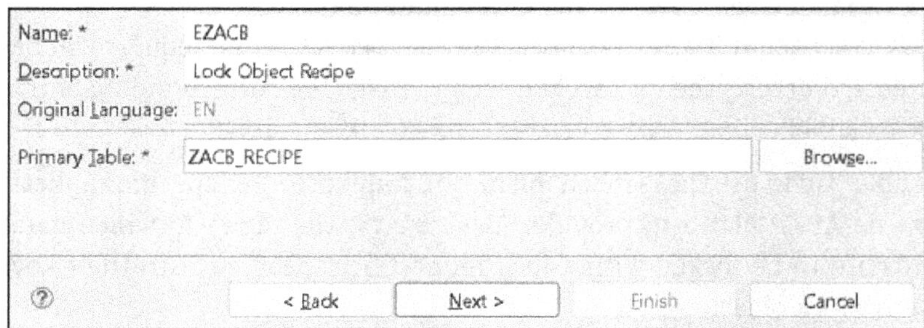

Figure 8.5 Creating a Lock Object

[«]

Name and Length of Lock Objects

The names of lock objects must always start with EY* or EZ* in the customer namespace. The name may have a maximum length of 16 characters.

4. Close the dialog box by clicking the **Next** button. This will create the lock object.
5. Before the object can be activated, the **Lock Mode** must be defined (see Figure 8.6). The most common lock modes are described in Table 8.1. For our example, we want to use a write lock.

8

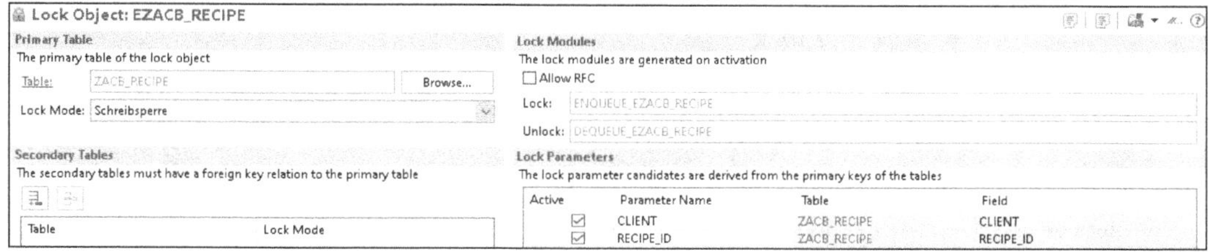

Figure 8.6 Creating the Lock Object

Lock Mode	Description
Read lock (S = Shared)	A read lock can be set by multiple users. If a read lock is set, a write lock will no longer be accepted.
Write lock (E = Exclusive)	A write lock locks the object against all other types of lock requests from other users. Only the same user can request an additional lock.
Optimistic lock (O = Optimistic)	This lock initially behaves like a read lock and can then be converted into a write lock.

Table 8.1 The Most Common Lock Modes

6. You should also pay attention to the following settings:
   - **Allow RFC**: If this indicator is enabled, the generated function modules are RFC-capable; that is, they can be called from other systems via the *remote function call* (RFC) protocol (can't be used in SAP S/4HANA Cloud Public Edition). In our example, we don't want to use it.
   - **Secondary Tables**: Additional tables with foreign key relationships can be added. This isn't necessary either for our example.

You can now activate the lock object. This will trigger the creation of the function modules. After that, you can use the generated lock object.

8.4 API for Lock Objects

Once the lock object has been activated, the system automatically generates the function modules for setting and releasing the locks. However, the direct use of these function modules is no longer permitted in ABAP Cloud. The function modules were therefore implemented here as a general application programming interface (API). You can use this API to set and release locks.

Possible API calls

In the following paragraphs, we present the API calls to execute the following actions:

- Instantiating a lock object
- Setting a lock
- Releasing a lock
- Releasing all locks

Instantiating lock objects

The API is provided via various ABAP classes. Using the GET_INSTANCE method of the CL_ABAP_LOCK_OBJECT class, it's possible to instantiate a lock object to perform further operations on the lock object. Listing 8.1 shows the implementation of this method for our sample application.

```
TRY.
    DATA(lock) = cl_abap_lock_object_factory=>get_instance(
      iv_name = 'EZACB_RECIPE' ).
  CATCH cx_abap_lock_failure
    INTO DATA(exception).
    RAISE SHORTDUMP exception.
ENDTRY.
```

Listing 8.1 The GET_INSTANCE Method

The name of the lock object (IV_NAME) is passed as a parameter, and the lock object is returned as an instantiated object of type IF_ABAP_LOCK_OBJECT. In addition to the parameter, the method can throw an error class (CX_ABAP_LOCK_FAILURE) if the lock object to be instantiated doesn't exist.

Setting a lock

Once the lock object has been created, a lock entry can be generated. The ENQUEUE method of the IF_ABAP_LOCK_OBJECT object is implemented for this purpose. Listing 8.2 shows the corresponding code for our example.

```
TRY.
    lock->enqueue(
      it_parameter  = VALUE #(
        ( name        = 'RECIPE_ID'
          value       = REF #( '12345' ) )
```

```
    )
      it_table_mode = VALUE #(
        ( table_name = 'ZACB_RECIPE'
          mode       =
        if_abap_lock_object=>cs_mode-write_lock ) )
    ).
  CATCH cx_abap_foreign_lock cx_abap_lock_failure
    INTO DATA(exception).
    RAISE SHORTDUMP exception.
ENDTRY.
```


Listing 8.2 The ENQUEUE Method

We transfer a table with values that are to be locked (IT_PARAMETER) to the method. We can also specify a lock mode (IT_TABLE_MODE), which determines whether a different lock mode should be assigned than was initially entered in the lock object.

In addition, a parameter can be added that specifies who owns the lock (_SCOPE) with the following values:

- **_SCOPE=1**
  The lock belongs only to the dialog owner.
- **_SCOPE=2**
  The lock belongs only to the owner of the update.
- **_SCOPE=3**
  The lock belongs to both of them.

The last parameter (_WAIT) can be specified to check whether a locking attempt can be repeated for a certain period of time. If this parameter isn't specified, the locking attempt will get canceled after the first failure.

In addition to the parameters, the following error classes can be used:

- **CX_ABAP_FOREIGN_LOCK**
  This error displays if there's already a lock on the entry.
- **CX_ABAP_LOCK_FAILURE**
  This error occurs if there's another error, for example, that the entry to be locked doesn't exist, and therefore no lock can be set.

Releasing a lock

Once the lock has been set, the lock can be removed again using the DEQUEUE method. A lock is normally removed by the SAP LUW (see Listing 8.3).

```
TRY.
    lock->dequeue(
      it_parameter  = VALUE #(
        ( name        = 'RECIPE_ID'
```

```
            value       = REF #( '12345' ) )
      )
        it_table_mode = VALUE #(
          ( table_name = 'ZACB_RECIPE'
            mode       =
          if_abap_lock_object=>cs_mode-write_lock ) )
      ).
    CATCH cx_abap_foreign_lock  cx_abap_lock_failure
      INTO DATA(exception).
      RAISE SHORTDUMP exception.
ENDTRY.
```

Listing 8.3 The DEQUEUE Method

The method also has the _SCOPE parameter, the lock mode, and a table that specifies the entries for which the locks are to be removed. However, the method doesn't have the _WAIT parameter, but the _SYNCHRONOUS parameter, which can be used to determine whether the locks should be lifted synchronously or asynchronously.

The method uses the CX_ABAP_LOCK_FAILURE class as the error class. This error occurs if there's an internal error, for example, that no unlocking function module exists for the lock object.

Releasing all locks In addition to these three main methods, the DEQUEUE_ALL method of the CL_ABAP_LOCK_OBJECT_FACTORY class is available. This method is used to remove all locks from the current LUW. It should only be used in an emergency and would be implemented as follows:

```
cl_abap_lock_object_factory=>dequeue_all(
  _synchronous = if_abap_lock_object=>cs_synchronous-yes ).
```

The only transfer parameter specified is whether the locks should be released synchronously or, if no value is transferred, asynchronously.

Constants In addition to these methods, the method provides constants for use via the IF_ABAP_LOCK_OBJECT interface, as you've already seen in various code sections in the previous listings. The most common constants are listed in Table 8.2.

Structure	Element	Meaning
CS_MODE	WRITE_LOCK	Write lock
CS_MODE	SHARED_LOCK	Read lock

Table 8.2 Constants in the IF_ABAP_LOCK_OBJECT Interface

Structure	Element	Meaning
CS_MODE	OPTIMISTIC_LOCK	Optimistic lock
CS_SYNCHRONOUS	YES	Synchronous
CS_SYNCHRONOUS	NO	Asynchronous
CS_WAIT	YES	Perform lock again
CS_WAIT	NO	Don't lock again

Table 8.2 Constants in the IF_ABAP_LOCK_OBJECT Interface (Cont.)

You've now become familiar with the various methods of the API for lock objects. In the following, we'll use both the pessimistic and the optimistic lock method.

Pessimistic lock mechanism

Listing 8.4 shows the implementation of the pessimistic lock mechanism.

```
TRY.
    DATA(lock) = cl_abap_lock_object_factory=>get_instance(
      iv_name = 'EZACB_RECIPE' ).

    lock->enqueue(
      it_parameter  = VALUE #(
        ( name        = 'RECIPE_ID'
          value       = REF #( '12345' ) )
    )
      it_table_mode = VALUE #(
        ( table_name = 'ZACB_RECIPE'
          mode       =
        if_abap_lock_object=>cs_mode-exclusive_lock ) )
    ).

    "Change operation on DB

  CATCH cx_abap_foreign_lock cx_abap_lock_failure
    INTO DATA(pessimistic_exception).
    RAISE SHORTDUMP pessimistic_exception.
ENDTRY.
```

Listing 8.4 Implementation of the Pessimistic Lock Mechanism

First, we instantiate the EZACB_RECIPE lock object and set an exclusive lock (if_abap_lock_object=>cs_mode-exclusive_) on the recipe with the ID 12345. Once this code has been executed, a lock will be set in the lock table. The fact that the lock is set can be checked in Transaction SM12 in the **Mode** column using the value **X** (see Figure 8.7). As an alternative to Transaction SM12, the lock can also be checked in the Maintain User Sessions app (app ID F6049).

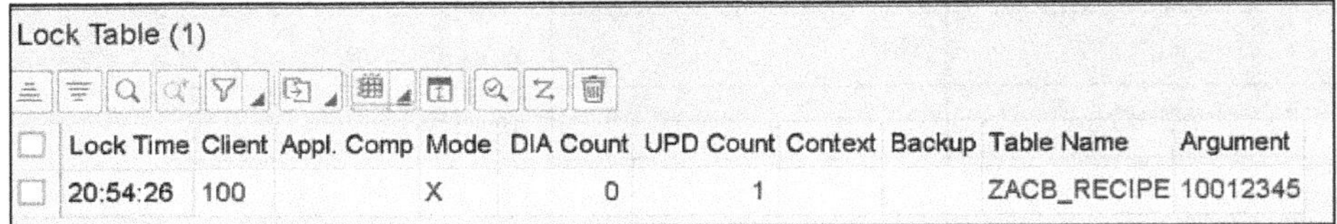

Figure 8.7 Checking the Exclusive Lock in Transaction SM12

Lock argument

The lock argument in the **Argument** column is made up of the client and the key. The lock entry contains the table name in addition to this.

[»]

Maintain User Sessions App

Information on the Maintain User Sessions app (app ID F6049) can be found in the SAP Fiori Apps Reference Library at *http://s-prs.de/v1064806*. First, you need to select your release. On the **IMPLEMENTATION INFORMATION** tab, you'll then find all the information you need to configure the SAP Fiori app in your system, for example, which standard roles are available and which Open Data protocol (OData) services need to be activated.

The exclusive lock is automatically removed by the SAP LUW. When that happens, the corresponding entry will no longer be displayed in Transaction SM12 (see Figure 8.8).

Figure 8.8 Removal of the Exclusive Lock in Transaction SM12

Optimistic lock mechanism

The optimistic lock mechanism is implemented in Listing 8.5.

```
TRY.
    DATA(lock) = cl_abap_lock_object_factory=>get_instance(
      iv_name = 'EZACB_RECIPE' ).

    " Read operation on DB
    lock->enqueue(
      it_parameter  = VALUE #(
        ( name        = 'RECIPE_ID'
          value       = REF #( '12345' ) )
    )
```

```
      it_table_mode = VALUE #(
        ( table_name = 'ZACB_RECIPE'
          mode       =
                     if_abap_lock_object=>cs_mode-optimistic_lock ) )
    ).

    " Make change to DB
    lock->enqueue(
      it_parameter  = VALUE #(
        ( name       = 'RECIPE_ID'
          value      = REF #( '12345' ) )
    )
      it_table_mode = VALUE #(
        ( table_name = 'ZACB_RECIPE'
          mode       =
            if_abap_lock_object=>cs_mode-promote_optimistic_lock ) )
    ).

    "Change operation on DB
  CATCH cx_abap_foreign_lock cx_abap_lock_failure
    INTO DATA(optimistic_exception).
    RAISE SHORTDUMP optimistic_exception.
ENDTRY.
```

Listing 8.5 Implementation of the Optimistic Lock Mechanism

This time, an exclusive lock (`if_abap_lock_object=>cs_mode-exclusive`) isn't set immediately after instantiating the lock object, but an optimistic lock (`if_abap_lock_object=>cs_mode-optimistic_lock`). In the lock table in Transaction SM12, the value **O** is set in the **Mode** column (see Figure 8.9).

Lock Table (1)

Lock Time	Client	Appl. Comp	Mode	DIA Count	UPD Count	Context	Backup	Table Name	Argument
20:56:58	100		O	0	1			ZACB_RECIPE	10012345

Figure 8.9 Setting the Optimistic Lock in a Lock Table

The optimistic lock is then converted into an exclusive lock (`if_abap_lock_object=>cs_mode-promote_optimistic_lock`). This can be recognized in the lock table by the value **E** in the **Mode** column (see Figure 8.10). Finally, the lock is removed again by the SAP LUW.

Lock Table (1)

Lock Time	Client	Appl. Comp	Mode	DIA Count	UPD Count	Context	Backup	Table Name	Argument
20:56:58	100		E	0	1			ZACB_RECIPE	10012345

Figure 8.10 Converted Lock in the Lock Table

8.5 Integration into the Sample Application

In this section, we'll first show you how the lock behavior is integrated into our SAP Fiori app by default.

Automatic lock mechanism in the ABAP RESTful application programming model

To use the automatic lock mechanism of the ABAP RESTful application programming model framework, you must specify the `lock master` annotation in the root entity (`ZR_ACB_RECIPE`) (see Listing 8.6).

```
define behavior for ZR_ACB_RECIPE alias Recipe
persistent table zacb_recipe
draft table ZACB_RECIPE_D
etag master LocalLastChangedAt
lock master
total etag LastChangedAt
authorization master( global )
```

Listing 8.6 Addition to the Lock Master

The dependent entities also require the `lock dependent` annotation as well as the specification as to which entity the lock is dependent on (`by _RECIPE`). The associated entities are supplemented as follows:

```
define behavior for ZR_ACB_INGREDIENT alias Ingredient
persistent table zacb_ingredient
lock dependent by _Recipe
```

Lock entry

In the lock table, you can see that there's an exclusive lock for the recipe (see Figure 8.11).

Lock Table (1)

Lock Time	Client	Appl. Component	Mode	DIA Count	UPD Count	Context	Backup	Table Name	Lock Argument
21:02:33	100	BC-ESI-RAP-CSP	E	0	1	BO		SCSP_S_SRV_ENQ_PRIMARY_KEY	100 ZACB_R_RECIPE 00002

Figure 8.11 Display of the Lock Table in Transaction SM12

However, as you can see in Figure 8.11, the lock entry looks different here than shown in the previous section for the implementation of the lock via the lock object API. The lock argument consists of the client, the entity, and the key. In addition, the values of the table name and the context differ from the values of a normal lock entry. This is due to the automatic lock implementation in the ABAP RESTful application programming model framework. If two users want to edit a recipe at the same time, they will receive the error message shown in Figure 8.12.

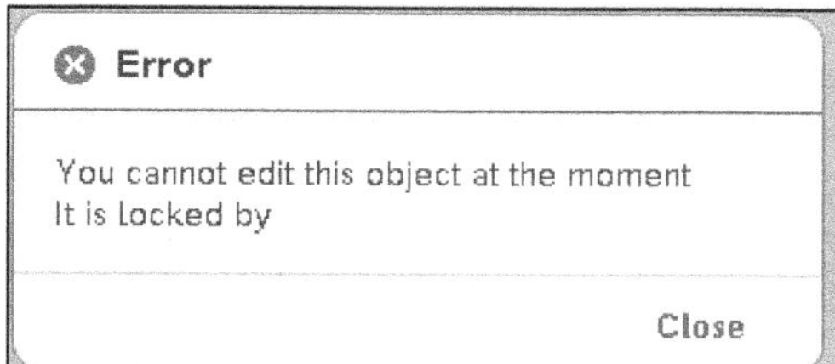

Figure 8.12 Error Message for a Locked Entry

Custom lock logic in an ABAP RESTful application programming model application

This is the default behavior for locking an ABAP RESTful application programming model application in the managed scenario. If you don't want to use the predefined scenario for locks, but instead want to lock client-independent tables, for example, you can develop a lock yourself in the unmanaged scenario. In this case, you must implement the lock behavior yourself. We also show this here for our sample application.

First, you need to add the `unmanaged` addition to the `lock master` command:

```
lock master unmanaged
```

Then, a warning gets displayed in the behavior definition stating that the `FOR LOCK` method must be implemented for the `lock master unmanaged` scenario (see Figure 8.13).

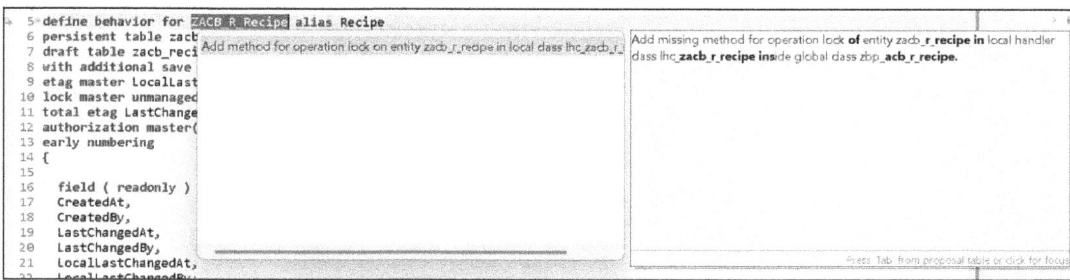

Figure 8.13 ABAP Development Tools Message on the FOR LOCK Method

By double-clicking on the method, the method will automatically be created in the class for the behavior definition. This method is automatically called by the framework before a change process, such as an update, is called.

Definition of the FOR LOCK method

The method definition of the predefined FOR LOCK method in the behavior definition reads as follows:

```
METHODS lock FOR LOCK
  IMPORTING keys FOR LOCK recipe.
```

The lock method provides implicit CHANGING parameters for this purpose:

- The failed parameter is used to log the causes when a lock fails.
- The reported parameter is used to save messages about the cause of the error.

Method implementation

The method implementation in Listing 8.7 shows one way in which the lock behavior can be developed in the unmanaged scenario.

```
TRY.
    DATA(lock) = cl_abap_lock_object_factory=>get_instance(
        iv_name = 'EZACB_RECIPE' ).

    LOOP AT keys ASSIGNING FIELD-SYMBOL(<recipe>).
      TRY.
          lock->enqueue(
              it_parameter  = VALUE #( (
                name = 'RECIPE_ID'
                value = REF #( <recipe>-recipeid ) ) )
          ).
        CATCH cx_abap_foreign_lock INTO DATA(foreign_lock).
          APPEND VALUE #( recipeid =
             <recipe>-recipeid ) TO failed-recipe.
          APPEND VALUE #( recipeid =
             <recipe>-recipeid
                %msg = new_message(
                    id = 'ZACB_COMMON'
                        number = '000'
                        v1 = <recipe>-recipeid
                        v2 = foreign_lock->user_name
                        severity = CONV #( 'E' ) )
         ) TO reported-recipe.
      ENDTRY.
    ENDLOOP.

  CATCH cx_abap_lock_failure INTO DATA(exception).
    RAISE SHORTDUMP exception.
ENDTRY.
```

Listing 8.7 FOR LOCK Method Implementation

Logic of the custom lock implementation

To set a lock in the code (Listing 8.7), the `EZACB_RECIPE` lock object must be instantiated first. This is done using the `GET_INSTANCE` method of the `CL_ABAP_LOCK_FACTORY_OBJECT` class. Then, the `ENQUEUE` method is called by the lock object and the value of the key field for executing the lock is specified. If the instance is already locked, a short dump (`CX_ABAP_FOREIGN_LOCK`) will be triggered. An individual error message to be displayed can be developed within the interception block of this short dump.

Using the custom lock logic

Now, we want to check how the implementation affects the behavior of the SAP Fiori app and the lock table. The lock table in Transaction SM12 now contains an entry for an exclusive lock (see Figure 8.14). As you can see, the table name and the argument from the client and key field are stored there again.

Lock Table (1)

Lock Time	Client	Appl. Comp	Mode	DIA Count	UPD Count	Context	Backup	Table Name	Argument
21:07:51	100		E	0	1	BO		ZACB_RECIPE	10000001

Figure 8.14 Display of the Lock Table for the Custom Lock Logic

The error message from Figure 8.15 will be displayed in the case of simultaneous editing. We've added it ourselves in the `lock` method as the value of the `reported` parameter (see Listing 8.7).

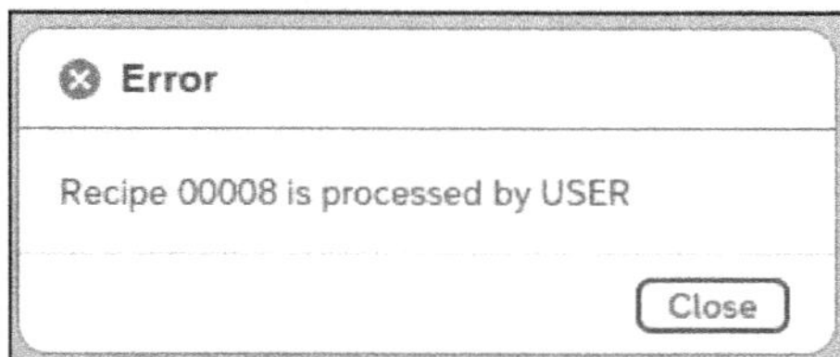
Error
Recipe 00008 is processed by USER
Close

Figure 8.15 Error Message in the Unmanaged Scenario

8.6 Summary

This chapter deals with the topic of lock objects. First, we explained the general problem of locking in databases. Locks are important so that the databases have a consistent status. There are various lock mechanisms available that can be applied to databases. With the pessimistic lock mechanism, an exclusive lock is requested immediately as soon as a data record gets read. Other programs that use this data record will immediately receive an error message, even though no database operation to be changed has yet been carried out. In contrast, the optimistic lock sets an exclusive lock on the data record shortly before a database operation to be changed is required.

Unlike older language versions, it's no longer permitted to call the appropriate ENQUEUE and DEQUEUE function modules in ABAP Cloud. They have been replaced by various classes of the API for lock objects. Table 8.3 summarizes the most important methods of the classes provided by this API.

Class	Method	Description
CL_ABAP_LOCK_OBJECT_FACTORY	GET_INSTANCE	Instantiating the lock object
IF_ABAP_LOCK_OBJECT	ENQUEUE	Locking the data record using the lock object
IF_ABAP_LOCK_OBJECT	DEQUEUE	Unlocking the data record using the lock object
CL_ABAP_LOCK_OBJECT_FACTORY	DEQUEUE_ALL	Unlocking all possible locks of an LUW

Table 8.3 Available Classes of the API for Lock Objects

Using these methods, we've presented the implementation of optimistic and pessimistic lock mechanisms in ABAP Cloud. We've also explained the integration of a lock mechanism in our application scenario. The ABAP RESTful application programming model provides an automatic lock mechanism in the managed scenario. With the lock master and lock dependent annotations, locks are automatically available without the need for custom coding. Finally, we've shown that it's possible to implement a custom lock mechanism. The ABAP RESTful application programming model provides the FOR LOCK method for this purpose. This method can be redefined if the lock master is unmanaged.

Chapter 9

Number Range Objects

You probably know the following scenario: You save a data record and receive an error message that the number has already been assigned. You're surprised and check the facts. You notice that the same number is being used again and again. In this chapter, you'll learn why this problem is gone in ABAP Cloud.

In the SAP environment, reference is often made to number ranges. However, different people often understand this to mean different things. One of them refers to the number range object, the other to the number status, and the next to the number range interval. In the context of number range objects, it's therefore important to explain exactly which function is being talked about.

A number range is used to assign unique numbers (e.g., financial documents) to individual database entries for business objects. You can also use number ranges to meet legal requirements such as continuous and consecutive numbering. In this chapter, you'll get to know some useful facts about number ranges.

Structure of this chapter

In Section 9.1, you'll learn about the basics of number ranges. In Section 9.2, we show you how to create a number range object in the ABAP development tools for Eclipse and maintain it via an SAP Fiori app. Section 9.3 presents the new application programming interface (API) provided by ABAP Cloud for number ranges. Section 9.4 describes the various numbering options in the ABAP RESTful application programming model.

Automatic numbering for the recipe portal

Up to now, it has only been possible to create data records in our recipe portal and assign a number manually. This is going to change in this chapter. We'll integrate automatic numbering into the sample application. To do this, we create a number range object based on the `ZACB_RECIPE_ID` domain. We then implement three different types of numbering for our application scenario.

Using the Sample Application

The objects created in this chapter are located in the `NUMBER` subpackage.

9.1 Number Ranges in SAP Systems

As mentioned in the introduction to this chapter, different people understand different things when talking about number ranges. In this section, we'll therefore define the various concepts and relate them to each other.

Number ranges, objects, and intervals

A *number range* is used to assign a number to individual database entries for business objects. To be able to assign a number range to a business object, a technical *number range object* must first be defined. The scope of a number range is determined by the *number range interval*. An interval is delimited by the **From number** and **To number** fields.

Numbering

The numbers from the number range are assigned by the SAP system (internal numbering), but can also be assigned by a user (external numbering).

[!]

> **The Term "Number"**
>
> The term *number* is actually not quite correct. A number range can contain special characters and letters along with numbers.

9.2 Maintaining a Number Range

In this section, we'll show you how to create and maintain a number range. To be able to create a number range, you first need a domain that determines the data type. We've described how to create a domain in the ABAP development tools in Chapter 2, Section 2.2. Note that only the `NUMC` or `CHAR` data types may be selected for number ranges. Otherwise, you can't create a number range for the domain. Once the domain has been created, it can be used as a template for a number range object. We use the `ZACB_RECIPE_ID` domain as the basis for our number range object. Its data type is `NUMC`. We can use this domain to determine the numbers for new recipes via the number range object.

Creating a number range object

A number range object can be created using the ABAP development tools as follows:

1. Right-click to open the context menu of our package, and select **New • Other ABAP Repository Object**.
2. Search for the **Number Range Object** Numbering: internal type, select it, and click the **Next** button (see Figure 9.1).
3. Enter the following information in the next window (see Figure 9.2):
   - **Name**: The name may have a maximum length of 11 characters. For our example, we entered "ZACP_RECIP".
   - **Description**: Enter a suitable description for the number range object.

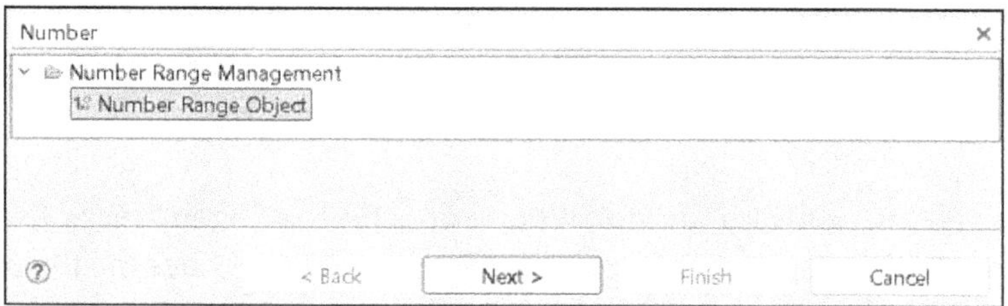

Figure 9.1 Opening the Number Range Object Creation Wizard

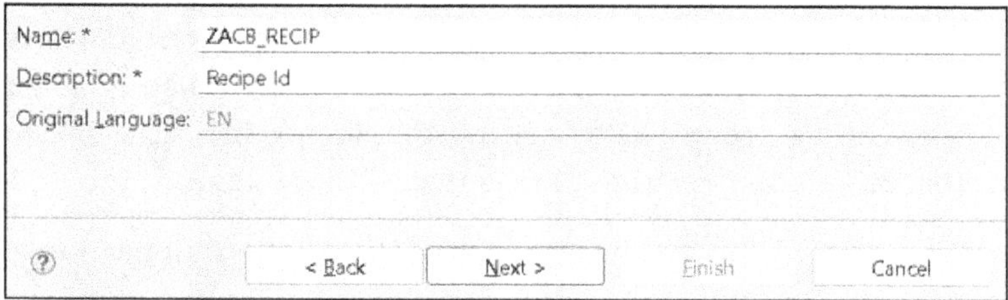

Figure 9.2 Creating a Number Range Object

4. Confirm your entries by clicking the **Next** button after which you can select a transport request. Exit the wizard by clicking the **Finish** button.

The number range object has now been created (see Figure 9.3).

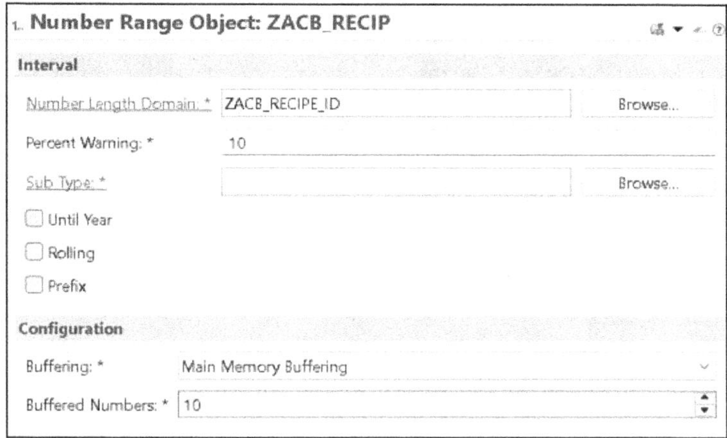

Figure 9.3 Created Number Range Object

Properties of the number range object

You can enter additional information on the number range object in the maintenance dialog shown in Figure 9.3:

- **Number Length Domain**
  You can enter the previously created domain from the ABAP Dictionary. It must be of type NUMC or CHAR and have a length of 1 to 20 characters.

- **Percent Warning**
  This number specifies the percentage of unassigned numbers per interval from which a warning will be issued during numbering.
- **Until Year**
  When you activate this indicator, it's also possible to differentiate by fiscal year. The prerequisite is that the application table provides the fiscal year as a key field. The specified fiscal year is regarded as the upper time limit.
- **Rolling**
  When the numbers of an interval have been used up, the system starts from the beginning by default, and the lowest number will be assigned again. This can be prevented by setting the **Rolling** indicator, which means that other numbers from the interval will be assigned.

Once you've made all the settings, you can activate the number range object.

Maintaining the number range interval

Once the number range object has been created, the number range interval must be maintained for the number range object. You can use the Manage Number Range Intervals app (app ID F4290) for this purpose.

The Manage Number Range Intervals App

All information on the Manage Number Range Intervals app can be found in the SAP Fiori apps reference library at *http://s-prs.de/1064807*. The **IMPLEMENTATION INFORMATION** tab lists all the prerequisites for making this SAP Fiori app available in your system.

Follow these steps for the maintenance:

1. To start the application, the SAP Fiori launchpad must first be started.
2. Click on the **Manage Number Range Intervals** tile there (see Figure 9.4).

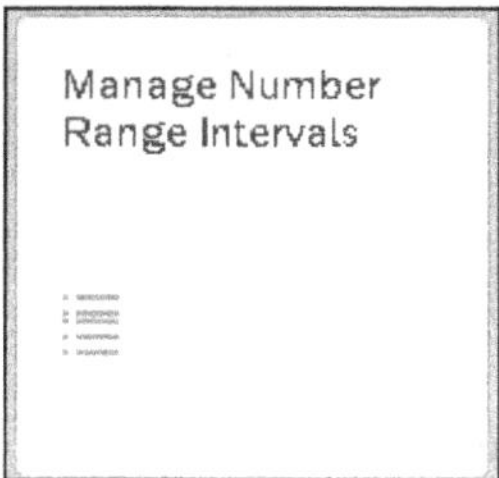

Figure 9.4 Manage Number Range Intervals App Tile

3. In the overview, select the previously created number range object **ZACB_RECIP** (see Figure 9.5).

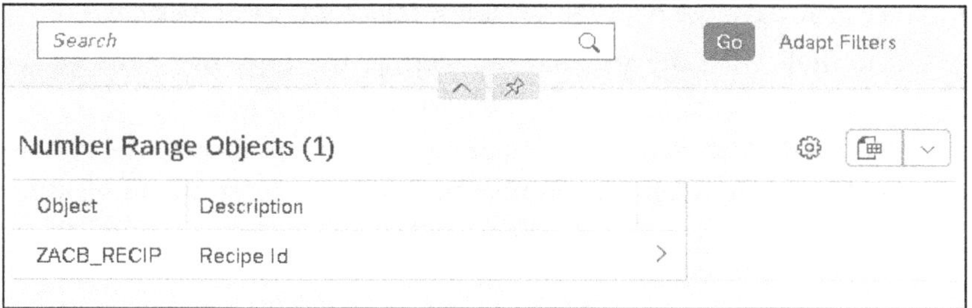

Figure 9.5 Overview of Number Range Objects

This takes you to the individual view of the number range object, which is where the information on the number range object is displayed, as shown in Figure 9.6.

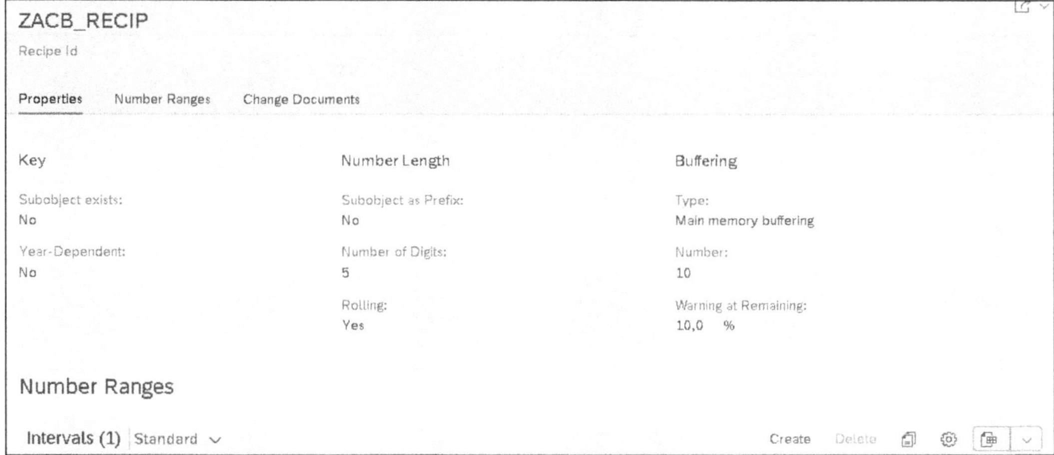

Figure 9.6 Number Range Object Created via a Class

4. Click on the **Create** button there to create a new number range interval.
5. Maintain the following values in the creation view (see Figure 9.7):
   - **Interval Number**: Enter a unique identification number here.
   - **Lower Limit/Upper Limit**: Define the interval in which the numbers may lie.

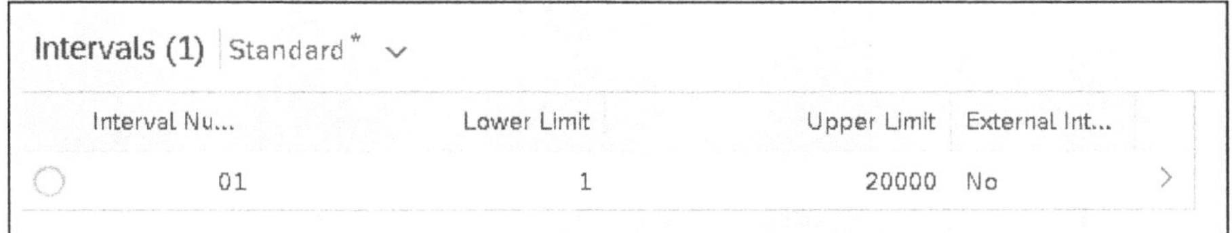

Figure 9.7 Maintaining a Number Range Interval

6. You can also choose between the following types of numbering:
   - **Internal**: With the internal numbering, the next free consecutive number in the corresponding interval is determined.
   - **External**: With external numbering, the user or an external system assigns the number. The number must be in the corresponding number range interval. The system doesn't check whether this number has already been used.
7. Click on the **Create** button to create the number range interval.

After saving, all important information about the number range interval, such as the current number status, the utilization (how many numbers have already been used), and the use of external numbering, is displayed in the corresponding row of the **Intervals** table. Click on the row to switch to change mode (see Figure 9.8).

Figure 9.8 Saved Number Range Interval

In change mode, you can also update the current value using the **Change Number Range Status** button (see Figure 9.9).

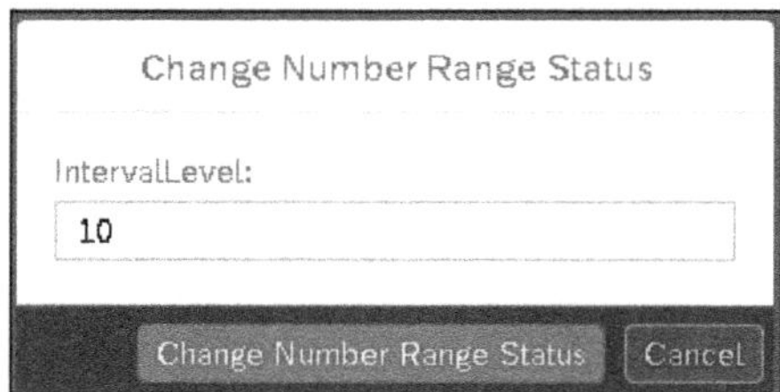

Figure 9.9 Change Number Range Status

9.3 API for Number Range Objects

In the previous section, we showed how you can manually maintain a number range. In this section, we present a new API that is available in ABAP Cloud as an alternative. This new API is object oriented and provides various classes. However, the old function modules are still technically used in the methods provided.

Classes and methods

Table 9.1 shows the most important classes and methods of the API.

Function	Class	Method
Create Number Range Object	CL_NUMBERRANGE_OBJECTS	CREATE
Change Number Range Object	CL_NUMBERRANGE_OBJECTS	CHANGE
Delete Number Range Object	CL_NUMBERRANGE_OBJECTS	DELETE
Create Number Range Interval	CL_NUMBERRANGE_INTERVALS	CREATE
Change Number Range Interval	CL_NUMBERRANGE_INTERVALS	CHANGE
Delete Number Range Interval	CL_NUMBERRANGE_INTERVALS	DELETE
Read Current Number	CL_NUMBERRANGE_RUNTIME	NUMBER_GET
Read Attributes of a Number Range Object	CL_NUMBERRANGE_OBJECTS	READ

Table 9.1 Important Classes and Methods of the API for Number Range Objects

Reading a number range object

In the following sections, we want to take a look at the use of the READ and NUMBER_GET methods of the CL_NUMBERRANGE_OBJECTS and CL_NUMBERRANGE_RUNTIME classes. The first method we present here is the READ method of the CL_NUMBERRANGE_OBJECTS class. This method can be used to read attributes of the number range object and its intervals. Listing 9.1 shows a possible implementation of this method, which we use to read the properties of the number range object.

```
TRY.
  cl_numberrange_objects=>read(
          EXPORTING
            language        = cl_abap_context_info=>
                get_user_language_abap_format( ).
            object          = 'ZACB_RECIPE'
          IMPORTING
            attributes      = DATA(attributes)
            interval_exists = DATA(interval_exists)
            obj_text        = DATA(obj_text)
        ).
```

```
  CATCH cx_nr_object_not_found cx_number_ranges INTO
    DATA(exception).
  RAISE SHORTDUMP exception.
ENDTRY.
```

Listing 9.1 READ Method: Reading Attributes of a Number Range

To read the attributes of a number range object, the static READ method of the CL_NUMBERRANGE_OBJECT class is called. The number range object (OBJECT) in which the attributes are to be read can be specified as an import parameter. The language (LANGUAGE) is also specified, as the texts are also returned. These are saved in the number range object depending on the language and must therefore also be read depending on the language. As just described, the object texts (OBJ_TEXT) are returned as return values. The attributes of a number range object listed in Table 9.2 are available in the ATTRIBUTES structure.

Attribute	Description
OBJECT	Name of the number range object
DTELSOBJ	Data element of the possible object type
YEARIND	Whether the number range object is dependent on the fiscal year
DOMLEN	Length of the underlying domain
PERCENTAGE	Percentage from which a warning is issued that the number range interval is narrowing
CODE	Obsolete for ABAP Cloud
BUFFER	Buffer type for the numbering process
NOIVBUFFER	Quantity of numbers in the buffer
NRSWAP	Notification of whether numbering should start again from the beginning at the end of an interval
NRCHECKASCII	Obsolete for ABAP Cloud

Table 9.2 Attributes of a Number Range Object

Retrieving a number

To retrieve new numbers from an interval, the CL_NUMBERRANGE_RUNTIME class can be used. This class contains the static NUMBER_GET method which determines the next number. We now want to call the next free number for a new recipe for our application. The code from Listing 9.2 can be used for this purpose.

```
TRY.
    cl_numberrange_runtime=>number_get(
  EXPORTING nr_range_nr        = '01'
            object             = 'ZACB_RECIP'
            quantity           = 1
  IMPORTING number             = DATA(number)
            returned_quantity  = DATA(returned_quantity)
            returncode         = DATA(rcode) ).

  CATCH cx_nr_object_not_found cx_number_ranges INTO DATA(exception).
    RAISE SHORTDUMP exception.
ENDTRY.
```


Listing 9.2 The NUMBER_GET Method

In Listing 9.2, you can see how the method can be called. The number range interval (NR_RANGE_NR) and the name of the number range object (OBJECT) must be specified as import parameters. Another important parameter is the quantity of numbers (QUANTITY) that are required. If the QUANTITY parameter isn't specified, a 1 is given by default. The numbers taken from the specified number range interval (NUMBER) and the quantity of numbers (RETURNED_QUANTITY) are returned. It's important that the numbers have a generic data type. This must be converted to the correct data element as follows:

```
DATA: number_recipe TYPE zacb_recipe_id.
number_recipe = number.
```

Once you've executed the NUMBER_GET method and the number has been converted into our data element, you can use it. The number range interval now shows that the number status has been increased.

[«]

Other Available Classes

There are two other methods in the CL_NUMBERRANGE_RUNTIME class. The first method is NUMBER_CHECK. This method can be used to check whether a number range interval converts an external numbering. The second method is GET_NUMBER_STATUS. This method can be used to determine the current number status from the number range.

9.4 Numbering in the ABAP RESTful Application Programming Model

In the previous sections, we introduced the concept of number range objects and the API for using number range objects. Now we come to the integration in our application scenario, that is, in the ABAP RESTful application programming model application for the recipe portal.

Numbering in ABAP RESTful application programming model applications

The ABAP RESTful application programming model follows a slightly different numbering concept than that used in classic ABAP programming. In ABAP RESTful application programming model applications, numbering means that values are assigned for the primary key of an entity. For our sample application, we'll determine a number for new recipes, for new ingredients, and for new reviews. The primary key can consist of one or multiple fields. If you develop an ABAP RESTful application programming model application, you must therefore ask yourself the central question of who should do the numbering and when. You have the following options:

- **Early numbering**
  With early numbering, the number is assigned as early as possible; that is the primary key is assigned a number as soon as the `create` method has been executed.
- **Late numbering**
  With late numbering, the number is assigned as late as possible, usually shortly before the data is saved.

In addition to these two options for when the numbers should be assigned, there are also the following options regarding who assigns the numbers:

- **External numbering**
  A number is assigned from outside.
- **Managed numbering**
  A Universally Unique Identifier (UUID) is automatically assigned by the ABAP RESTful application programming model framework during the `create` request.
- **Unmanaged numbering**
  A separate logic is implemented as part of the development.

Based on these options, Table 9.3 lists the possibilities as to when and how the numbers can be assigned.

Who/When	Early Numbering	Late Numbering
External	Yes	No
Managed	Yes	Yes
Unmanaged	Yes	Yes

Table 9.3 Possible Combinations of Options

We'll present the following three scenarios for our application scenario:

- Unmanaged early numbering for the recipes
- External early numbering for the ingredients
- Managed early numbering for the reviews

9.4.1 Unmanaged Early Numbering

We now want to implement unmanaged early numbering for the Recipe entity.

Behavior definition

The first extension that needs to be made to implement early numbering is an adjustment to the `ZACB_R_Ingredient` behavior definition. You need to add the `early numbering` addition here. At the same time, it's now necessary to set the `readonly` property for the `RecipeID` key field so that the user can no longer make any entries in it (see Listing 9.3).

```
define behavior for ZACB_R_Recipe alias Recipe
persistent table zacb_recipe
draft table zacb_recipe_d
with additional save
etag master LocalLastChangedAt
lock master unmanaged
total etag LastChangedAt
authorization master ( global )
early numbering
{

  field ( readonly )
  CreatedAt,
  CreatedBy,
  LastChangedAt,
  LocalLastChangedAt,
  LocalLastChangedBy;
```

```
field ( readonly )
RecipeID;
```

Listing 9.3 Extending the Behavior Definition to Include Information on Unmanaged Early Numbering

The earlynumbering_create method

After activating the behavior definition, the editor tells us that we should implement the earlynumbering method for the create addition (see Figure 9.10).

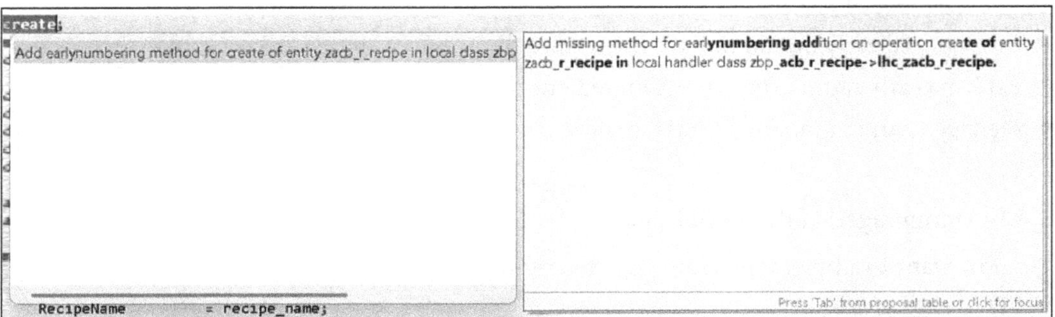

Figure 9.10 Message on the Necessity of Implementing the earlynumbering_create Method

Select this entry. The method definition and a template for the method implementation are then added to the behavior definition class. The following method definition was generated:

```
earlynumbering_create FOR NUMBERING
  IMPORTING entities FOR CREATE Recipe.
```

The signature of the earlynumbering method has the following parameters:

- **IMPORTING parameter entities**
  Includes all entities for which keys must be assigned.
- **Implicit CHANGING parameters:**
  - mapped: Used to provide mapping information.
  - failed: Used to identify the data record in which an error has occurred.
  - reported: Used to return messages in the event of an error.

Implementing the method

The earlynumbering_create method must now be implemented. Listing 9.4 shows the implementation for our sample application.

```
METHOD earlynumbering_create.
  DATA recipe_id_max TYPE zacb_recipe_id.
  DATA entity TYPE STRUCTURE FOR CREATE zacb_r_recipe.
```

```
TRY.
  LOOP AT entities INTO entity
  WHERE recipeid IS NOT INITIAL.
    APPEND CORRESPONDING #( entity )
    TO mapped-recipe.
  ENDLOOP.
DATA(entities_wo_recipeid) = entities.
DELETE entities_wo_recipeid
WHERE recipeid IS NOT INITIAL.
  IF entities_wo_recipeid IS NOT INITIAL.
    cl_numberrange_runtime=>number_get(
      EXPORTING nr_range_nr  = '01'
                Object   = 'ZACB_RECIP'
                Quantity = CONV # ( lines( entities_wo_recipeid ) )
      IMPORTING number = DATA(number_range_key)
        returncode = DATA(number_range_return_code)
        returned_quantity = DATA(number_range_returned_quantity) ).
  ENDIF.
    recipe_id_max = number_range_key -
      number_range_returned_quantity.
    LOOP AT entities INTO entity.
      recipe_id_max += 1.
      entity-recipeid = recipe_id_max.
      APPEND VALUE #( %cid      = entity-%cid
                      %key      = entity-%key
                      %is_draft = entity-%is_draft )
             TO mapped-recipe.
    ENDLOOP.
    CATCH cx_number_ranges INTO DATA(error).
      LOOP AT entities INTO entity.
        APPEND VALUE #( %cid      = entity-%cid
                        %key      = entity-%key
                        %is_draft = entity-%is_draft
                        %msg      = error )
             TO reported-recipe.
      APPEND VALUE #( %cid      = entity-%cid
                      %key      = entity-%key
                      %is_draft = entity-%is_draft )
             TO failed-recipe.
    ENDLOOP.
  ENDTRY.
ENDMETHOD.
```

Listing 9.4 Implementation of the earlynumbering_create Method

9

You must first check whether an ID has been set for a recipe entity. This check is required if the draft method is used for the entity, which is the case in our example. If no recipe ID has been assigned, we remove all instances with a set recipe ID from the `entities` import parameter. Then, we retrieve new numbers for the new entities and save them in the internal `number_range_key` table.

The `ZACB_RECIP` number range object is used by all users. If an entry isn't saved but discarded in draft mode, there will be gaps in the numbering. The new numbers determined are then assigned to the entities. If errors occur within the implementation, these will also be processed.

Testing the application

Now that we've made these changes, let's take a closer look at the behavior in the application:

1. Open the application, and click on **Create** to add a new recipe (see Figure 9.11).

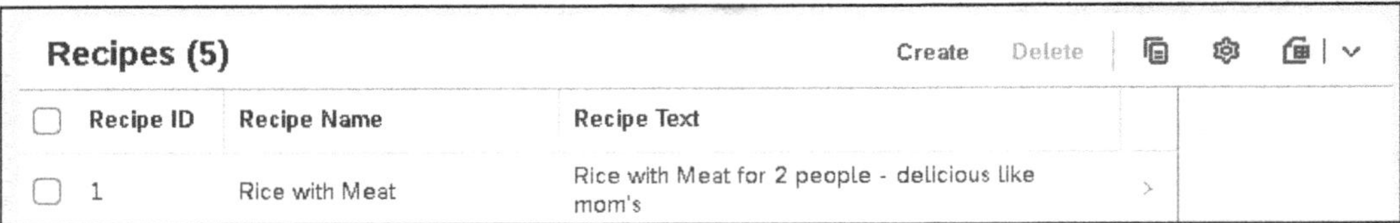

Figure 9.11 Creating a Data Record

2. Enter the values for the new data record in the creation window, and you'll see that a recipe ID has already been created (see Figure 9.12). Once you've entered the recipe data, click on the **Create** button.

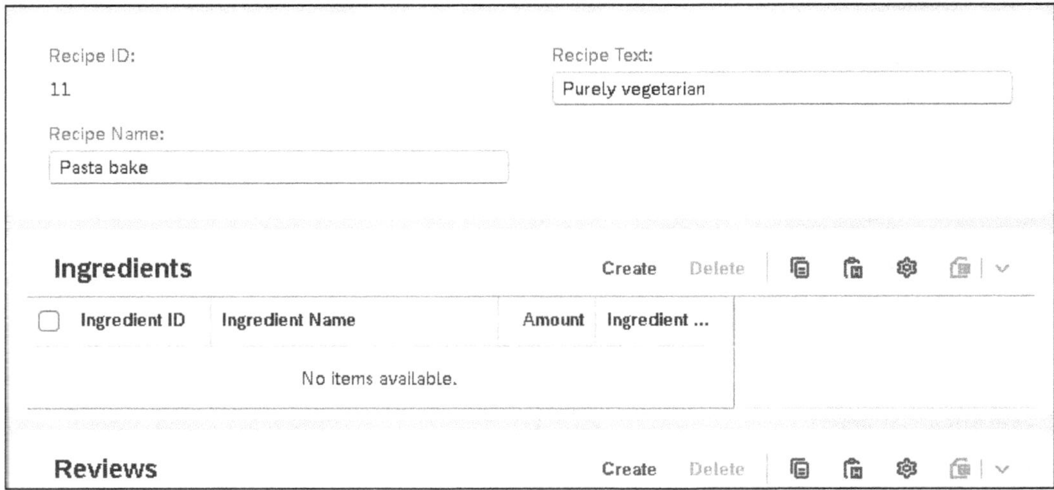

Figure 9.12 Unmanaged Early Numbering When Creating a New Recipe

3. Return to the recipe overview. The data record still has a recipe ID, but it's not yet stored in the database. It won't be saved in the database until you click on the **Save** button.

[«]

Unmanaged Late Numbering

The unmanaged late numbering scenario is implemented in a similar way. Instead of the `early numbering` addition, the `late numbering` addition is defined at the entity level. In addition, the key fields must be set to `readonly`. The implementation takes place in an `adjust_numbers` method, which is called during the save process. The generated number must be assigned to the internal temporary key via the `mapped` export parameter.

9.4.2 External Early Numbering

Manual numbering

The next numbering option we want to present in more detail is external early numbering. The numbers are assigned externally at an early stage, for example, by the user. This is the default scenario for a generated ABAP RESTful application programming model application. In our case, we use it to assign the ingredient ID.

Testing the default behavior

To test this default behavior, open our sample application. Then, add a new ingredient to a recipe using the **Create** button (see Figure 9.13).

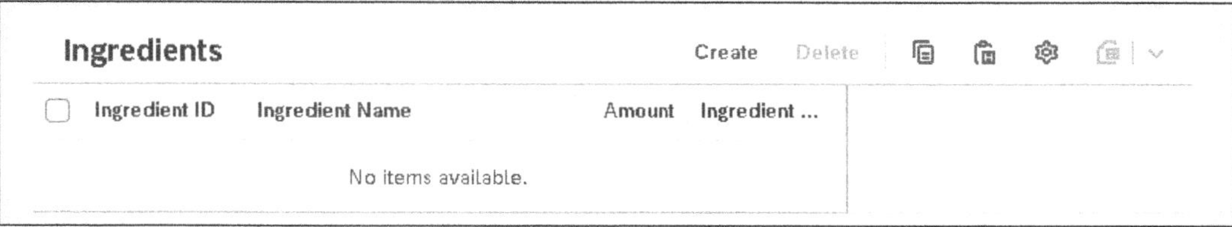

Figure 9.13 Sample Application: Adding a New Ingredient

In the dialog box for creating an ingredient, the **Ingredient ID** is a mandatory entry and is maintained via a popup window (see Figure 9.14).

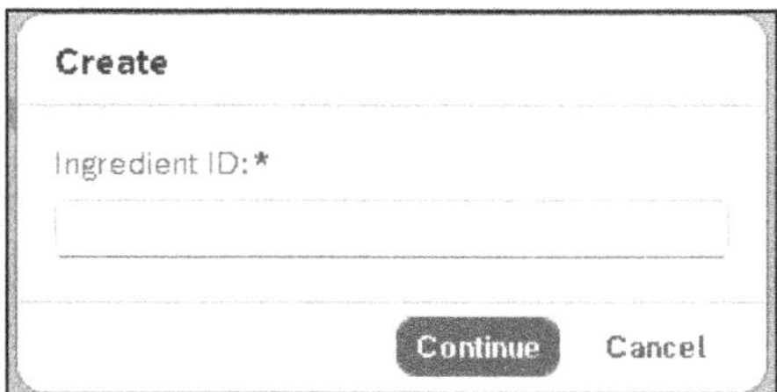

Figure 9.14 Manual Entry of an Ingredient ID

The specified ingredient ID will be used immediately and is reserved for this entry (see Figure 9.15). Once you've entered all other information about this ingredient, click on the **Create** button.

Ingredient

Ingredient ID: 1

Amount: 200

Ingredient Name: Noodles

Ingredient Unit: g

Figure 9.15 Using the ID in the Ingredient Data Record

After saving the data record, the number entered is used for this data record and saved in the database (see Figure 9.16).

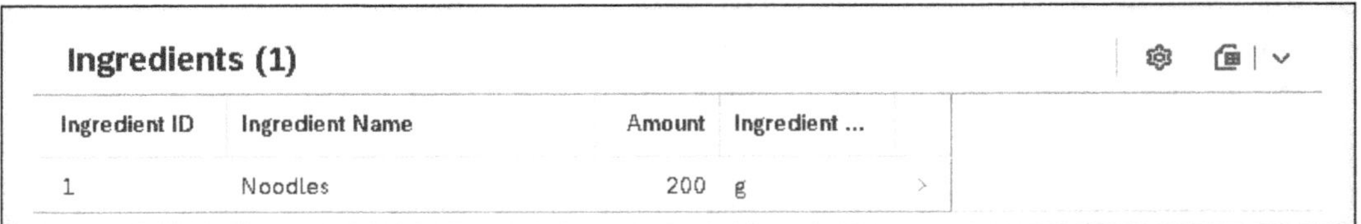

Figure 9.16 Saving the Ingredient ID

However, if you've entered a number that has already been used, you'll be notified automatically (see Figure 9.17).

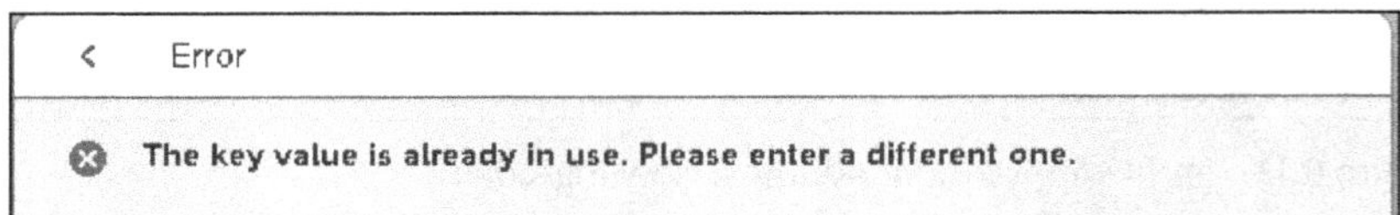

Figure 9.17 Message for Duplicate Key Values

Even if you enter a key value that is longer than permitted, you'll receive a corresponding message (see Figure 9.18).

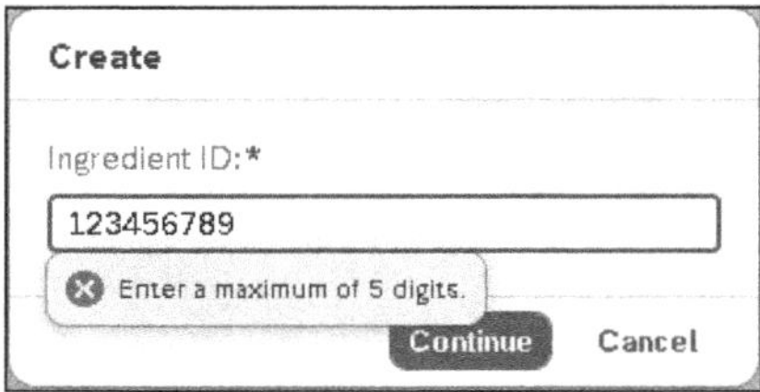

Figure 9.18 Message When the Maximum Field Length Is Exceeded

9.4.3 Managed Early Numbering

Behavior definition

The third scenario is the automatic assignment of a review ID by the ABAP RESTful application programming model framework using managed early numbering. To implement the behavior, you must first extend the `ZACB_R_Review` behavior definition for the Review entity with the `field ( numbering : managed ) ReviewId;` addition (see Listing 9.5). This addition tells the ABAP RESTful application programming model framework that it should take care of the automatic numbering process.


```
define behavior for ZACB_R_Review alias Review
...
{
...
  field ( readonly ) RecipeId;
  field ( readonly ) ReviewId;
  field ( numbering : managed ) ReviewId;
  association _Recipe { with draft; }
  mapping for zacb_review
```

Listing 9.5 Extension of the Behavior Definition

Once you've inserted this addition into the code, the editor notifies you that the `ReviewId` field must be compatible with the `ABP_BEHV_PID` type (see Figure 9.19).

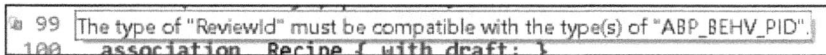

The type of "ReviewId" must be compatible with the type(s) of "ABP_BEHV_PID".

Figure 9.19 Message from the Editor About the Data Type Used

For this to be the case, the field must be assigned to a compatible type in both the database table and the draft table. You can see the corresponding extension of the table definitions in Listing 9.6. We use the `SYSUUID_X16` type here.

```
define table zacb_review {
  key client          : abap.clnt not null;
  key review_id       : sysuuid_x16 not null;
...
define table zacb_review_d {

  key client        : abap.clnt not null;
  key reviewid      : sysuuid_x16 not null;
```

Listing 9.6 Extension of the Database Tables by the SYSUUID_X16 Type

After the subsequent activation of the two tables, the error message in the editor of the behavior definition disappears and can be activated.

[»]

Table Conversion

After changing the field in the database table, the database must be converted. This can be done using the quick fix function in the ABAP development tools or in Transaction SE14. During a table conversion, existing database contents are adapted to the modified table.

Testing the application

After successfully activating the behavior definition, you can check the behavior of the automatic numbering process in our SAP Fiori app. In the editing mode of the recipe detail view, you can add a new review using the **Create** button. When the review is created, a number is automatically generated in UUID format and assigned to the review before it's saved (see Figure 9.20).

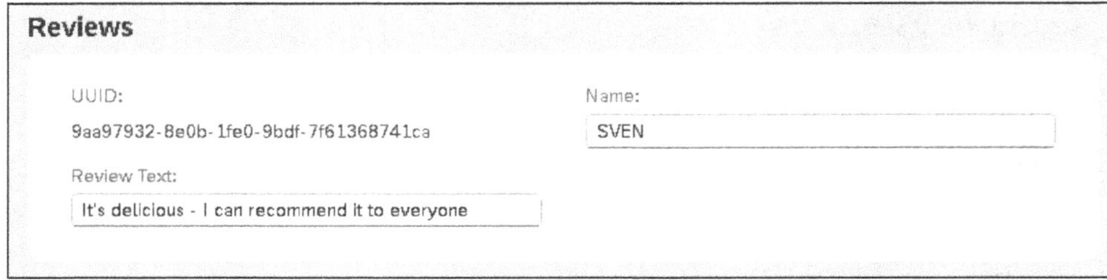

Figure 9.20 Assignment of a UUID by the SAP Fiori App

[»]

Universally Unique Identifier (UUID)

A UUID is a 128-bit number that is used to individually identify data. It's designed to be unique in the world. This minimizes the risk of accidental duplication.

Accept the review by clicking on the **Apply** button. After saving the recipe, the review will be transferred with its UUID, and the data record is saved in the database (see Figure 9.21).

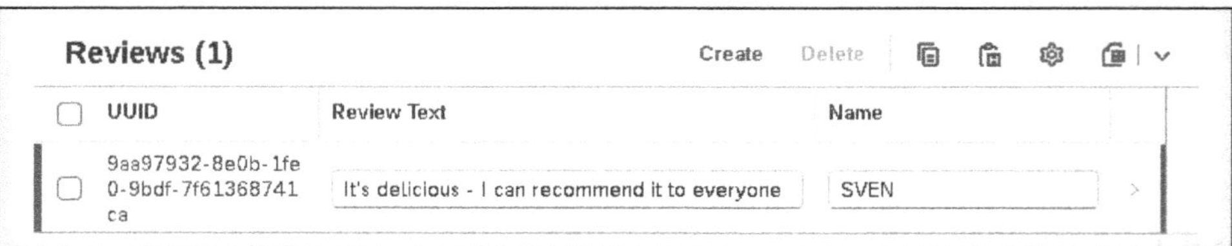

Figure 9.21 Saving the UUID in the SAP Fiori App

9.5 Summary

In this chapter, we've explained which objects exist in the context of number ranges. The following objects were presented:

- Number range object
- Number range interval
- Number status

You've also learned how to create a number range object in the ABAP development tools. In addition, you got to know the new SAP Fiori app for editing number range intervals and number statuses.

Then, we introduced you to the new API for number ranges. Instead of using function modules as in previous ABAP versions, you can now use classes to work with number ranges.

You've also learned how number ranges are implemented in the ABAP RESTful application programming model environment. There are various options here for early and late numbering or a numbering process that's either managed or unmanaged by the ABAP RESTful application programming model framework.

Chapter 10
Background Processing

You're executing a program in a dialog and an error occurs because you've exceeded the maximum dialog time. In such a case, your program is too runtime intensive. One solution is to run the program in the background. In this chapter, you'll learn how you can schedule programs in the background in ABAP Cloud and execute them automatically.

Background processing enables the system to perform tasks automatically without any manual intervention. The *background jobs* used for this can be run once or recurrently, such as every Wednesday at midnight. This allows you to move resource-intensive processing tasks to night times when few or no users are logged into the system. Thanks to background processing, the dialog processes don't get overloaded. Reports running in the background are also not subject to the runtime restrictions of the dialog modes.

Up until now, you've been familiar with the following classic ABAP procedure for executing a job:

1. Develop a program with a selection screen.
2. Create a selection variant for the previously developed program.
3. Define a job flow in Transaction SM36.
4. Define the times at which the job is to be processed (also in Transaction SM36).
5. Execute the program in a background process (batch process).
6. Determine the job status and output in a log via Transaction SM37.

In this chapter, you'll learn how to develop, schedule, and monitor such jobs in ABAP Cloud. You'll get to know the successor to the background jobs: the *application jobs*.

Background processing for our example

In this chapter, we implement the first preparatory measure for the requirement to send newsletters about new recipes by email. To do this, we develop a logic for how a background job can be created and schedule its execution. In Chapter 11, the logic executed by this job will then be developed to generate a newsletter for the new recipes.

[»]

Using the Sample Application

The objects presented in this chapter are located in the JOB subpackage.

10.1 Developing an Execution Logic

Classes instead of programs

Previously, you developed a repository object with the PROG object type (for program) if you wanted to execute logic for processing in the background. With ABAP Cloud, you need to develop the logic for the program in a class. This is because the PROG object type is no longer available and the execution logic can no longer be mapped in selection screens and variants due to the unavailability of SAP GUI.

Creating an execution class

We now want to create a class in the ABAP development tools for Eclipse to provide our job with parameters that serve as starting points for the execution and the actual function. Follow these steps:

1. Click on **New • ABAP Class** in the context menu of the package.
2. Enter the following values in the dialog box that opens (see Figure 10.1):
   - **Name**: Enter a unique name for the class. For our example, we entered "ZCL_ACB_DET_RECIPE".
   - **Description**: Enter a suitable description for the class.
   - **Interfaces**: These must be selected manually via the **Browse** button:
     - IF_APJ_DT_EXEC_OBJECT: The parameters can be defined using this interface.
     - IF_APJ_RT_EXEC_OBJECT: This interface serves as the starting point for background processing.
3. Click on the **Next** button.

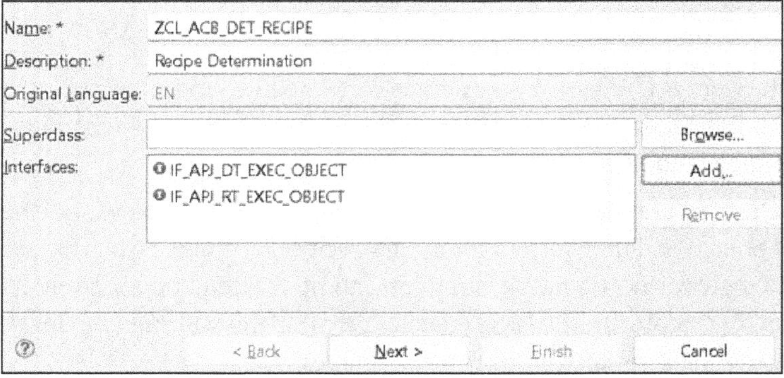

Figure 10.1 Defining Parameters for a New Class

4. Select a transport request, and complete the processing step by clicking the **Finish** button.

Generated class

Now the class will be created. In this context, the code shown in Listing 10.1 gets generated.

```
CLASS zcl_acb_recipe_background DEFINITION
  PUBLIC
  FINAL
  CREATE PUBLIC.

  PUBLIC SECTION.

    INTERFACES if_apj_dt_exec_object.
    INTERFACES if_apj_rt_exec_object.
  PROTECTED SECTION.
  PRIVATE SECTION.
ENDCLASS.
CLASS zcl_acb_recipe_background IMPLEMENTATION.

  METHOD if_apj_dt_exec_object~get_parameters.
  ENDMETHOD.

  METHOD if_apj_rt_exec_object~execute.
  ENDMETHOD.
ENDCLASS.
```

Listing 10.1 Generated Code for the Execution Logic

The GET_PARAMETERS method

You can now edit the generated code. First, we add two parameters to the program logic so that we can recreate the classic selection screen (see Listing 10.2). For this purpose, we define the GET_PARAMETERS method.

```
METHODS get_parameters
  EXPORTING
   !et_parameter_def TYPE tt_templ_def
   !et_parameter_val TYPE tt_templ_val.
```

Listing 10.2 Extension of the Program Logic with Parameters

The two parameters serve the following purposes:

- **ET_PARAMETER_DEF**
  This parameter is used to define the appearance of the input options and their behavior.

- **ET_PARAMETER_VAL**
  This parameter is used to preset the input options with values.

The ET_PARAMETER_DEF parameter

First, we define the input options in ET_PARAMETER_DEF in Listing 10.3.

```
et_parameter_def = VALUE #(
        changeable_ind = abap_true
        (
        datatype       = 'C'
        selname        = para_lastdays
        kind           = if_apj_dt_exec_object=>parameter
        param_text     = 'Determination days'
        length         = 2
        mandatory_ind = abap_true )
        (
        datatype       = 'C'
        selname        = para_lastfourweeks
        kind           = if_apj_dt_exec_object=>parameter
        param_text     = 'Last 4 weeks'
        length         = 1
        checkbox_ind  = abap_true ) ).
```

Listing 10.3 Extension of the Execution Logic to Include Parameter Presetting

In Listing 10.3, you can see that two parameters are generated, PARA_LASTDAYS with the length 2 and PARA_LASTFOURWEEKS with the length 1. They both have the CHAR data type and can be changed. The PARA_LASTFOURWEEKS parameter is displayed as a checkbox in the user interface (UI).

Table 10.1 shows the possible fields for defining parameters.

Fields	Description
SELNAME	Name of the field (maximum eight characters)
KIND	A parameter or a selection option: ■ Parameter: if_apj_dt_exec_object=>parameter ■ Selection option: if_apj_dt_exec_object=>select_option
PARAM_TEXT	Display name of the parameter
DATATYPE	Basic type of the field
LENGTH	Maximum input length of the field in the display screen
DECIMALS	Number of decimal places if the data type is Decimal

Table 10.1 Possible Elements for the Parameter Definition

Fields	Description
LOWERCASE_IND	Support for uppercase and lowercase letters in the input field
CHECKBOX_IND	Display of the field as a checkbox (ABAP_TRUE)
RADIO_GROUP_IND	Display of the field as a radio button (ABAP_TRUE)
RADIO_GROUP_ID	All fields in a radio button group
MANDATORY_IND	Mandatory input field (ABAP_TRUE)

Table 10.1 Possible Elements for the Parameter Definition (Cont.)

The ET_PARAMETER_VAL parameter

The values of ET_PARAMETER_VAL are preset with values in Listing 10.4.

```
et_parameter_val = VALUE #(
                    sign   = 'I'
                    option = 'EQ' (
                    selname = para_lastfourweeks
                    low     = abap_true ) ).
```

Listing 10.4 Definition of Parameter Values

We want the PARA_LASTFOURWEEKS checkbox to be selected by default. The PARA_DAYS parameter isn't preset with a value.

Table 10.2 shows the possible fields for the parameter values.

Fields	Description
SIGN	Inclusion (I) in or exclusion (E) from the total result set
OPTION	Condition in the form of comparison operators, such as EQ
SELNAME	Name for which the values are to be set
LOW	Value for PARAMETER or lower interval limit for SELECT-OPTIONS
HIGH	Upper interval limit of SELECT-OPTIONS

Table 10.2 Possible Elements for Parameter Values

Execution logic in the EXECUTE method

Now we add the most important component of the execution logic in the EXECUTE method to the execution class, as shown in Listing 10.5.

```
METHOD if_apj_rt_exec_object~execute.
  DATA days TYPE zacb_days.
  DATA test TYPE abap_bool.
```

```
  LOOP AT it_parameters INTO DATA(parameter).
    CASE parameter-selname.
      WHEN para_lastfourweeks.
        days = fourweeks.
      WHEN para_lastdays.
          days = parameter-low.
    ENDCASE.
  ENDLOOP.

  DATA(det_date) = EXACT timestamp(   xco_cp=>sy->moment(
    xco_cp_time=>time_zone->utc
      )->subtract( iv_day = CONV #( days )
      )->as( xco_cp_time=>format->abap
      )->value ).
      SELECT FROM zacb_recipe
        FIELDS *
          WHERE created_at > @det_date
          INTO TABLE @DATA(recipes).
        ...
ENDMETHOD.
```

Listing 10.5 The EXECUTE Method

In the first step, we parse the parameters contained in the import IT_PARAMETERS parameter. Now we can parse the contents of the table and distribute them to the various fields. We've now distributed all content to variables and can start with the actual implementation logic. This logic is used to determine the recipes that were created in the last *x* days based on the determination date.

Second Part of the Execution Logic for Sending Newsletters

The execution logic described here is the first step. The second step is described in Chapter 11. An email is sent as a newsletter based on the selected newly created recipes.

10.2 Creating Application Jobs

In the following sections, we explain how this execution class becomes a background job. You'll get to know the following objects:

- Application job catalog entry
- Application job template
- Application job

Figure 10.2 illustrates the interaction of these objects.

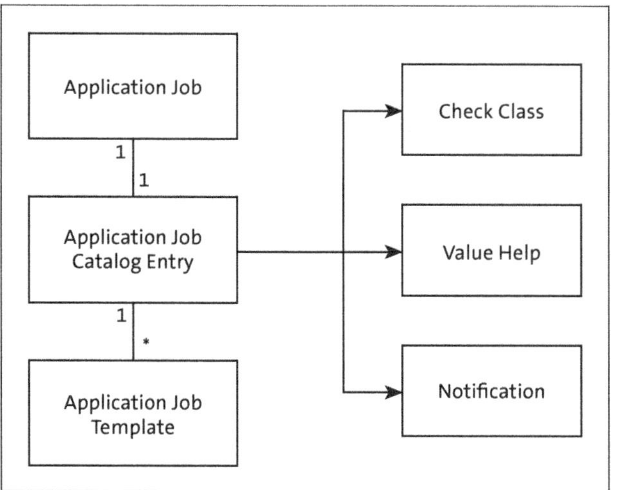

Figure 10.2 Objects for Application Jobs

Application job catalog entry

Once the execution class has been created, the *application job catalog entry* can be created as well. An application job catalog entry contains the information required for planning and executing an application job. The entry in the application job catalog contains the name of the class that implements the business logic as well as information on the display of selection fields. In addition, the application job catalog entry defines other settings that are used during the scheduling and execution of the application job.

10

10.2.1 Creating an Application Job Catalog Entry

Creating a catalog entry

The following steps are required to create an application job catalog entry:

1. Right-click on the package in the project explorer, and select the **New • Other ABAP Repository Object** path from the context menu.
2. In the dialog box that opens, select the **Application Job Catalog Entry** item (see Figure 10.3).
3. In the dialog that opens next, make the following settings for the application job catalog entry (see Figure 10.4):
   - **Name**: Enter a unique name. For our example, we entered "ZACB_DET_RECIPE".
   - **Description**: Enter a suitable description.
   - **Class with Execute Method**: Select the execution class for the job that we created in Section 10.1.

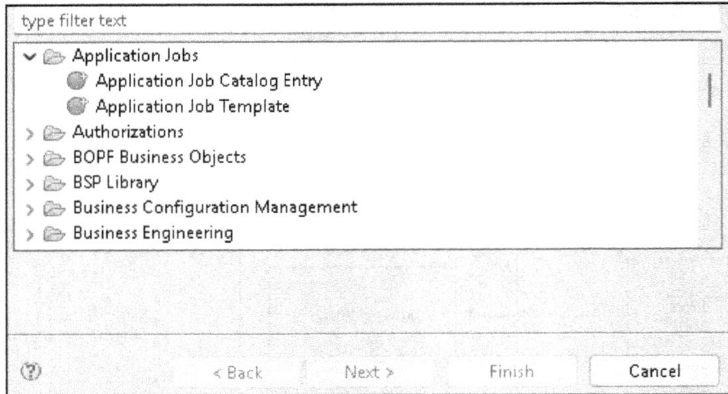

Figure 10.3 Selecting the Application Job Catalog Entry Object Type

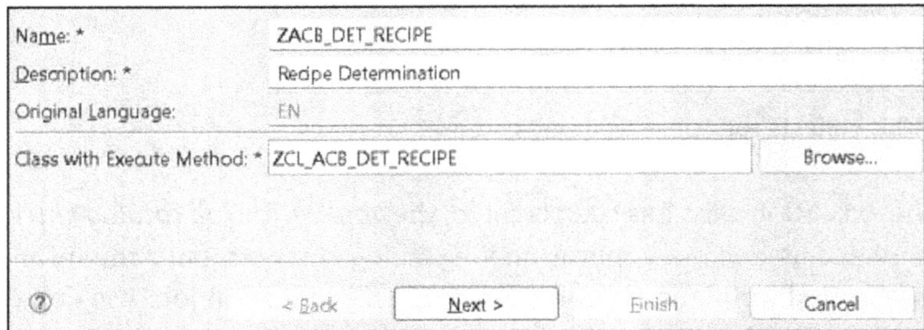

Figure 10.4 Settings for the Application Job Catalog Entry

4. Confirm your input by clicking the **Next** button.
5. Select a transport request, and close the dialog by clicking the **Finish** button.

Now the application job catalog entry gets created. Figure 10.5 shows the maintenance dialog.

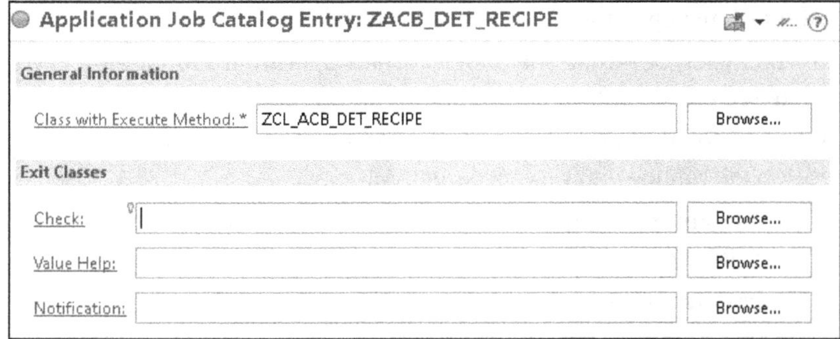

Figure 10.5 Created Application Job Catalog Entry

Exit classes

Three different exit classes can be made available for the application job catalog entry:

- **Check**
  We explain the exit class for checks in Section 10.3.
- **Value Help**
  You can find out more about the exit class for value help in the following information box.
- **Notification**
  We explain the exit class for notifications in Chapter 11.

[!]

Value Helps

At the time this book went to press, the exit classes for value helps weren't yet supported in SAP S/4HANA 2023, although a field was already available for them in the application job catalog entry (refer to Figure 10.5).

The value help will be available for the first time from SAP S/4HANA Cloud Public Edition 2502. Further information on this can be found in a blog post in the SAP Community (see *http://s-prs.de/v1064807*).

[«]

Notifications

We'll go into the topic of notifications in more detail in Chapter 11. There, we describe how you can send notifications by email in the event of errors within an application job.

10.2.2 Creating an Application Job Template

Application job template

Based on the application job catalog entry, the next step is to create an *application job template*. An application job template contains values for some or all selection fields. It can be regarded as a variant of an application job catalog entry. An application job catalog entry can contain more than one application job template. An application job is then created by scheduling an application job template.

Creating the job template

The following steps are required to create an application job template:

1. Open the context menu of the package, and select **New • Other ABAP Repository Object**.
2. In the dialog box that opens, select the **Application Job Template** object type. Confirm your selection by clicking the **Next** button to proceed to the next dialog box.

3. Enter the following information there (see Figure 10.6):
   - **Name**: Enter a unique name for the job template. For our example, we entered "ZACB_DET_RECIPE".
   - **Description**: Enter a suitable description.
   - **Job Catalog Entry**: Here, you specify the referenced object for which the job template is to be created. In our example, this is the previously created job catalog entry "ZABC_DET_RECIPE".

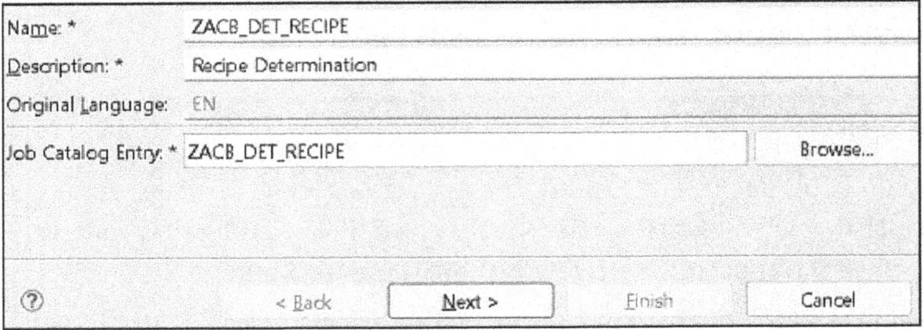

Figure 10.6 Entering Information on the Application Job Template

4. Confirm your entries by clicking the **Next** button.
5. In the next step, you can select a transport request. The wizard can then be closed by clicking the **Finish** button.

The application job template has now been created and preset with the selection data from the execution class (see Figure 10.7).

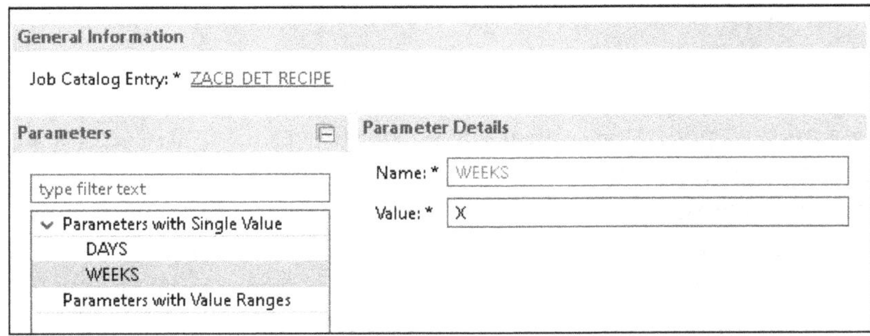

Figure 10.7 Created Application Job Template

10.2.3 Creating an Application Job

Application Jobs app To be able to execute an application job, it must first be configured. To do this, you can use the Application Jobs app (app ID F1240). Open the SAP Fiori launchpad and start the app via the corresponding tile (see Figure 10.8).

Figure 10.8 Tile for Calling the Application Jobs App

> **Application Jobs App**
>
> All important information on the Application Jobs app (app ID F1240) can be found in the SAP Fiori apps reference library at *http://s-prs.de/v1064808*. Among other things, you can find out which Open Data protocol (OData) service and which roles need to be configured for the version available in your system.

Overview of application jobs

The app starts with an overview of the application jobs (see Figure 10.9). All jobs are listed there with their status, names, and times.

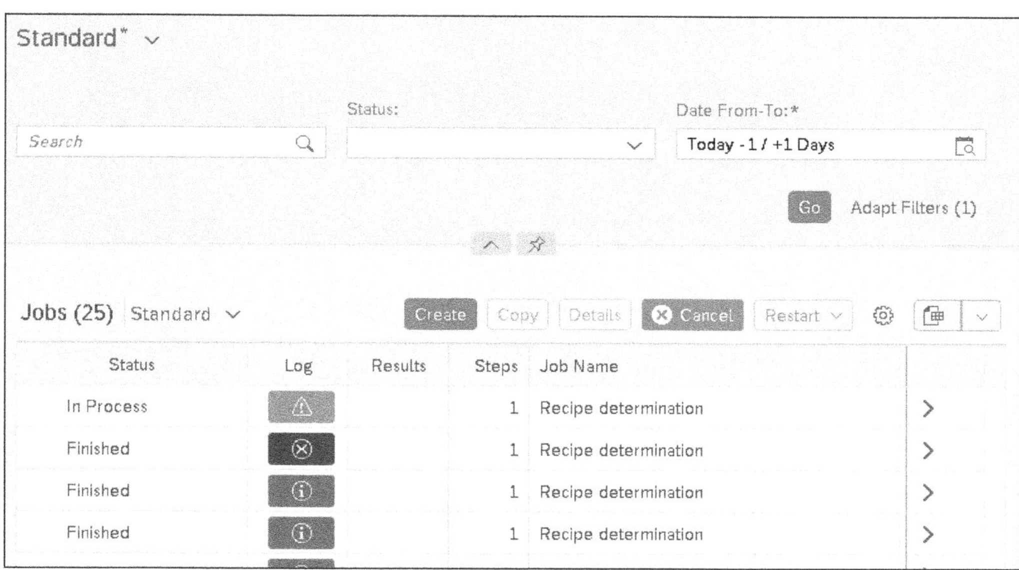

Figure 10.9 Overview of Application Jobs

Creating an application job

To create a new application job here, follow these steps:

1. Click the **Create** button.
2. A creation wizard opens (see Figure 10.10), in which you must first make the following entries:

   - **Job Template**: The name of the job template can be entered manually or selected from the input help. Select the application job template created in Section 10.2.2.
   - **Job Name**: The job name is automatically determined from the name of the job template. It can be overwritten manually.
3. Click on the **Step 2** button.
4. Configure the scheduling options in the next window (see Figure 10.11). The application job can be scheduled as a single run or as a repeat pattern.

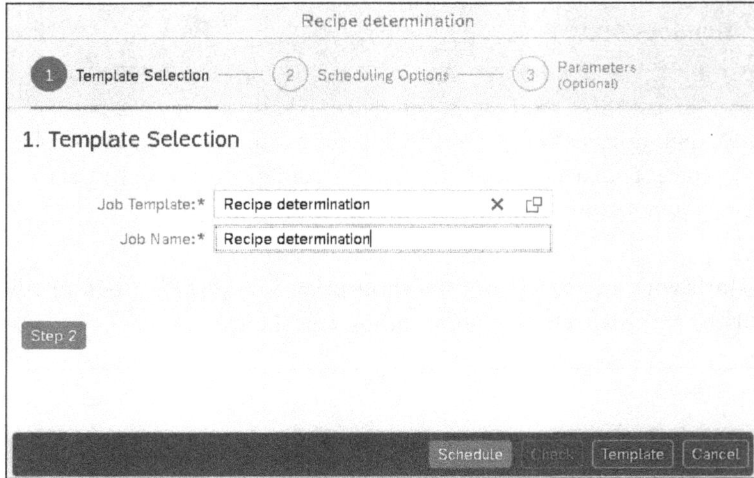

Figure 10.10 Specification of the Application Job Template

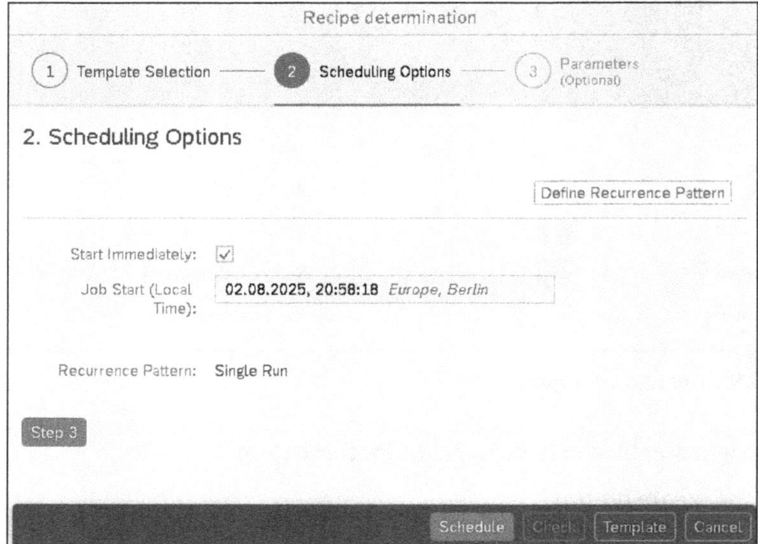

Figure 10.11 Scheduling Options

5. Because we want to use our newsletter on a regular basis, we define a recurrence pattern. To do this, click on the **Define Recurrence Pattern** button. The following settings can be made there (see Figure 10.12):
   - **Start immediately**: Specify whether the job should be started immediately or at a specific time.
   - **Job Start (Local Time)**: The start date and time can be defined here. You can also select the time zone to which the start date and time belong. The local time of your SAP user is calculated and displayed from the previously defined time and time zone.
   - **Recurrence Pattern**: The recurrence pattern can be defined here. The options range from **Monthly** to **Every minute**. As soon as an option other than a single run has been selected, a fixed interval can be specified as well, such as every five minutes.
   - **On Non-Working Day**: If the execution date is a nonworking day, you can specify how this should be handled. Here, you can select, for example, whether the job should be carried out on the next or previous working day.
   - **Calendar**: The calendar can be used to determine whether it's a work day or a day off, such as a Saturday or public holiday.

   For our sample application, we define that the job should be started immediately. It should also be performed once a day.

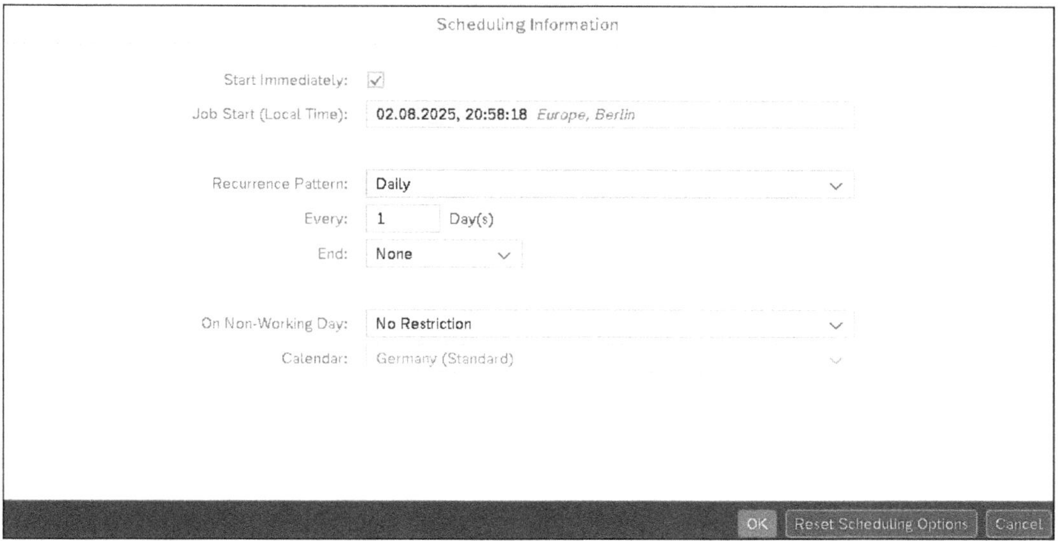

Figure 10.12 Defining a Recurrence Pattern

6. As soon as the recurrence pattern has been defined, you can confirm your entries by clicking the **OK** button. Then click on the **Step 3** button to continue (see Figure 10.13).

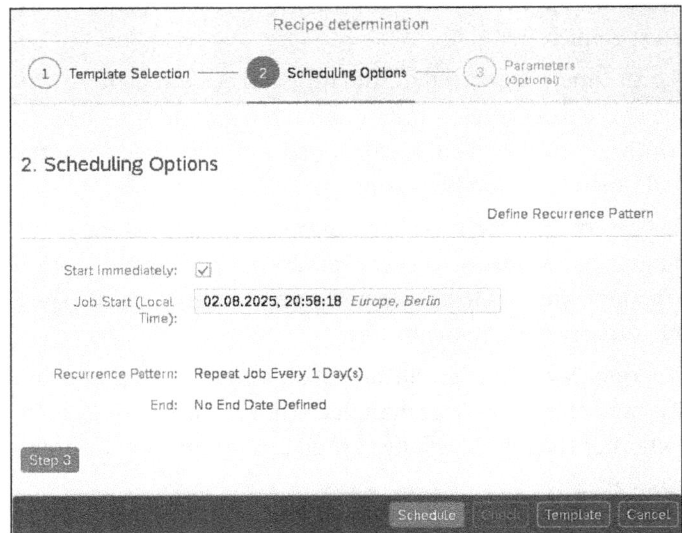

Figure 10.13 Overview of the Selected Scheduling Options and the Recurrence Pattern

7. In step 3, the parameters defined in the execution class can optionally be checked with the set values (see Figure 10.14). The parameter values can be manually overwritten or supplemented.

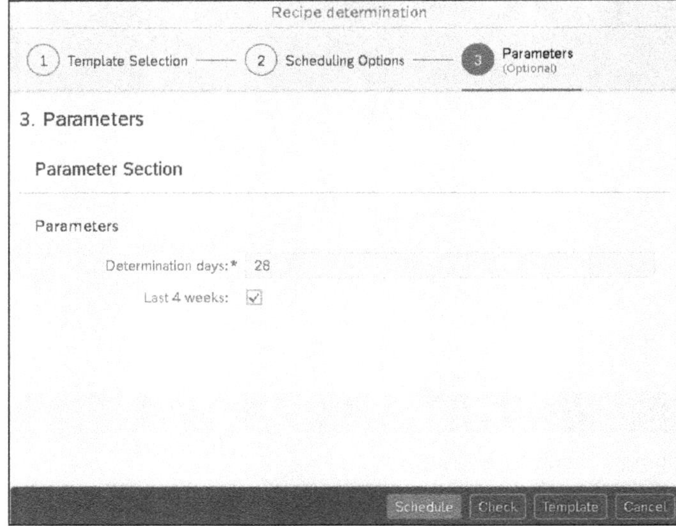

Figure 10.14 Checking the Parameters (Optional)

8. You can now schedule the application job by clicking the **Schedule** button.

Scheduled application job

The application job is then displayed with the selected scheduling options in the Application Jobs app (app ID F1240), as shown in Figure 10.15.

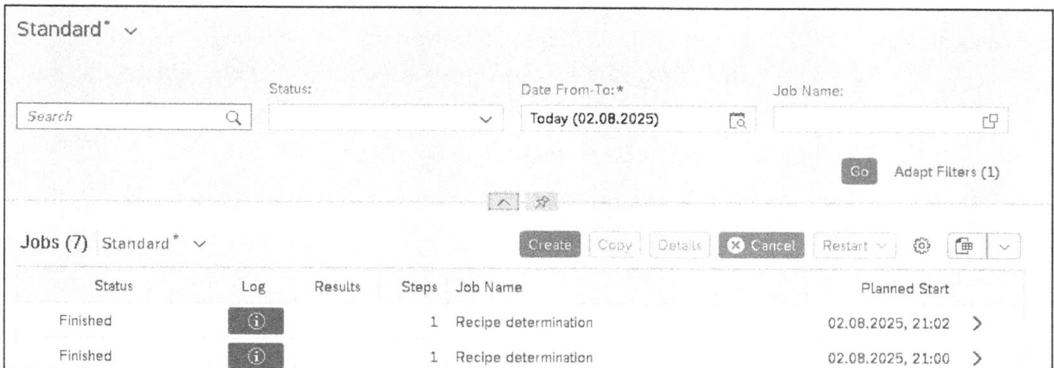

Figure 10.15 Application Job Overview After Scheduling

Job execution log

Once the job has been executed, you can switch to the detailed view by clicking on the corresponding entry with the **Finished** status (see Figure 10.16).

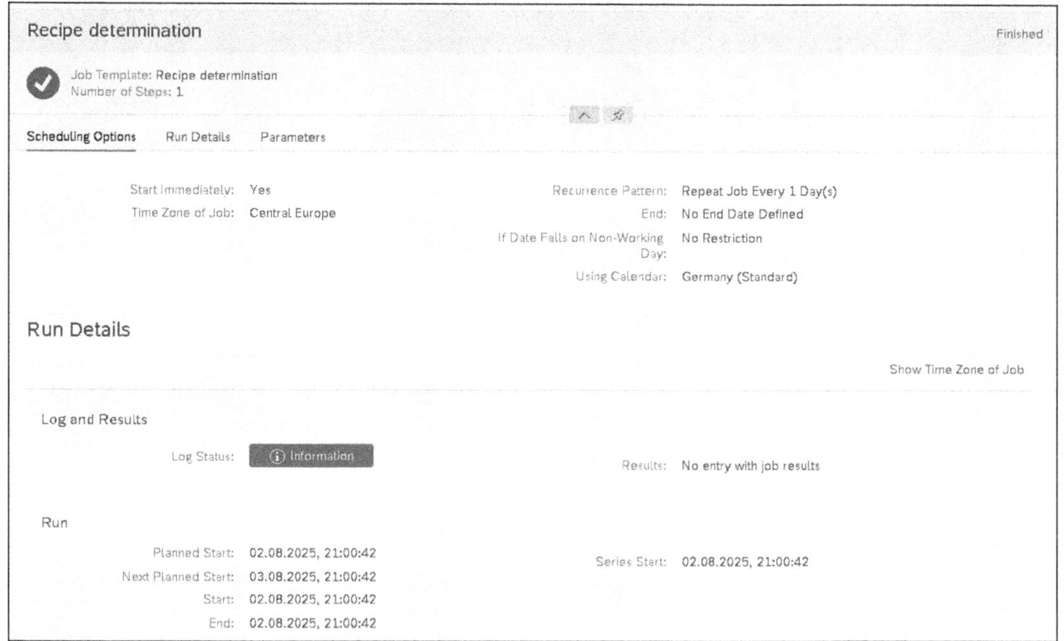

Figure 10.16 Detailed View of the Executed Application Job

You can also display the output log via the **Information** button in the **Log and Results** section.

10.3 Checks

Validating the input parameters

With the help of checks, the application jobs can be made more robust and the various functions can be better used. The checks validate the input parameters and provide feedback in the form of messages.

The checks can be implemented in two ways:

- Storing an existing class
- Creating a new check class

Creating a check class

In the following sections, we'll show you how to create a check class. This is necessary for our example because there's no existing class with a corresponding logic available yet. For this purpose, you need to click on the **Check** link in the maintenance view of the application job catalog entry (see Figure 10.17).

Figure 10.17 Creating a Check Class via the Application Catalog Entry

The CHECK_AND_ADJUST method

The CHECK_AND_ADJUST method of our check class checks the inputs and can output messages that indicate problems. In addition, the user's entries can be overwritten or corrected. The job catalog name, the parameters, and the user name are provided as transfer parameters. Listing 10.6 shows the implementation of this method.

```
METHOD if_apj_jt_check_20~check_and_adjust.
  DATA det_days TYPE zacb_days.
  det_days = ct_value[ KEY param parameter_name =
    zcl_acb_det_recipe=>para_lastdays ]-low.

  IF det_days <> initial.
    ev_successful = abap_true.
```

```
  ELSE.
    INSERT VALUE #( id         = 'ZACB_JOB'
                    type       = 'E'
                    number     = '001'
                    message_v1 = det_days )
    INTO TABLE et_msg.
  ENDIF.
ENDMETHOD.
```

Listing 10.6 Implementation of the CHECK_AND_ADJUST Method

The method first reads the PARA_LASTDAYS parameter. If its value is initial or not a number, an error message will be returned. It's important that the value of the ev_successful parameter is set to abap_true. This informs you that the check was successful.

We now enter the value "0" as an example and receive an error message because the value is initial (see Figure 10.18).

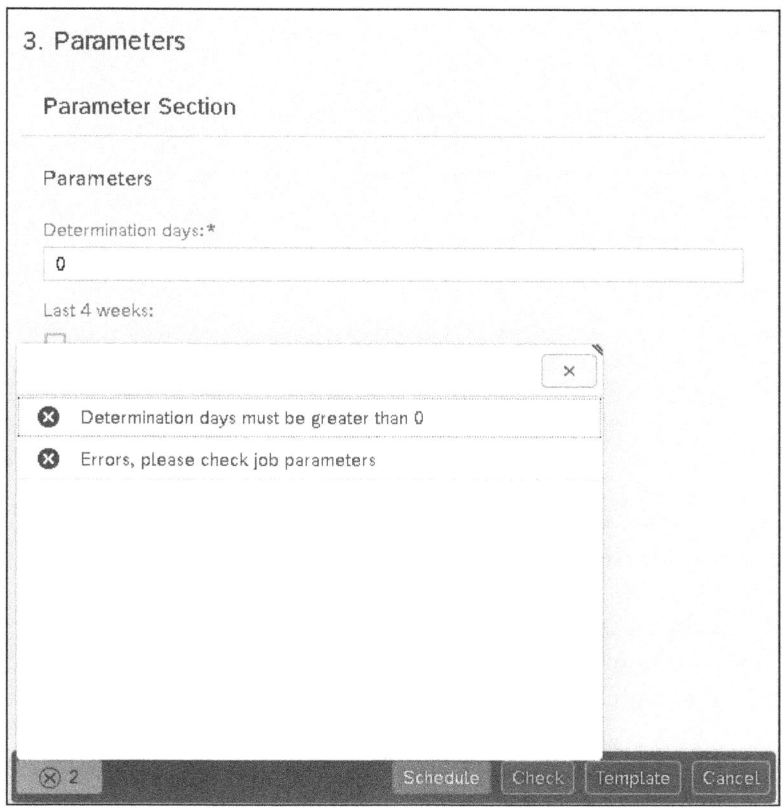

Figure 10.18 Testing the CHECK_AND_ADJUST Method

The INITIALIZE method

The next method is INITIALIZE. It's only called once, namely, when the parameters are loaded for the first time for display. It's possible to overwrite default values from the job template. The method has the same parameters as CHECK_AND_ADJUST. In addition, the parameter values set in the CT_VALUE variable can be changed.

In Listing 10.7, you can see that the PARA_LASTDAYS parameter is preset with the value 28. When the job is started, the default value of the **Determination days** parameter is therefore set to **28** (see Figure 10.19).

```
METHOD if_apj_jt_check_20~initialize.
  TRY.
    ct_value[ KEY param parameter_name = zcl_acb_det_recipe=>
      para_lastdays ]-low = '28'.
  CATCH cx_sy_itab_line_not_found.
  INSERT VALUE #( parameter_name = zcl_acb_det_recipe=>para_lastdays
    sign = 'I' option = 'EQ'
    low = '28' )
    INTO TABLE ct_value.
  ENDTRY.
ENDMETHOD
```

Listing 10.7 Implementation of the INITIALIZE Method

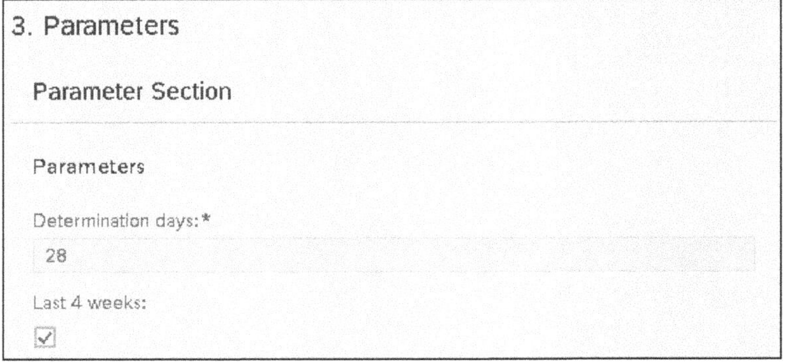

Figure 10.19 Testing the INITIALIZE Method

The CHECK_AUTHORIZATIONS method

Another method is CHECK_AUTHORIZATIONS. This method enables you to perform authorization checks. In Listing 10.8, we define the check as to whether the executing user is an admin.

```
METHOD if_apj_jt_check_20~check_authorizations.
  SELECT SINGLE FROM zacb_user
  FIELDS *
  WHERE username = @iv_username AND admin = @abap_true
  INTO @DATA(user).
```

```
  IF user IS INITIAL.
    INSERT VALUE #( id = 'ZACB_JOB' type = 'E' number = '000'
      message_v1 = iv_username )
    INTO TABLE et_msg.
  ELSE.
    ev_successful = abap_true.
  ENDIF.
ENDMETHOD.
```

Listing 10.8 Implementation of the CHECK_AUTHORIZATIONS Method

If you're authorized to execute the job according to this method, it's important to set the value of the ev_successful parameter to abap_true. If that isn't the case, a corresponding error message will display when the job execution is tested (see Figure 10.20).

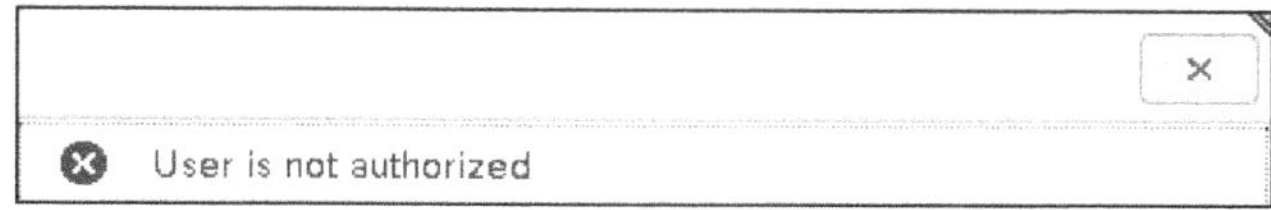

Figure 10.20 Testing the CHECK_AUTHORIZATIONS Method

The CHECK_START_CONDITIONS method

The CHECK_START_CONDITIONS method checks the start conditions of the job. This method can be used to check the scheduling criteria. The implementation in Listing 10.9 checks whether recurring scheduling is planned in the IS_SCHEDULE_INFO parameter (periodic_granularity IS NOT INITIAL). Otherwise, the parameter is set and the application job returns an error message. We also need to set the ev_incorrect parameter to abap_true so that the method knows that scheduling isn't allowed.

```
METHOD if_apj_jt_check_20~check_start_condition.
  IF is_schedule_info-periodic_granularity IS INITIAL.
    INSERT VALUE #( id = 'ZACB_JOB' type = 'E' number = '002' )
    INTO TABLE et_msg.
    ev_incorrect = abap_true.
  ELSE.
    ev_incorrect = abap_false.
  ENDIF.
ENDMETHOD.
```

Listing 10.9 Implementation of the CHECK_START_CONDITIONS Method

When checking the application job, corresponding messages are now displayed if the job doesn't have a recurrence pattern (see Figure 10.21).

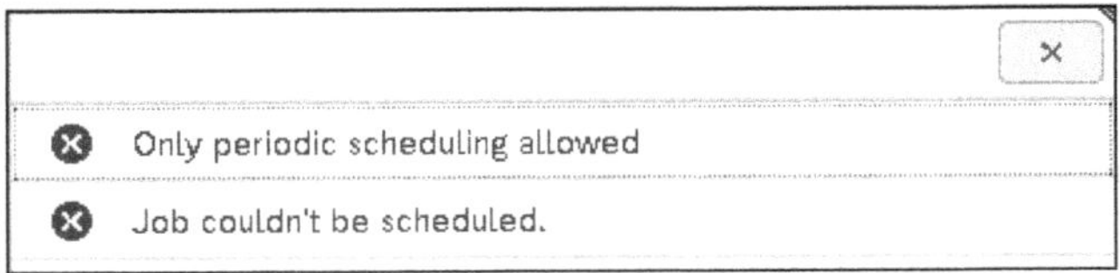

Figure 10.21 Testing the CHECK_START_CONDITIONS Method

GET_DYNAMIC_PROPERTIES

Using the GET_DYNAMIC_PROPERTIES method, it's possible to dynamically change parameters in the application. The visibility (HIDDEN) and editability (READ_ONLY) of the parameters can be changed. In the implementation in Listing 10.10, the PARA_LASTFOURWEEKS parameter is read. If this parameter is set, the PARA_LASTDAYS parameter can't be edited. If the TEST parameter isn't set, the DATE parameter is ready for input again.

```
METHOD if_apj_jt_check_20~get_dynamic_properties.
  IF it_value[ KEY param parameter_name =
      zcl_acb_det_recipe=>para_lastfourweeks ]-low = abap_true.
    INSERT VALUE #( job_parameter_name =
      zcl_acb_det_recipe=>para_lastdays read_only_ind = abap_true )
    INTO TABLE rts_dynamic_property.
  ELSE.
    INSERT VALUE #( job_parameter_name =
      zcl_acb_det_recipe=>para_lastdays read_only_ind = abap_false )
    INTO TABLE rts_dynamic_property.
  ENDIF.
ENDMETHOD.
```

Listing 10.10 Implementation of the GET_DYNAMIC_PROPERTIES Method

The following behavior now occurs when the job gets scheduled: If the **Last 4 weeks** indicator is ticked, the **Determination days** field can't be edited (see Figure 10.22). If the **Last 4 weeks** indicator isn't checked and the **Check** button is clicked, the **Determination days** field can be filled with content.

Parameter Section

Parameters

Determination days:* 28

Last 4 weeks: ☑

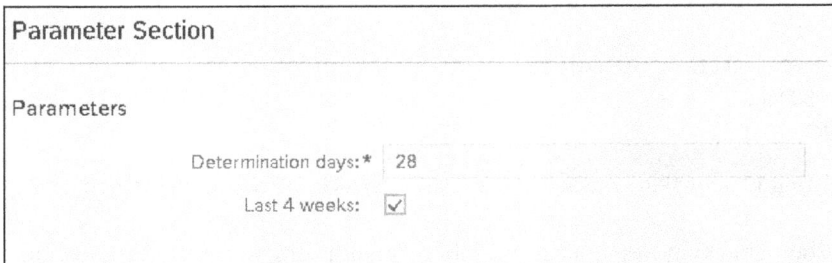

Figure 10.22 Testing the GET_DYNAMIC_PROPERTIES Method

In addition to the methods described, the following methods are also available:

Other methods

- CHECK_AND_ADJUST_PARAMETER
  This method is called when a user changes parameters. The values can then be checked and adjusted if necessary.
- ADJUST_READ_ONLY
  This method enables the adjustment of values for fields that are set to read only.
- ADJUST_HIDDEN
  This method enables the adjustment of values for fields that are hidden.

10.4 Logging

During the execution of a job, various different pieces of information are collected. These can be passed on to users if it's necessary to output these messages. To generate this kind of output, it's necessary to generate messages. The messages can be generated using the application log.

[«]

Application Log

Further information on the application log can be found in Chapter 6.

Application log object

To use corresponding messages, an application log object must first be created. You can create the corresponding object in the ABAP development tools as described in Chapter 6, Section 6.2.

Extending an execution class

In the next step, we want to extend the execution class ZCL_ACB_DET_RECIPE created in Section 10.1. We define a new method that creates messages and adds them to the application log. In the if_apj_rt_exec_object~execute execution method, we create a log object, add some messages, and save the log in the database (see Listing 10.11).

```
DATA(log) = cl_bali_log=>create( ).
DATA(lo_header) = cl_bali_header_setter=>
    create( object      = 'ZACB_JOB_LOG'
            subobject   = 'JOB'
            external_id =
...
cl_system_uuid=>create_uuid_c32_static( ) ).
log->set_header( lo_header ).
...
log_messages( severity = if_bali_constants=>c_severity_status
      message_text     = 'Start' ).
```

```
...
cl_bali_log_db=>get_instance( )->save_log( log = log
                    assign_to_current_appl_job = abap_true ).
```

Listing 10.11 Extension of the EXECUTE Execution Method

[!]

Assigning the Application Log to the Application Job

When saving the job, the value of the `assign_to_current_appl_job` indicator must be set to `abap_true`. This saves the application log for the current application job.

Executing the application job

The application job is now executed again. After the run, a corresponding identification for the status display will be stored in the **Log** column (see Figure 10.23). This is also used for navigation.

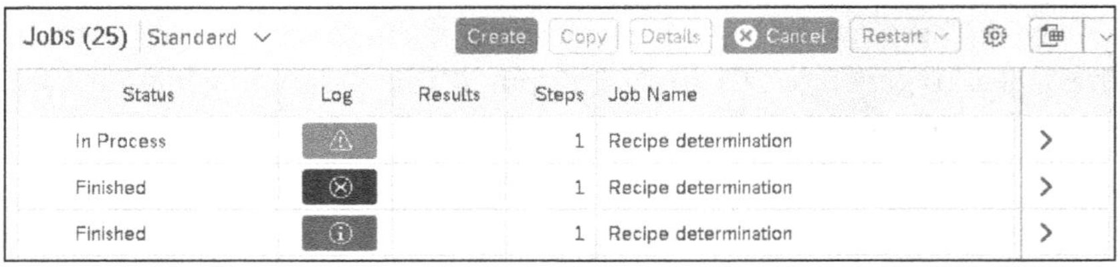

Jobs (25) Standard ∨ Create Copy Details Cancel Restart

Status	Log	Results	Steps	Job Name	
In Process	⚠		1	Recipe determination	>
Finished	⊗		1	Recipe determination	>
Finished	ⓘ		1	Recipe determination	>

Figure 10.23 Display of the Application Log in the Application Job

If you click on the button in the **Log** column, the application log will be displayed. There, you can filter by message texts and the severity of the message (see Figure 10.24). The overall status of the job is determined by the highest severity level of the messages output for this job.

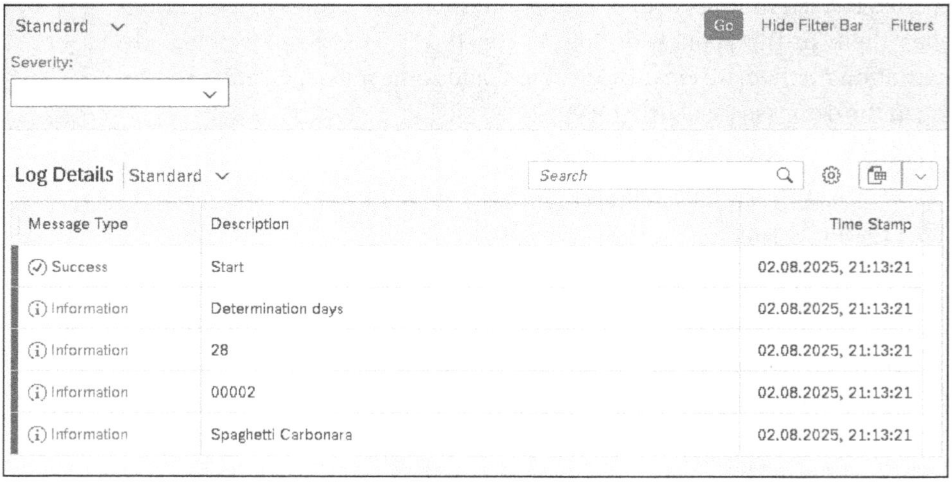

Standard ∨ Go Hide Filter Bar Filters

Severity:

Log Details Standard ∨ Search

Message Type	Description	Time Stamp
Success	Start	02.08.2025, 21:13:21
Information	Determination days	02.08.2025, 21:13:21
Information	28	02.08.2025, 21:13:21
Information	00002	02.08.2025, 21:13:21
Information	Spaghetti Carbonara	02.08.2025, 21:13:21

Figure 10.24 Display of the Application Log

10.5 Summary

In this chapter, you learned how to develop and schedule application jobs in ABAP Cloud.

You've learned how to configure an application job with its parameters and values. When scheduling jobs, you have the option of adding additional classes for checks, value helps (not yet available in SAP S/4HANA 2023), and notifications. You've also learned about logging for application jobs.

[«]

Announced Further Development of the Application jobs

As of SAP S/4HANA Cloud Public Edition 2411, a new version of the application jobs for creating background jobs is available: Application Jobs V2. Further information on this can be found in the SAP Help Portal at *http://s-prs.de/1064809*.

New API for application jobs

Application jobs are an important part of an SAP system. They can perform time-consuming tasks in the background. Application jobs are currently under constant development. While we were writing this book, SAP introduced a new API for application jobs. They also offer the support of value helps. In addition, the parameters can be created and set in a dialog box when the application job catalog is created. Furthermore, there's also an option for debugging application jobs.

Table 10.3 shows the differences between background processing in classic ABAP and in ABAP Cloud.

Aspect	Classic ABAP	ABAP Cloud
Program logic	Classic ABAP program	ABAP class implements the `IF_APJ_RT_EXEC_OBJECT` interface.
Definition of parameters	Definition of parameters as a selection screen in the program and subsequent saving of the variant	Parameters are defined in the implementation class by implementing the `IF_API_DT_EXEC_OBJECT` interface and then filling the application job template.
Job scheduling	Transaction SM36 (Job Definition)	Maintain Application Job app (app ID F1240).
Job overview	Transaction SM37 (Job Overview)	Maintain Application Job app (app ID F1240).

Table 10.3 Comparison of Background Processing in Classic ABAP and in ABAP Cloud

Chapter 11
Email Dispatch

Would you like to send other members of your team important information from the SAP system that they don't have access to? In the past, written media would have been used for this, but, as we know, today it's email.

11

Emails have long been regarded as one of the most effective and reliable means of communication in a professional environment. They can be sent to recipients anywhere in the world within seconds. Emails can also be used to reach multiple people at the same time, and information can be forwarded quickly.

In the SAP system, for example, orders or error reports can be sent automatically by email. Whereas previously the CL_DOCUMENT_BCS class or Transaction SOST was used, the CL_BCS_MAIL_MESSAGE class and a new SAP Fiori app are now used to monitor the dispatch of emails. In ABAP Cloud, the entire application programming interface (API) for emails has also been revised.

Structure of this chapter

In this chapter, you'll first learn how to configure this API in Section 11.1. In Section 11.2, we explain how you can create simple formatted emails and emails with attachments. Section 11.3 then describes the integration into the application job created in Chapter 10 for an automatic dispatch via email.

[«]

Did You Know?

Queen Elizabeth II was the first royal person to send an email on March 26, 1976. This happened during a visit to a telecommunications center. The email was sent within ARPANET, the predecessor of the current internet.

Extending the application job

In this chapter, we use the ZCL_ACB_MAIL_DEMO class to explain how emails can be sent in ABAP Cloud. We also add two functions to the application job from Chapter 10:

- An extension of the execution logic to include the dispatch of a newsletter
- Notifications in the event of an error during job execution

To enable you to try out both the example from Chapter 10 and the extension from this chapter on your own, you should copy the application job from Chapter 10 for this chapter and extend it to include the requirements mentioned.

[»]

Using the Sample Application

The objects created in this chapter are located in the `MAIL` folder.

11.1 Configuring and Monitoring the Email Dispatch

Email configuration

To be able to use the new ABAP Cloud API for emails, some email-specific settings must be made. Formerly, Transaction SCOT was used for this purpose. However, the email settings for the new API in the on-premise SAP S/4HANA solution are no longer configured via this transaction, but via Transaction SBCS_MAIL_CONFIGSMTP. Figure 11.1 shows the initial screen of this transaction.

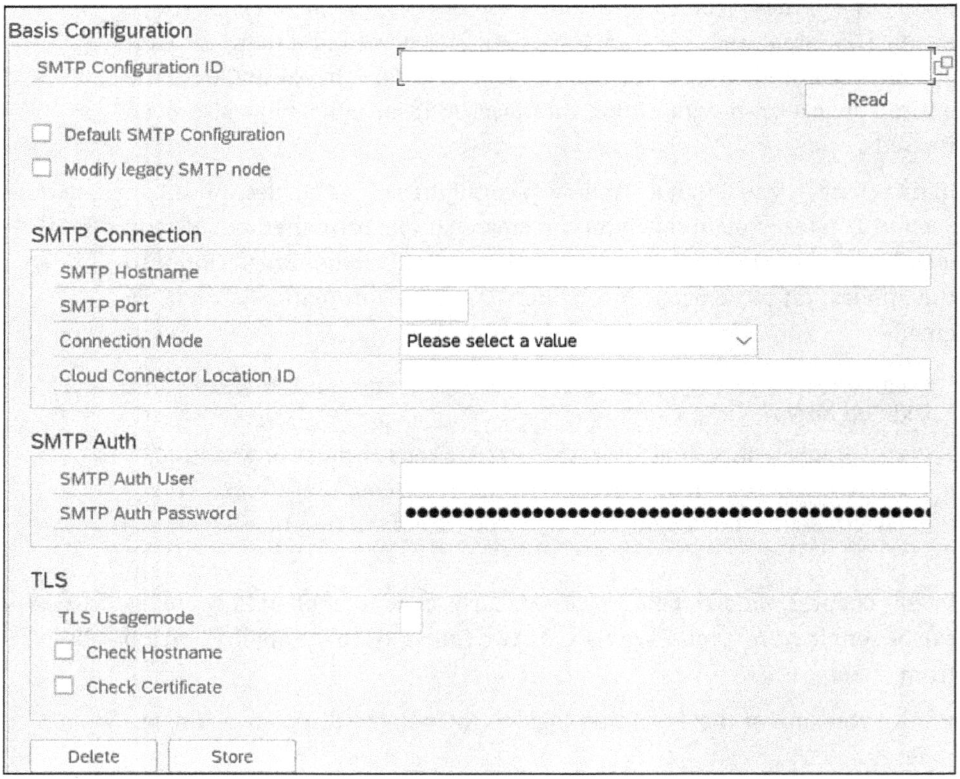

Basis Configuration
SMTP Configuration ID
Read
Default SMTP Configuration
Modify legacy SMTP node
SMTP Connection
SMTP Hostname
SMTP Port
Connection Mode
Please select a value
Cloud Connector Location ID
SMTP Auth
SMTP Auth User
SMTP Auth Password
TLS
TLS Usagemode
Check Hostname
Check Certificate
Delete
Store

Figure 11.1 Email Configuration in Transaction SBCS_MAIL_CONFIGSMTP

All information about the mail server must be entered there. Remember to confirm your changes via the **Read**, **Delete**, and **Save** buttons. Otherwise, the configuration won't be saved.

[!]

Network Release

You should implement these settings in consultation with your system administrator. Firewall activations may need to be set up between the SAP system and the mail server. The certificates of the mail server may also have to be stored in the SAP system. The email trace can be an important aid here. It's used to log the transmission process in detail.

Email monitoring

To test the configuration, you need to switch to the email monitoring application. Monitoring also no longer takes place in Transaction SOST, but in the Monitor Email Transmissions app (app ID F5442).

[«]

Monitor Email Transmissions App

Information on the Monitor Email Transmissions app (app ID F5442) can be found in the SAP Fiori apps reference library at *http://s-prs.de/v1064802*. First select your release. On the **IMPLEMENTATION INFORMATION** tab, you'll find all the information you need to configure the SAP Fiori app in your system and find out, for example, which standard roles are available and which Open Data (OData) protocol services need to be activated.

All emails sent via the new API in ABAP Cloud are displayed in the SAP Fiori app. Important information such as status, subject, or creation date is displayed (see Figure 11.2). In contrast to Transaction SOST, content and attachments aren't available.

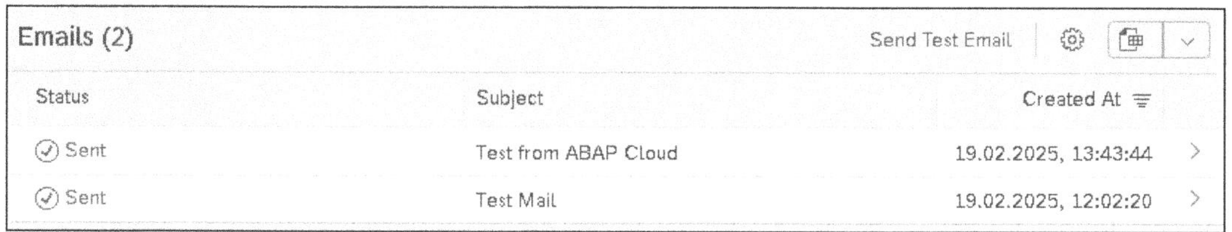

Figure 11.2 Monitor Email Transmissions App

Test email

In addition to the overview, you also have the option of sending a test email. Follow these steps:

1. Click on the **Send Test Email** button on the right above the email overview.
2. In the popup window that appears, select a recipient.

3. Click on **Send Test Email**. If the email configuration is correct, a test email will now be sent.

After that, the test email is in the recipient's mailbox. It contains a subject, the email content (body), and an attachment (see Figure 11.3).

Figure 11.3 Test Email Received

Structure of an email

The most important components of a standard email can be briefly shown using this test email as an example. You can see the structure of an email in Figure 11.4:

- **Header**
  The header provides information about the route of a message, the recipients, and the subject of the email.

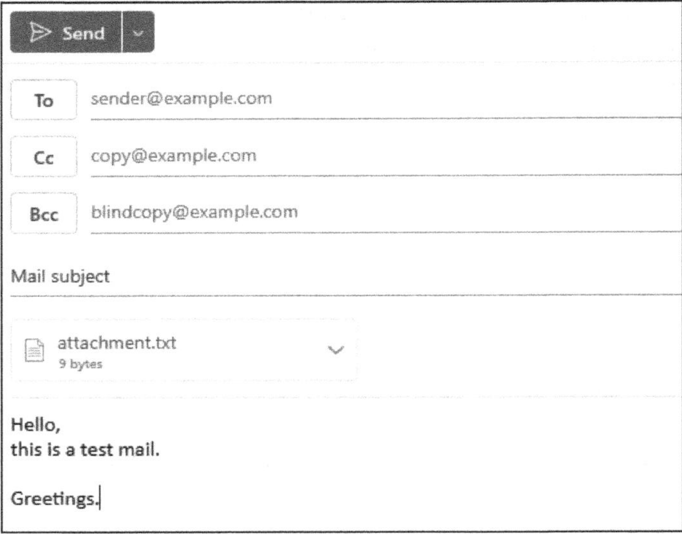

Figure 11.4 Structure of an Email

- **Attachment**
  The attachment is optionally located below this. It can contain any type of file such as images or documents.
- **Body**
  The body contains the actual text of the email. It can consist of plain text or formatted text (e.g., HTML).
- **Footer**
  The email footer below the body also contains text, such as notes or advertising.

11.2 ABAP Cloud API for the Email Dispatch

API classes

Let's now move on to the new API for emails in ABAP Cloud. The main class of this API for creating and sending emails is the CL_BCS_MAIL_MESSAGE class. The CL_BCS_MAIL_TEXTPART and CL_BCS_MAIL_BINARYPART classes are used to format texts and add attachments. In this section, we'll show you how to use these classes and their methods to send an email. We want to show you three examples that build on each other:

- Sending a simple email
- Sending a formatted email
- Sending a formatted email with attachment

Sending a simple email

To send a simple, unformatted email, you can implement the SEND_TEXT_MAIL method of the CL_BCS_MAIL_MESSAGE class in a separate method, as shown in Listing 11.1. You can find our sample implementation in the download material in our ZCL_ACB_MAIL_DEMO class with the SEND_NORMAL_MAIL method.

```
METHOD send_text_mail.
  DATA content TYPE  string.
  TRY.
      DATA(mail) = cl_bcs_mail_message=>create_instance( ).

      mail->set_sender( address ).

      mail->add_recipient(
           iv_address = address
           iv_copy    = cl_bcs_mail_message=>to ).
      mail->add_recipient(
           iv_address = address
           iv_copy    = cl_bcs_mail_message=>cc ).
```

```
      mail->set_subject(
           'Test new API - sample text' ).

      content = 'This is a normal text.
      mail->set_main(
           cl_bcs_mail_textpart=>create_instance(
           iv_content           = content
           iv_content_type      = 'text/plain' ) ).

      mail->send( IMPORTING
           et_status      = DATA(status)
           ev_mail_status = DATA(mail_status) ).

      out->write( 'Test: Sende normal method' ).
      out->write( status ).
      out->write( |Status of the mails: { mail_status }| ).

    CATCH cx_bcs_mail INTO DATA(exception).
      out->write( exception->get_text( ) ).
  ENDTRY.
ENDMETHOD.
```

Listing 11.1 Sending a Simple Email

In Listing 11.1, you can see the program sequence for sending an email:

1. The CL_BCS_MAIL_MESSAGE class is instantiated.
2. A sender address is added using the set_sender method.
3. The recipient addresses (iv_address) are added using the add_recipient method. You can also specify the type of reception (iv_copy). The possible reception types are listed in Table 11.1.
4. The subject of the email is specified (set_subject method).

Constant	Reception Type
cl_bcs_mail_message=>to	This is the direct contact person.
cl_bcs_mail_message=>cc	CC stands for carbon copy and enables a copy to be sent to someone who isn't the main recipient of an email.
cl_bcs_mail_message=>bcc	BCC stands for blind carbon copy. The other recipients don't see that the recipients listed have received the email.

Table 11.1 Possible Reception Types and Constants

5. The actual content of the email is then included in the body. First, the content text is assigned (`content = test`). The `set_main` method is used to communicate the content. The type of text is also indicated. In this case, it's a simple text (`cl_bcs_mail_textpart=>create_text_plain( iv_content = content)`).
6. To send the email, the `send` method is executed. After sending the email, you'll receive the transmission status for each recipient address (`et_status`) as well as the general status (`ev_status`).
7. If an error occurs in one of the methods presented, the `CX_BCS_MAIL` error class will be thrown.

As a result, the email shown in Figure 11.5 is now in the recipient's mailbox.

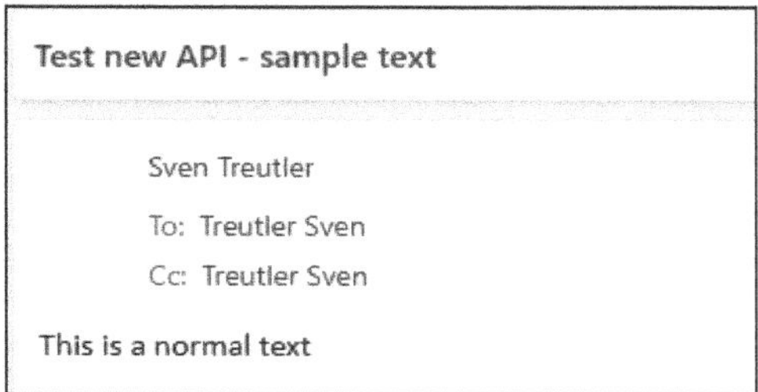

Figure 11.5 Received Email with Sample Text

Sending a formatted email

To make an email more visually appealing, it can be formatted with HTML code. Listing 11.2 shows the implementation of the `SEND_FORMAT_MAIL` method.

```
METHOD send_format_mail.
  DATA content TYPE  string.
  TRY.
      DATA(mail) = cl_bcs_mail_message=>create_instance( ).

      mail->set_sender( address ).

      mail->add_recipient(
        iv_address = address
        iv_copy     = cl_bcs_mail_message=>to ).
      mail->add_recipient(
        iv_address = address
        iv_copy     = cl_bcs_mail_message=>cc ).

      mail->set_subject(
            'Test new API - sample formatting' ).
      content = '<h1>Hello,</h1>' &&
```

```
                '<p>I have been formatted.</p>' &&
                '<p><strong>Have a nice day</strong></p>'.
      mail->set_main( cl_bcs_mail_textpart=>create_text_html(
          iv_content = content ) ).

      mail->send( IMPORTING
          et_status      = DATA(status)
          ev_mail_status = DATA(mail_status) ).

      out->write( 'Test: Send formatting' ).
      out->write( status ).
      out->write( |Status of the mails: { mail_status }| ).

    CATCH cx_bcs_mail INTO DATA(exception).
      out->write( exception->get_text( ) ).
  ENDTRY.
ENDMETHOD.
```

Listing 11.2 Email Formatting Example

The content of the email (content) is transferred as HTML code. This time, the CREATE_TEXT_HTML method of the CL_BCS_MAIL_TEXTPART class must also be called because HTML content is also provided. The results are displayed in Figure 11.6.

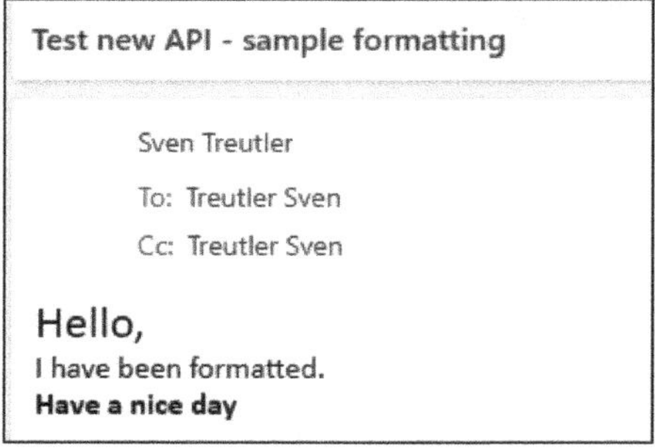

Figure 11.6 Example of a Formatted Email

Sending an email with attachment

The new API provides the following options for attaching a file to the email:

- **Simple text content as an attachment (MIME type text/plain)**
  The CL_BCS_MAIL_TEXTPART class can also be used to attach a text file to the email. It provides the CREATE_TEXT_PLAIN method for text files. The content and file name of the attachment must be passed to it. Listing 11.3 shows how this method must be called.

```
mail->add_attachment( cl_bcs_mail_textpart=>create_text_plain(
          iv_content      = 'Text attachment'
          iv_filename     = 'TestAttachmentText.txt' ) ).
```

Listing 11.3 Transferring a Simple Text File as Attachment

- **Text content as attachment**
  To send a file that has a MIME type other than text/plain, the create_instance method of the CL_BCS_MAIL_TEXTPART class can be called. In addition to the text content (iv_content_type) and the filename (iv_filename), the MIME type (iv_content_type) must also be specified (see Listing 11.4).

```
mail->add_attachment( cl_bcs_mail_textpart=>create_instance(
  iv_content = '<mail><an>Sven</an><von>Fabian</von>' &&
               '<content>My XML file!</content></mail>'
  iv_content_type = 'text/xml'
  iv_filename     = 'TestAttachmentXml.xml'
) ).
```

Listing 11.4 Transferring Text Content as Attachment

11

[«]

MIME Type

A *MIME type* consists of the media type and the subtype. These details are separated by a slash. The media type specifies the group of the file, such as text for text file or image for an image.

Table 11.2 lists some of the most common MIME types.

MIME Type	File Type
text/plain/html	HTML files
application/pdf	PDF files
image/jpeg	JPEG files
text/plain	Plain text files
video/mpeg	MPEG files

Table 11.2 Commonly Used MIME Types

- **Binary files as attachment**
  Binary files are files that aren't saved in text format and must be interpreted by another external program. Unlike when transferring text content as an attachment, the CL_BCS_MAIL_BINARYPART class is used for this

purpose. The binary content (`iv_content_type`), the filename (`iv_filename`), and the MIME type (`iv_content_type`) are given to the `CREATE_INSTANCE` method (see Listing 11.5).

```
mail->add_attachment( cl_bcs_mail_binarypart=>create_instance(
  iv_content = pdfcontent
  iv_content_type = 'application/pdf'
  iv_filename = 'Testpdf.pdf'
) ).
```

Listing 11.5 Transferring a Binary File as PDF

After the sample code has been executed, an email with attachments gets sent, and the recipients can open these files with the sample content (see Figure 11.7).

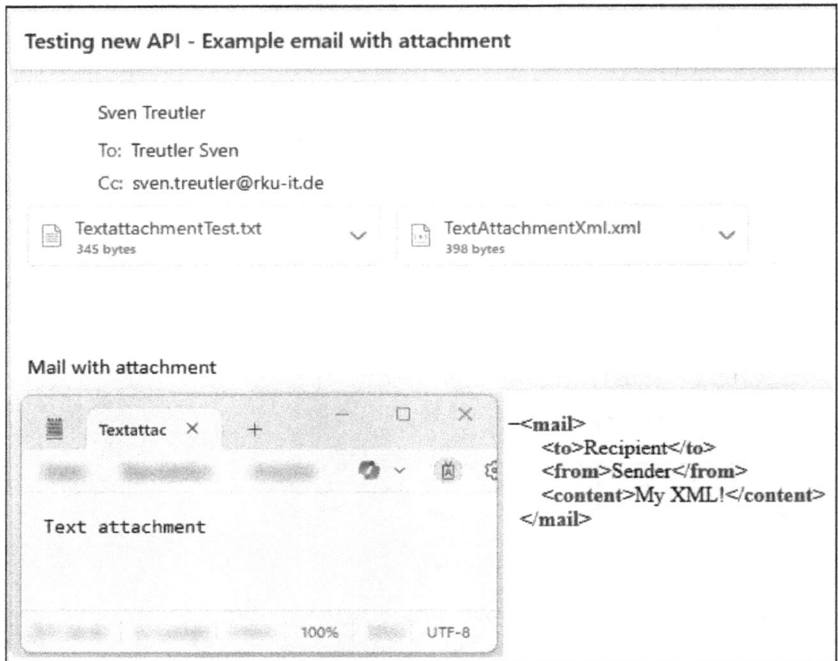

Figure 11.7 Email with Attachment

Validating email addresses

Another important method is used to validate email addresses. The `CL_MAIL_ADDRESS` class checks whether an email address is syntactically correct. Listing 11.6 shows an example of such a validation.

```
TRY.
        DATA mail_address TYPE cl_mail_address=>ty_address_string.
        mail_address = 'test@example.com'.
```

```
    DATA(mail_check) = cl_mail_address=>create_instance(
      iv_address_string = mail_address ).

    DATA(address_valid) = mail_check->validate( ).

    out->write( |Email address { mail_address
      } is valid: { address_valid }| ).

    mail_address ='test-example.com'.

    mail_check = cl_mail_address=>create_instance(
      iv_address_string = mail_address ).

    address_valid = mail_check->validate( ).

    out->write( |Email address { mail_address
      } is valid: { address_valid }| ).
  CATCH cx_bcs_mail INTO DATA(exception).
    out->write( exception->get_text( ) ).
ENDTRY.
```

Listing 11.6 Sample Code for Validating an Email Address

First, an instance of the CL_MAIL_ADDRESS class is created. The CREATE_INSTANCE method is transferred along with the email address to be validated (iv_address_string). Then, the VALIDATE method is used. It returns the information as to whether an email address is syntactically valid or not (see Figure 11.8).

mail address	test@example.com	is valid:	X
mail address	test-example@com	is valid:	

Figure 11.8 Output of the Sample Code

11.3 Integrating the Newsletter Dispatch into the Recipe Portal

Sending an email as a newsletter

For our recipe portal, we want to dispatch a newsletter by email. The newsletter contains the names of the recipes that have been added in the past few days. The number of days after a new email will be sent is stored in the system.

11.3.1 Creating an Application Job

To implement this requirement scenario, we need to create a job. For this purpose, all the objects required for an application job must first be created. These objects are listed in Table 11.3.

Object	Application Jobs
ZCL_ACB_NEWSLETTER	Execution class for the application job
ZCL_ACB_NEWSLETTER_CHK	Test class for the application job
ZACB_NEWSLETTER	Application job catalog entry
ZACB_NEWSLETTER	Application job template
ZACB_NEWSLETTER	Application log

Table 11.3 Objects Created for the Application Job

[»]

The Application Job from Chapter 10

In Chapter 10, we described how you can create and use an application job. Now, you can copy the execution logic developed there into a new application job for the present scenario.

Extending the execution logic

Now, you want to extend the execution method (EXECUTE) of the execution class of the application job, as shown in Listing 11.7.

```
METHOD if_apj_rt_exec_object~execute.
  ...
      log = cl_bali_log=>create( ).

      DATA(newsletter) = NEW zcl_acb_mail_newsletter(
                               days ).
      newsletter->send(  ).
...
ENDMETHOD.
```

Listing 11.7 The if_apj_rt_exec_object~execute Method

This application job evaluates the period for which the recipes created are to be determined and sent by newsletter. The result gets transferred to the ZCL_ACB_MAIL_NEWSLETTER class. If an error occurs during the execution of this class, it will be intercepted.

The recipes are first determined within the ZCL_ACB_MAIL_NEWSLETTER class (see Listing 11.8).

Determining data within the application job

```
METHOD constructor.
  days = det_days.
  det_recipes( ).
ENDMETHOD.

METHOD det_recipes.
  DATA(det_date) = EXACT timestamp(
    xco_cp=>sy->moment( xco_cp_time=>time_zone->utc )
      ->(subtract( iv_day = CONV #( days ) )
      ->as( xco_cp_time=>format->abap )
      ->value ).

  SELECT FROM zacb_recipe
    FIELDS *
    WHERE created_at > @det_date
    INTO TABLE @recipes.

  IF sy-subrc <> 0.
    RAISE EXCEPTION NEW zcx_acb_newsletter_error( ).
  ENDIF.
ENDMETHOD.
```

Listing 11.8 Determination of Recipes in the ZCL_ACB_MAIL_NEWSLETTER Class

All information such as the sender, the formatted HTML content, and the attachments are then added to the email and sent (see Listing 11.9).

Sending the formatted email with attachment

```
METHOD send.
  mail = cl_bcs_mail_message=>create_instance( ).
  add_sender( ).
  add_recipients(  ).
  add_content( ).
  add_attachment( ).
  mail->send( ).
ENDMETHOD.
```

Listing 11.9 Creation of an Email Newsletter

The received email, including attachment, is shown in Figure 11.9.

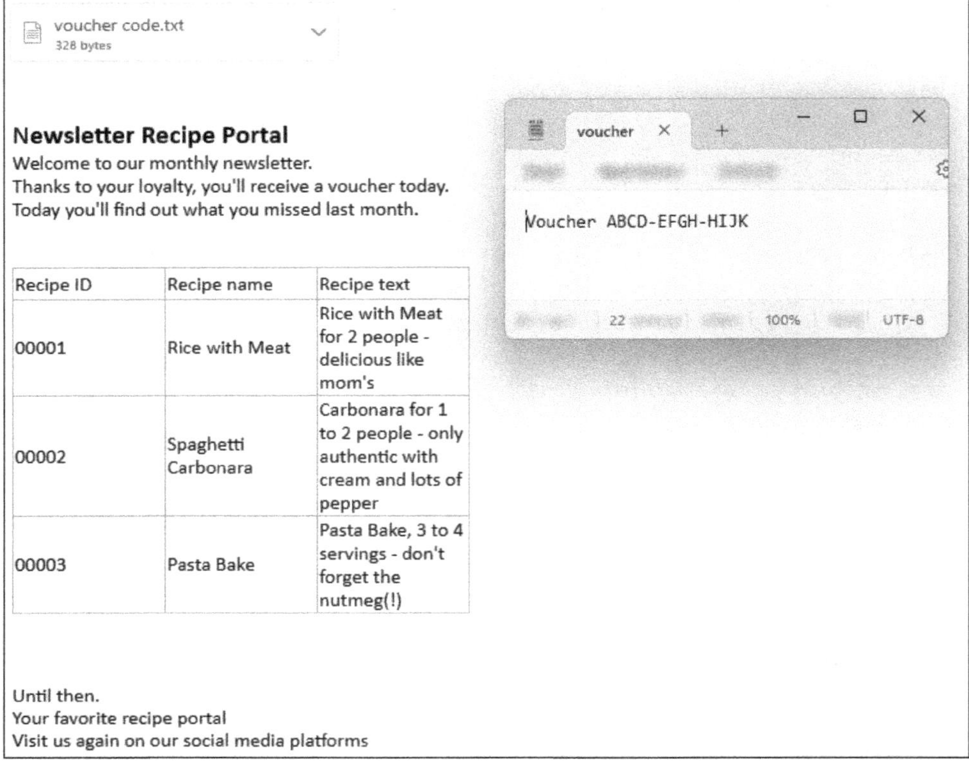

voucher code.txt
328 bytes

Newsletter Recipe Portal

Welcome to our monthly newsletter.
Thanks to your loyalty, you'll receive a voucher today.
Today you'll find out what you missed last month.

Recipe ID	Recipe name	Recipe text
00001	Rice with Meat	Rice with Meat for 2 people - delicious like mom's
00002	Spaghetti Carbonara	Carbonara for 1 to 2 people - only authentic with cream and lots of pepper
00003	Pasta Bake	Pasta Bake, 3 to 4 servings - don't forget the nutmeg(!)

Until then.
Your favorite recipe portal
Visit us again on our social media platforms

Figure 11.9 Sample of an Email Newsletter

11.3.2 Email with Error Message

Notification class

If the job gets canceled in the system, there's a way to inform the admins about the cancelation. As described in Section 10.2.1, a notification class can also be stored in the job catalog in addition to the execution class and the check class (see Figure 11.10).

Application Job Catalog Entry: ZACB_NEWSLETTER

General Information

Class with Execute Method:* ZCL_ACB_NEWSLETTER Browse...

Exit Classes

Check: ZCL_ACB_NEWSLETTER_CHK Browse...

Value Help: Browse...

Notification: Browse...

Figure 11.10 Application Job Catalog Entry

You can either specify an existing class as the class for notifications or you can create a new class by clicking on the **Notification** link. The creation wizard for classes then opens (see Figure 11.11).

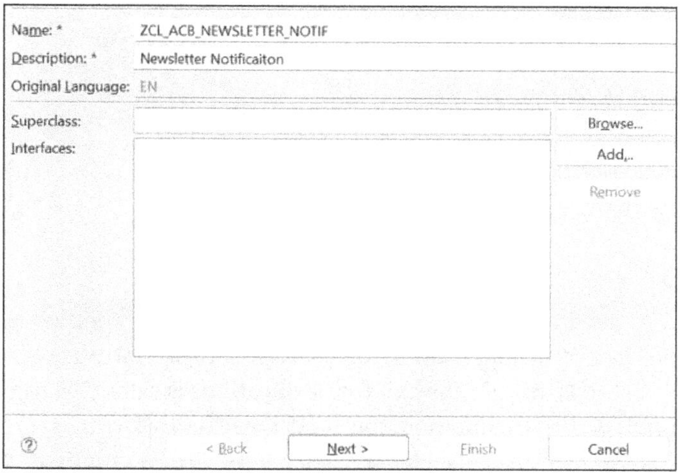

Figure 11.11 Creation Wizard for Classes

Name the class for our example ZCL_ABC_NEWSLETTER_NOTIF. The IF_APJ_RT_JOB_NOTIF_EXIT interface must be added manually. To do this, click on the **Add** button. Then, complete the creation wizard by clicking on the **Finish** button. The class has now been created and entered in the job catalog. Listing 11.10 shows the automatically generated code.

```
CLASS zcl_acb_newsletter_notif DEFINITION
  PUBLIC
  FINAL
  CREATE PUBLIC.
  PUBLIC SECTION.
    INTERFACES if_apj_rt_job_notif_exit.
  PROTECTED SECTION.
  PRIVATE SECTION.
ENDCLASS.
CLASS zcl_acb_newsletter_notif IMPLEMENTATION.
  METHOD if_apj_rt_job_notif_exit~notify_jt_end.
  ENDMETHOD.

  METHOD if_apj_rt_job_notif_exit~notify_jt_start.
  ENDMETHOD.
ENDCLASS.
```

Listing 11.10 Generated Code for Notification Class ZCL_ACB_NEWSLETTER_NOTIF

Methods of the notification class

As shown in Listing 11.10, two methods can be implemented here:

- NOTIFY_JT_START
  This method is currently not yet supported. You'll find this information in the class outline (see Figure 11.12).

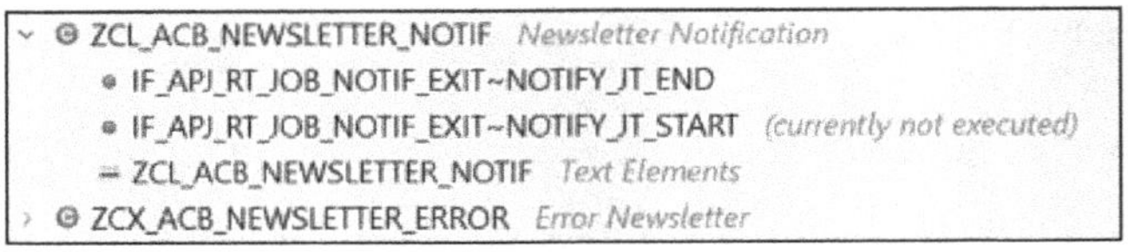

```
⌄ ZCL_ACB_NEWSLETTER_NOTIF  Newsletter Notification
    IF_APJ_RT_JOB_NOTIF_EXIT~NOTIFY_JT_END
    IF_APJ_RT_JOB_NOTIF_EXIT~NOTIFY_JT_START  (currently not executed)
    ZCL_ACB_NEWSLETTER_NOTIF  Text Elements
› ZCX_ACB_NEWSLETTER_ERROR  Error Newsletter
```

Figure 11.12 Nonexecution of the NOTIFY_JT_START Method

- NOTIFY_JT_END
  This method is executed if an error is triggered in the EXECUTE method in the execution class of the application job catalog. The job status is then automatically set to **Error**, and the exit for notifications is called. To trigger an exception within the method, the RAISE EXCEPTION NEW cx_apj_rt_content( ) statement must be executed. The NOTIFY_JT_END method is used to transfer the job information listed in Table 11.4.

Parameter	Description
IS_JOB_INFO	General information about the job such as job name or job runs
IV_JOB_CATALOG_ENTRY_NAME	Job catalog name
IV_JOB_TEMPLATE_NAME	Job template name
IV_STEP_STATUS	Status of the step
IV_JOB_STATUS	Status of the job

Table 11.4 Parameters of the NOTIFY_JT_END Method

The NOTIFY_JT_END method thus contains the logic shown in Listing 11.11.

```
METHOD if_apj_rt_job_notif_exit~notify_jt_end.
  DATA(mail) = cl_bcs_mail_message=>create_instance( ).

  mail->set_sender( 'error@example.com' ).
  mail->add_recipient( 'test@example.com' ).

  mail->set_subject( CONV #( | Job error | ) ).
```

```
  mail->set_main( cl_bcs_mail_textpart=>create_text_plain(
     iv_content = | Der Job { is_job_info-job_name
     } has an error. Please check | ) ).

  mail->send( ).
ENDMETHOD.
```

Listing 11.11 Implementation of the NOTIFY_JT_END Method

As shown in Listing 11.11, the following logic has been developed to send a notification email from the system. First, the CL_BCS_MAIL_MESSAGE class is instantiated. The sender (set_sender), recipient (set_recipient), and subject (set_subject) are then added. The actual content of the email is next added using the set_content method. In the final step, the email gets sent with the previously added information (mail->send).

11

Extending the execution logic

To trigger an exception for the execution of the notification class, the code from Listing 11.7 shown earlier for the EXECUTE execution method must be extended by the CX_APJ_RT_CONTENT exception class (see Listing 11.12).

```
CATCH   cx_bali_runtime
        cx_uuid_error
        cx_bcs_mail
        zcx_acb_newsletter_error
    INTO DATA(error).
  RAISE EXCEPTION TYPE cx_apj_rt_content
    MESSAGE ID 'ZACB_NEWSLETTER' NUMBER 003
    WITH error->get_text( ).
  ENDTRY.
```

Listing 11.12 Extension by the CX_APJ_RT_CONTENT Exception

As soon as the job is canceled, the email shown in Figure 11.13 gets sent.

Job error

The job ZACB_NEWSLETTER has an error. Please check.

Figure 11.13 Email Notification in the Event of an Error During Job Execution

11.4 Summary

In this chapter, we've shown you how the topic of sending emails in ABAP Cloud is handled. We've introduced you to the new configuration Transaction SBCS_MAIL_CONFIGSMTP and the new options for monitoring email transmissions using the Monitor Email Transmissions app (app ID F5442).

Next, you got to know the new email API. In ABAP Cloud, there's now a central class that can add the recipient and sender data as well as the content to an email. The CL_BCS_MAIL_TEXTPART class can be used to add text files as attachments, while the CL_BCS_MAIL_BINARYPART class adds binary files. The CL_MAIL_ADDRESS class can also be used to validate email addresses.

Finally, the email dispatch was integrated into an application job to implement a newsletter dispatch for our recipe portal. First, we've explained how an email can be sent from an application job. We've also revisited the topic of notifications from Chapter 10 and explained it in more detail. In this context, we've shown how an email can be generated that informs about an error in the execution of the application job.

Table 11.5 compares the management and execution of email dispatch in classic ABAP and in ABAP Cloud.

Topic	Classic ABAP	ABAP Cloud
Email configuration	Transaction SCOT	Transaction SBCS_MAIL_CONFIGSMTP
Email monitoring	Transaction SOST	Monitor Email Transmissions app (app ID F5442)
Email API	Function modules: SO_NEW_DOCUMENT_ATT_SEND_API1 SO_NEW_DOCUMENT_SEND_API1 Classes: CL_BCS CL_DOCUMENT_BCS CL_CAM_ADDRESS_BCS	Classes: CL_BCS_MAIL_MESSAGE CL_BCS_MAIL_TEXTPART CL_BCS_MAIL_BINARYPART

Table 11.5 Email Dispatch in Classic ABAP and Modern ABAP

Chapter 12
Parallelizing Application Logic

To optimize performance, application logic can be parallelized using multiple work processes. For this purpose, ABAP Cloud provides a special class. In this chapter, you'll learn how to use this class using the recipe portal as an example, and you'll also get to know the respective technical details.

Background processing

In Chapter 10, you gained an initial insight into the application jobs. As the successor to classic background jobs, they allow programs to be executed in ABAP Cloud without any active user interaction. They enable periodic scheduling as well as status monitoring and log analysis. Application jobs are one of multiple options in ABAP Cloud for implementing background processing if, for example, a process is to be executed automatically on a time-controlled basis or takes too long to allow users to wait for a certain status.

Parallelization

An alternative approach to moving a process with a long runtime to background processing is to shorten the runtime. In addition to the usual performance optimization measures, the parallelization of application logic and thus the division into multiple simultaneously running work processes is another option.

Of course, this approach can also be combined with background processing techniques. Traditionally, modules with remote function call (RFC) capabilities were used for this purpose in ABAP development in various forms and sometimes using frameworks. In ABAP Cloud, there's an encapsulating class that significantly simplifies the implementation and allows you to remain completely in the object-oriented context instead of having to create a function group.

Structure of this chapter

In Section 12.1, we first explain some basic concepts of parallelization before describing the specific class-based application programming interface (API) in ABAP Cloud in Section 12.2. Section 12.3 first revises the existing code of our sample application to create a technical basis before we implement the parallelization logic using the example of the recipe portal in Section 12.3.1. Section 12.3.2 then describes the runtime behavior and debugging of parallelized applications.

Improvement of the user data import

In Chapter 6, we presented the application log using the example of user data import. An HTTP service received and processed mass data for user master records in CSV format. For this purpose, there was a conversion method available that converted the content of a CSV file into the format of the ZACB_USER database table. We now assume that this logic isn't sufficiently performant and want to improve the performance by means of parallelization.

[!]

Disadvantages of Parallelization

The example in this chapter is designed to demonstrate the technique of parallelization. As a matter of fact, with the volume of user data in our example, the parallelized application logic will ultimately run more slowly than the previous sequential processing with its comparatively simple processing logic. Parallelization is always accompanied by overhead and performance losses for the serialization and deserialization of data. You should therefore only use parallelization in scenarios in which no other performance parameters can be optimized while being aware of the disadvantages you're accepting.

[»]

Using the Sample Application

In the download material, you'll find the objects created in this chapter in the PARALLEL subpackage. In the ABAP development tools for Eclipse, you can use the F9 function key to run a demo application via the ZCL_ACB_USER_CONVERSION_DEMO class and set a breakpoint in the main method beforehand to analyze this application in the debugger. The parallelized logic can only be executed via the demo class and isn't actively integrated into the HTTP handler of the user data import component so that this component continues to describe the example from Chapter 6.

12.1 Parallelization on the ABAP Platform

To make it easier for you to understand the following explanations, in this section we'll first take a look at some relevant terms in the context of parallelization.

Threads

In many other programming languages, parallelization is an essential part of the language. You may therefore be familiar with the term *thread*, which has no relevance in ABAP development. This involves the parallel execution of statements within a process. Threads therefore also share the memory area, including the memory area of local variables. As a result, there's the problem of making applications *thread-safe* so that the application logic takes parallel execution into account. This is a complex problem in software development that fortunately doesn't exist in ABAP: ABAP doesn't support parallelization via threads.

Work processes

Instead, the ABAP platform offers parallel processes as a parallelization technique. Multiple *work processes* simultaneously execute processing logic separately from each other. They each use their own memory areas with separate life cycles for the variables. Once a process has completed its subtask, it returns the partial result to the calling process. Such partial results can be data objects or just a status. The caller collects the partial results of all started processes and hopes for a performance gain compared to the sequential execution.

Tasks

To start a second process, RFC-capable function modules are used in classic ABAP. You can use the `CALL FUNCTION ... STARTING NEW TASK` syntax to start a second process via an asynchronous RFC (aRFC) that is executed in parallel to your current process. You can use `RECEIVE RESULTS FROM FUNCTION` and `WAIT FOR` to query the results of the parallel process or wait for it to finish.

For our purposes, such an active, specific work process can be described as a task. Work processes exist regardless of whether they are currently busy executing a program. For this reason, the term *task* is suitable for a process in execution. You can even name it with a freely selectable identifier when calling it. You don't need to memorize the syntax for starting tasks and interacting with tasks because the `CL_ABAP_PARALLEL` class encodes precisely this syntax and thus also makes the described function available in ABAP Cloud.

12.2 The CL_ABAP_PARALLEL Class

Your starting point for parallelization is the global `CL_ABAP_PARALLEL` class. It provides various methods for starting, waiting for, and terminating parallel processes. An overview can be found in the documentation of the class, which you can call up in the ABAP development tools using the F2 function key and the ABAP element information (see Figure 12.1).

API for parallelization

From a technical perspective, the class doesn't provide any functions you couldn't already have implemented using RFC-capable function modules. However, there are some powerful arguments in favor of using this class. In contrast to previous technologies, it's available in ABAP Cloud and encapsulates the function modules, which are sometimes very difficult to use correctly, using methods in a comparatively lean API. Even if the class definition may seem extensive at first, you'll probably only have to actively deal with a fraction of the available functions during parallelization.

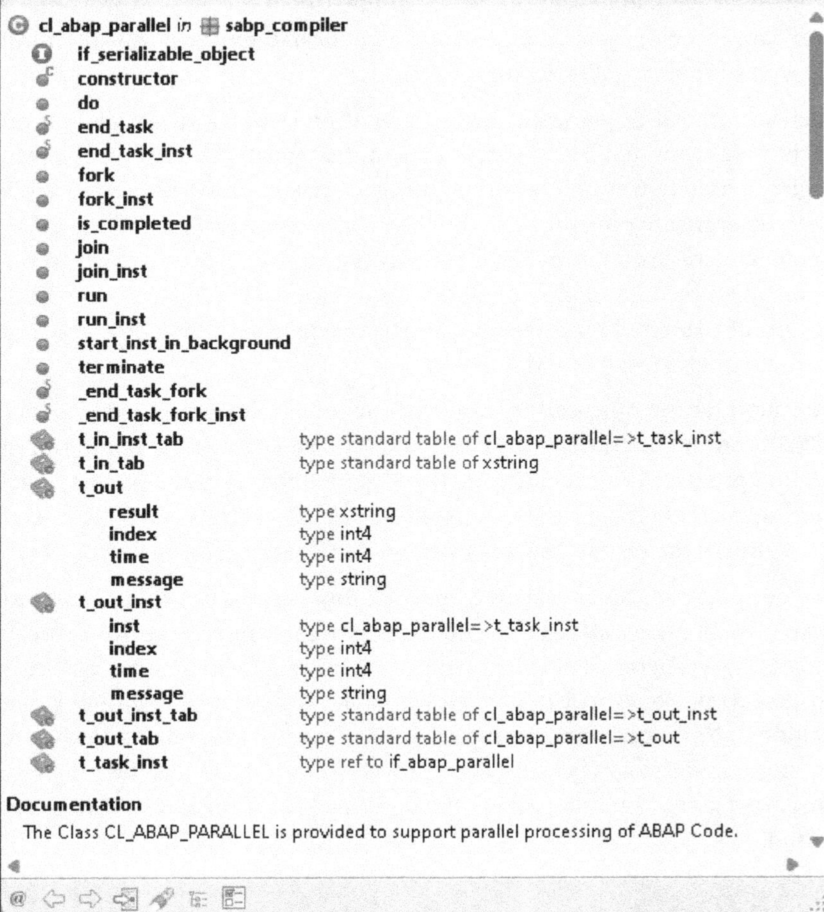

Figure 12.1 Documentation of the CL_ABAP_PARALLEL Class

Creating the instance

The methods of the class you can use are instance methods. To call them, you must first instantiate the class directly via its constructor, which provides some optional parameters that can be used for configuration (see Table 12.1).

Parameter	Usage
P_NUM_TASKS	Maximum number of simultaneous tasks
P_TIMEOUT	Maximum runtime of a task in seconds
P_PERCENTAGE	Maximum percentage of assignable work processes

Table 12.1 Parameters of the Constructor of the CL_ABAP_PARALLEL Class

Parameter	Usage
P_NUM_PROCESSES	Obsolete; use P_NUM_TASKS instead to control how many tasks can be run simultaneously
P_LOCAL_SERVER	Only uses the application server of the caller
P_ABORT_ON_ERROR	Terminates all started tasks if a task is terminated due to an error
P_KEEPING_TASKS	Don't delete task sessions after completion (see also terminate method)

Table 12.1 Parameters of the Constructor of the CL_ABAP_PARALLEL Class (Cont.)

Instance methods of CL_ABAP_PARALLEL

Once you've created an instance, you can use the following instance methods:

- **run**
  The run method starts parallel processing and waits until all tasks are completed or the timeout is reached.
- **fork**
  The fork method starts parallel processing and then allows the current process to continue as normal. You can use is_completed to query whether parallel processing has been completed. You can receive the results via join or wait synchronously for completion if parallel processing hasn't yet been completed.
- **join**
  The join method lets you receive the results of parallel processing started via fork or wait synchronously for this parallel processing until it has been completed.
- **is_completed**
  The is_completed method allows you to query whether parallel processing started via fork has now been completed.
- **terminate**
  The terminate method allows you to terminate processes started in parallel or to terminate the sessions of those processes. If you've started parallelization via the fork method, you can terminate it prematurely. If you've used the p_keeping_tasks parameter, you can use this method to delete the sessions.

A method we haven't mentioned yet is start_inst_in_background. We've deliberately omitted it at this point because it uses background RFC (bgRFC), a technology that isn't available in ABAP Cloud.

Parallel processing using "run"

For a better understanding, the differences between the `run` and `fork` methods are illustrated in the following sequence diagrams. Figure 12.2 shows the parallel execution using `run`. It starts three additional processes in parallel and then waits for them to finish.

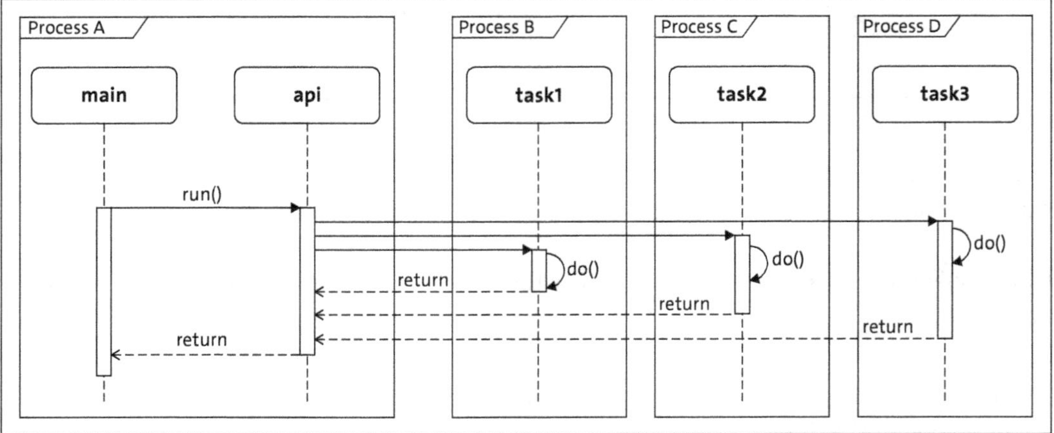

Figure 12.2 Sequence of the "run" Method in the CL_ABAP_PARALLEL Class

Parallel processing using "fork"

If, on the other hand, you want to run through other statements in the main process while the processes started in parallel are still being processed, the `fork` method is a good choice. In Figure 12.3, a `calculate` method is called during this period.

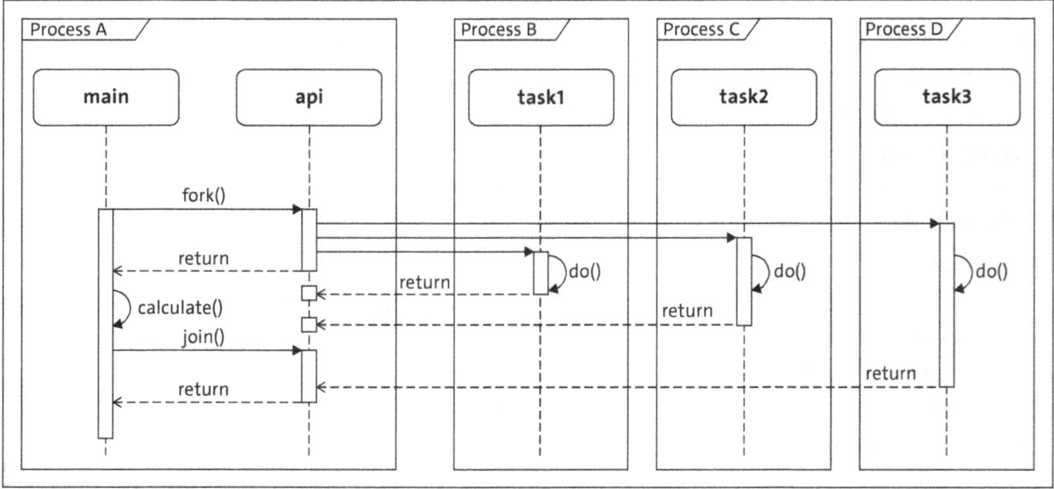

Figure 12.3 Sequence of the "fork" and "join" Methods in the CL_ABAP_PARALLEL Class

You can then use the join method to wait for the other processes or receive their partial results once the work in the main process has been completed. You can use the is_completed method to query whether or not parallel processing has been completed.

Base class or interface

The run, fork, and join methods are available in two versions: the regular version and the one with the _inst suffix. You have two options for implementing the logic that is to be executed in parallel:

- **CL_ABAP_PARALLEL as base class**
  The first option is to implement a class that inherits from the CL_ABAP_PARALLEL class. In the inheriting class, you want to redefine the do method and implement the logic to be parallelized there. Then you can use the run, fork, and join methods of your previously created instance of the CL_ABAP_PARALLEL class.
- **Implementing the IF_ABAP_PARALLEL interface**
  Alternatively, you can create a class that implements the IF_ABAP_PARALLEL interface and implement the do method there. Then, you use the run_inst, fork_inst, and join_inst methods of your previously created instance of the CL_ABAP_PARALLEL class.

In the following sections, we'll restrict ourselves to the second variant, that is, the implementation of the interface. Both approaches have their benefits and drawbacks. A decisive advantage of the interface variant is that you can define the data you need in the parallel process directly via attributes of your class. In the variant with the base class, you have to take care of serializing and deserializing the required data yourself. The interface variant is therefore easier for us to implement. In real life, the serialization provided in the standard system could be slower than a serialization you implement yourself when using the inheritance variant.

12.3 Refactoring the Sample Application

Up to this point, the conversion logic for user master data has been implemented as the private convert_csv_to_users method directly in the ZCL_ACB_USER_IMPORT_HANDLER handler class. In the course of introducing parallelization, we first redesign the existing logic in this section so that it's possible to decide between sequential and parallel execution without increasing the complexity of the handler class. The code to be parallelized must be implemented in a separate class anyway, or at least be called from there. It's therefore worth extracting the logic beforehand.

Extracting the conversion logic

We first define a new interface, which describes a conversion algorithm. We call the interface ZIF_ACB_USER_CONVERTER. It contains the definition of the convert_csv_to_users method, which was previously defined and implemented directly in the handler class (see Listing 12.1).

```
INTERFACE zif_acb_user_converter PUBLIC.
  TYPES users TYPE SORTED TABLE OF zacb_user
              WITH UNIQUE KEY username.
  METHODS convert_csv_to_users
    IMPORTING csv           TYPE string
              delimiter     TYPE string
    RETURNING VALUE(result) TYPE users
    RAISING   zcx_acb_user_conversion_error.
ENDINTERFACE.
```

Listing 12.1 Definition of the ZIF_ACB_USER_CONVERTER Interface

Creating an implementation class

For the implementation, we create the global ZCL_ACB_USER_CONVERTER class and implement the previously created interface in this class. We move the implementation of the convert_csv_to_users method from the handler class to the newly created class—now with reference to the method defined in the interface (see Listing 12.2).

```
CLASS zcl_acb_user_converter DEFINITION
  PUBLIC FINAL
  CREATE PUBLIC.
  PUBLIC SECTION.
    INTERFACES zif_acb_user_converter.
ENDCLASS.

CLASS zcl_acb_user_converter IMPLEMENTATION.
  METHOD zif_acb_user_converter~convert_csv_to_users.
    ...
    SPLIT text AT |\n| INTO TABLE DATA(lines).
    LOOP AT lines ASSIGNING FIELD-SYMBOL(<line>).
      TRY.
          SPLIT <line> AT delimiter
                INTO TABLE DATA(fields).
          INSERT VALUE #(
            username   = EXACT #( fields[ 1 ] )
            first_name = EXACT #( VALUE #(
                              fields[ 2 ] OPTIONAL ) )
```

```
            last_name  = EXACT #( VALUE #(
                              fields[ 3 ] OPTIONAL ) )
                              ...
                ) INTO TABLE result.
        CATCH cx_sy_itab_line_not_found
              cx_sy_conversion_error
              INTO DATA(format_error).
          RAISE EXCEPTION
                TYPE zcx_acb_user_conversion_error
                MESSAGE e004(zacb_user)
                EXPORTING previous = format_error.
      ENDTRY.
    ENDLOOP.
  ENDMETHOD.
ENDCLASS.
```

Listing 12.2 Implementation of the ZCL_ACB_USER_CONVERTER Class

We then change the call in the ZCL_ACB_USER_IMPORT_HANDLER handler class as shown in Listing 12.3.

```
METHOD if_http_service_extension~handle_request.
  ...
  DATA converter TYPE zif_acb_user_converter.
  ...
  converter = NEW zcl_acb_user_converter( ).
  ...
  DATA(users) = converter->convert_csv_to_users(
    csv       = request->get_text( )
    delimiter = delimiter ).
  ...
ENDMETHOD.
```

Listing 12.3 Modified Call of the Conversion Logic from Within the Handler Class

To ensure that the sample code in this chapter doesn't conflict with that in Chapter 6, we'll restrict ourselves in the following to the presentation of the adaptation in Listing 12.3. In the sample objects provided, the call is made exclusively via the ZCL_ACB_USER_CONVERSION_DEMO demo class.

Adapted architecture after refactoring

This results in the architecture shown in Figure 12.4.

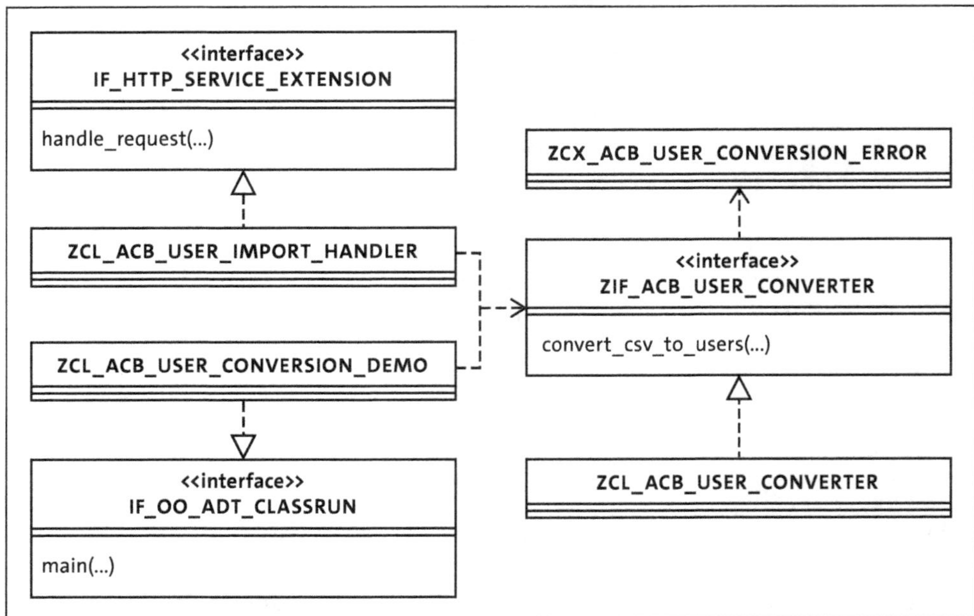

Figure 12.4 Architecture of the User Data Import After Refactoring

12.3.1 Implementing the IF_ABAP_PARALLEL Interface

Creating a task class

Now, we decide to parallelize the user data conversion in our sample application. To implement code that can be run in parallel, we need to create a new class. We opt for the ZCL_ACB_USER_CONVERSION_TASK identifier. We choose the TASK suffix because the implementation represents the starting point of a parallel process. Within the class, we implement the IF_ABAP_PARALLEL interface with the do method (see Listing 12.4).

```
CLASS zcl_acb_user_conversion_task DEFINITION
  PUBLIC FINAL
  CREATE PUBLIC.
  PUBLIC SECTION.
    INTERFACES if_abap_parallel.
ENDCLASS.

CLASS zcl_acb_user_conversion_task IMPLEMENTATION.
  METHOD if_abap_parallel~do.
    " TODO
  ENDMETHOD.
ENDCLASS.
```

Listing 12.4 Implementation of the IF_ABAP_PARALLEL Interface

Transferring parameters

The do method has no parameters. To transfer the necessary data from the calling process, attributes are used in the parallelization variant with the IF_ABAP_PARALLEL interface. We'll create instances of our ZCL_ACB_USER_CONVERSION_TASK class later when calling the API, which we parameterize directly. Technically, these instances can't be transferred to other processes. For this reason, the instance and its attribute values are serialized into a text format and deserialized from the text format into an object in the started process. You don't have to worry about this conversion because it's done for you by the API. You can simply add those parameters to your class that are required to implement the do method. In the other direction, you also receive the result of the parallel execution via the object instance. It therefore makes sense to add a method such as get_result.

[!]

Restriction Through CALL TRANSFORMATION ID

The API uses the ID transformation provided in the standard version to serialize and deserialize your object. To be able to use this transformation, classes must implement the IF_SERIALIZABLE_OBJECT interface. In our example, this is automatically the case because IF_ABAP_PARALLEL includes the IF_SERIALIZABLE_OBJECT interface. By implementing this *tag interface*, some stricter rules apply. For example, no objects can be passed as attributes to the second process that don't themselves implement the interface.

Transferring conversion parameters

To perform the conversion, we need the user data in CSV format and the separator (i.e., the delimiter), which introduces the next cell. We add attributes and a constructor to be able to initialize them during instantiation. We also add a get_result method that returns the result stored in another attribute, including possible errors. You can see the modified class in Listing 12.5.

```
CLASS zcl_acb_user_conversion_task DEFINITION
  PUBLIC FINAL
  CREATE PUBLIC.
  PUBLIC SECTION.
    INTERFACES if_abap_parallel.
    METHODS constructor IMPORTING csv       TYPE string
                                  delimiter TYPE string.
    METHODS get_result
      RETURNING VALUE(result) TYPE
        zif_acb_user_converter=>users
      RAISING   zcx_acb_user_conversion_error.
  PRIVATE SECTION.
    DATA csv                TYPE string.
```

```
    DATA delimiter         TYPE string.
    DATA conversion_result TYPE
      zif_acb_user_converter=>users.
    DATA conversion_error TYPE REF TO
      zcx_acb_user_conversion_error.
ENDCLASS.

CLASS zcl_acb_user_conversion_task IMPLEMENTATION.
  METHOD constructor.
    me->csv       = csv.
    me->delimiter = delimiter.
  ENDMETHOD.
  METHOD get_result.
    IF conversion_error IS BOUND.
      RAISE EXCEPTION conversion_error.
    ENDIF.
    RETURN conversion_result.
  ENDMETHOD.
  ...
ENDCLASS.
```

Listing 12.5 Addition of Parameters and Methods in the Task Class

Implementing the "do" method

The only thing missing now is the implementation of the actual logic. We don't want to duplicate it, so we simply call the existing ZCL_ACB_USER_CONVERTER class in the do method, transfer the parameters from the attributes, receive the result, and write it into other attributes (see Listing 12.6).

```
METHOD if_abap_parallel~do.
  TRY.
      conversion_result = NEW zcl_acb_user_converter(
        )->zif_acb_user_converter~convert_csv_to_users(
          csv       = csv
          delimiter = delimiter ).
    CATCH zcx_acb_user_conversion_error
      INTO conversion_error ##NO_HANDLER.
  ENDTRY.
ENDMETHOD.
```

Listing 12.6 Implementation of the "do" Method

Package creation

Technically, we've now created the option of running the user data conversion in parallel as often as required. In the next step, we want to do this in a smart manner and create small packages so that each parallel process is assigned a subtask of the conversion. We therefore no longer want to process the entire CSV file sequentially, but rather divide it into smaller packages and process these in parallel. With the CSV file format, this is easy to do because each row stands alone, requiring no context information from other rows.

We'll create a second implementation of the ZIF_ACB_USER_CONVERTER interface and name it ZCL_ACB_USER_CONV_PARALLEL for the purpose of the package creation. This second implementation no longer executes the conversion logic directly, but splits the transferred file and starts the parallel processes with their subtasks. As soon as these have been completed, it collects the partial results and combines them into an overall result.

Wrapping the parallelization

None of this is apparent to the caller. The interface turns parallelization into an implementation detail that is hidden in the ZCL_ACB_USER_CONV_PARALLEL class. The sequential implementation is still available unchanged in the ZCL_ACB_USER_CONVERTER class and is even simply reused in the parallel processes.

Grouping lines

The implementation of the ZCL_ACB_USER_CONV_PARALLEL class is described in greater detail later in this section. The convert_csv_to_users method starts with the package creation. We want to pass a few rows of CSV data to each parallel process for conversion. For this reason, we first divide the CSV variable into lines by using the line break (see Listing 12.7). We also harmonize possible different line breaks in advance.

```
DATA(text) = csv.
REPLACE ALL OCCURRENCES OF
  cl_abap_char_utilities=>cr_lf(1)
        IN text WITH ''.
SPLIT csv AT |\n| INTO TABLE DATA(lines).
```

Listing 12.7 Splitting the CSV File into Lines

We then merge the lines back into groups (see Listing 12.8). These are collected in the internal groups table. We decided on 10 lines per process. This value is primarily chosen for display reasons and would be far too small for a production application to achieve a performance gain through parallelization.

```
DATA groups TYPE string_table.
FIELD-SYMBOLS <group> TYPE string.
LOOP AT lines ASSIGNING FIELD-SYMBOL(<line>).
  IF sy-tabix = 1 OR sy-tabix MOD 10 = 0.
    INSERT <line> INTO TABLE groups
           ASSIGNING <group>.
  ELSE.
```

```
    <group> &&= |\n{ <line> }|.
  ENDIF.
ENDLOOP.
```

Listing 12.8 Grouping Lines

Creating the task instances

Our parameters have now been set and the packages created. We now create instances of our ZCL_ACB_USER_CONVERSION_TASK class and pass the parameters that form the subtasks. The CL_ABAP_PARALLEL class later expects an internal table with the instances of type IF_ABAP_PARALLEL and provides a corresponding table type. A VALUE constructor expression can be used at this point to turn the existing internal table with the line groups into an internal table of objects (see Listing 12.9).

```
DATA(tasks) = VALUE cl_abap_parallel=>t_in_inst_tab(
  FOR g IN groups
  ( NEW zcl_acb_user_conversion_task(
      csv       = g
      delimiter = delimiter ) ) ).
```

Listing 12.9 Creation of the Task Instances

Calling the parallelization

We've now created all the prerequisites for starting the user data conversion. We use the API presented earlier in Section 12.1 and instantiate the CL_ABAP_PARALLEL class. Then, we decide to wait for the processes started in parallel and therefore select the run_inst method. This means that no other method calls of the class are required. For this reason, we also skip the assignment of the created instance to a local variable.

We transfer the p_num_tasks parameter and limit the maximum number of tasks in our example to three. Then, we call the run_inst instance method and transfer our created and parameterized task instances via the p_in_tab parameter. We receive the results of the processing via EXPORTING parameter p_out_tab (see Listing 12.10).

```
NEW cl_abap_parallel( p_num_tasks = 3
  )->run_inst(
    EXPORTING p_in_tab  = tasks
    IMPORTING p_out_tab = DATA(processing_results) ).
```

Listing 12.10 Start of the Parallelization via the run_inst Method

Processing the partial results

Simultaneously, the run_inst method waits until all started processes have finished their work or have been terminated due to the timeout or other technical issues. We can therefore process the partial results in the next statement and provide the caller with the overall result.

The processing_results variable is typified as a structure type defined in the CL_ABAP_PARALLEL class. It contains the INST, INDEX, TIME, and MESSAGE components as described here:

- INST contains an object reference to the instance of our task class if processing was successful.
- INDEX identifies the task in relation to the p_in_tab input table.
- TIME contains the number of seconds the task took to execute.
- MESSAGE receives the error message if the task wasn't terminated regularly.

We iterate over the result table and query whether the task instance is available. If not, we'll raise an exception. If we do, we access the specific type of our task class via a downcast and add the partial result of the task to our overall result by calling the get_result method and adding the result to the result variable. If the user data conversion in the parallel process results in an exception, this is triggered by the get_result method and automatically propagated upward, as it's contained in the signature of the current method. You can find this implementation in Listing 12.11.


```
LOOP AT processing_results
     ASSIGNING FIELD-SYMBOL(<task>).
  IF <task>-inst IS NOT BOUND.
    IF <task>-message IS NOT INITIAL.
      cl_message_helper=>set_msg_vars_for_clike(
        <task>-message ).
      RAISE EXCEPTION TYPE zcx_acb_user_conversion_error
        USING MESSAGE.
    ELSE.
      RAISE EXCEPTION NEW
        zcx_acb_user_conversion_error( ).
    ENDIF.
  ENDIF.
  DATA(task) = CAST zcl_acb_user_conversion_task(
    <task>-inst ).
  INSERT LINES OF task->get_result( ) INTO TABLE result.
ENDLOOP.
```

Listing 12.11 Processing of Partial Task Results

Integrating the components in the overall architecture

At this point, the implementation is completed. We created an integration with the existing architecture and enabled the parallel execution of tasks. You can see the modified architecture in Figure 12.5.

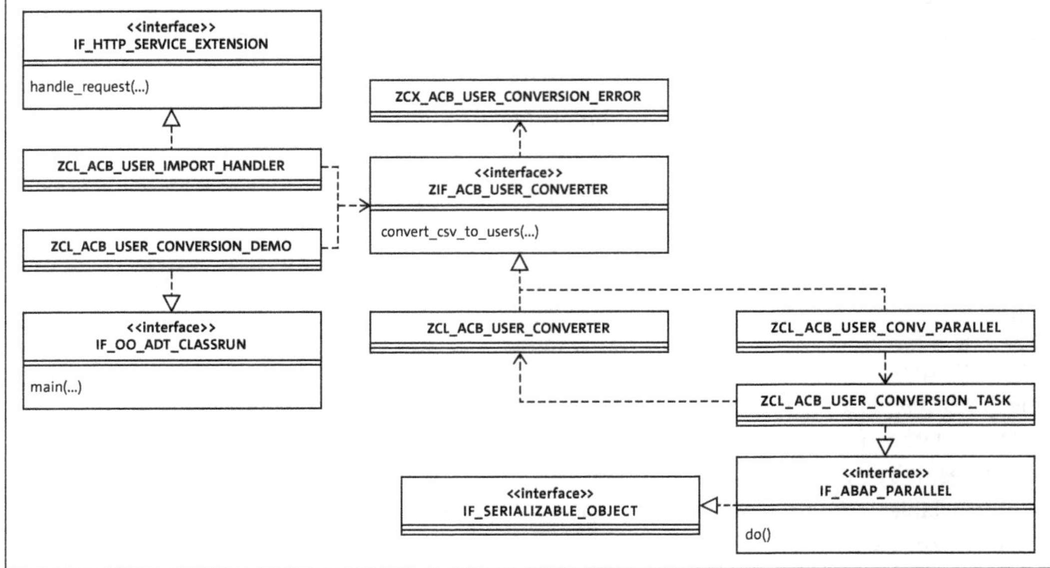

Figure 12.5 Integration of Parallelization into the Architecture of the User Data Import

In the following section, we'll take a look at the parallelized application at runtime.

12.3.2 Sequence and Debugging of Parallel Processes

Debugging applications that use parallelization can be somewhat difficult because you're debugging multiple sessions at the same time. Fortunately, the ABAP development tools provide the option of conveniently doing this in a single window, which would usually exceed the maximum number of SAP GUI windows in the new ABAP Debugger.

User data conversion process

Figure 12.6 shows what the user data conversion process now looks like using parallelization.

Execution in the debugger

In our example, 43 data records are transferred to the converter in CSV format in the ZCL_ACB_USER_CONVERSION_DEMO class. The converter then creates packages with 10 data records each and a maximum of three processes. This means that a total of five processes are started. When the first three processes have been completed, two more will be added as soon as the first ones have finished their work. You can easily reproduce this process via a breakpoint in the do method of the ZCL_ACB_USER_CONVERSION_TASK class. Sessions 2, 3, and 4 are the processes started in parallel, while session 1 waits until all processes have finished (see Figure 12.7).

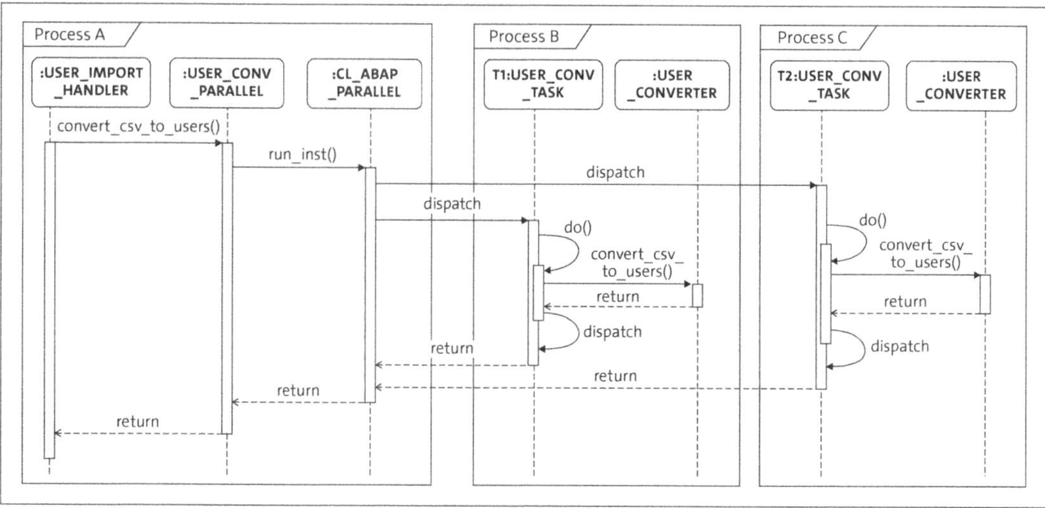

Figure 12.6 Sequence Diagram of the Parallelized User Data Conversion Process

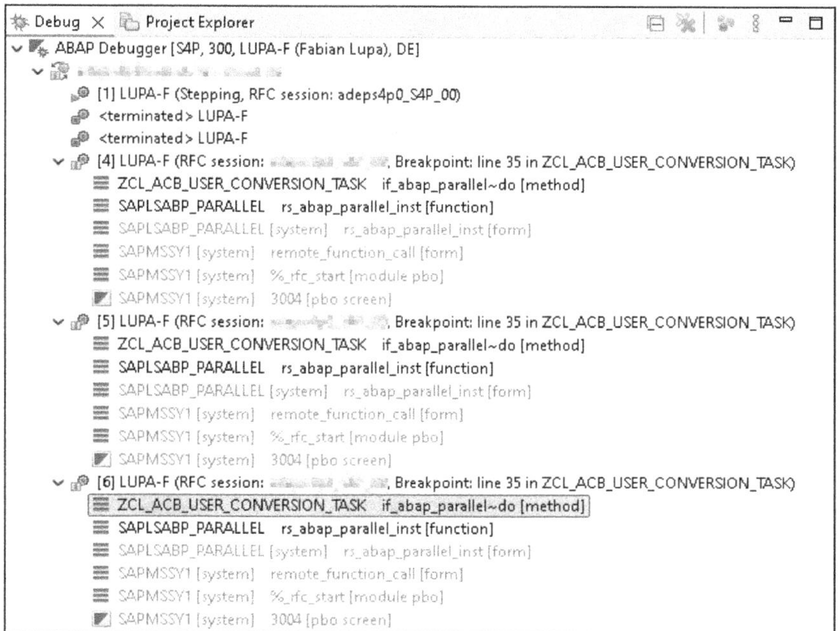

Figure 12.7 Debugging Multiple ABAP Sessions in the ABAP Development Tools

After completing session 2, 3, or 4, session 5 moves up, followed by session 6. Once all sessions have ended, the execution run continues in session 1 (see Figure 12.8).

This is where the partial results are processed. In Figure 12.9, you can see the status in the debugger before the method is completed.

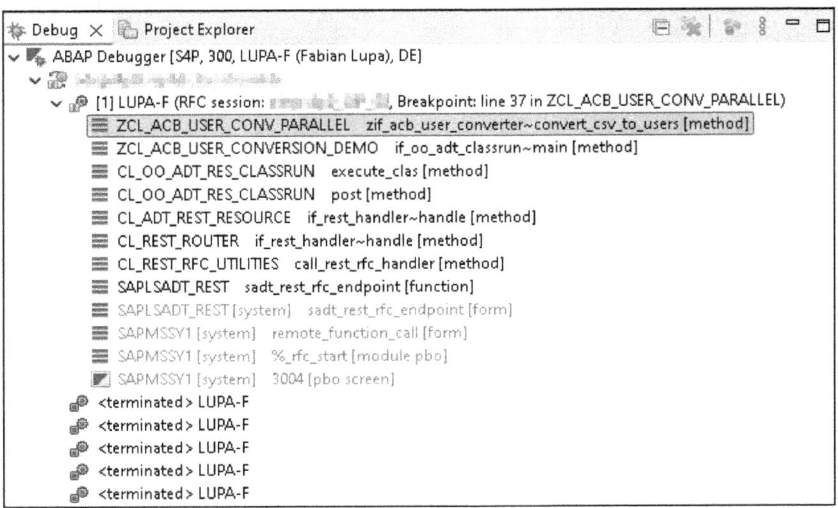

Figure 12.8 Continuation of the Calling Session After Completion of the Parallel Processes

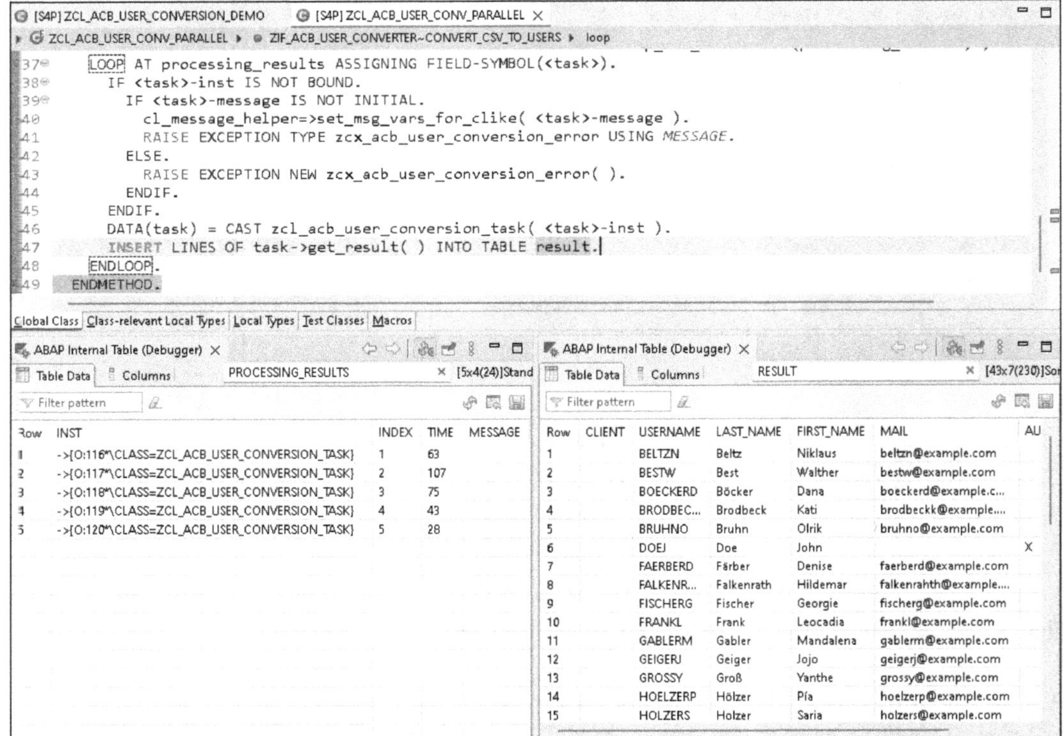

Figure 12.9 Partial Result Processing in the Debugger

The contents of the internal processing_results and result tables are displayed in the lower part of the screen. Note that the order in which the processes are terminated isn't fixed. In the processing_results table, you can see that tasks with a lower index were completed later and that there's no correlation whatsoever. For this reason, you should make sure that your implementation doesn't depend on the processes being started or ended in a specific order. Otherwise, you would catch a *race condition*, which is very difficult to debug because the analysis in the debugger itself, as well as the configuration and load of the respective system, can lead to a different sequence.

[+]

Option for Sequential Processing

To simplify the error analysis, we recommend that you provide an option in your implementation to deactivate parallelization and execute the code sequentially. This option can be provided as a Customizing option or simply as a variable in the code that can be changed in the debugger of the development system with the appropriate authorizations. In our example, the user was offered this choice via two different interface implementations.

The CL_ABAP_PARALLEL class already provides a function for this purpose. In the run, run_inst, fork, and fork_inst methods, the optional p_debug parameter is defined, which forces sequential processing.

12.4 Summary

CL_ABAP_PARALLEL class

In this chapter, you've learned how to use the CL_ABAP_PARALLEL class and the associated IF_ABAP_PARALLEL interface. The aim of their use is to optimize performance by parallelizing subtasks across multiple work processes. We've explained the technical background and how parallelization can be integrated into the architecture of your application. Finally, we've shown the runtime behavior and how to use the debugger in this context.

Technically, this option is highly interesting because parallelization is a software development discipline that isn't often used in ABAP development. The CL_ABAP_PARALLEL class simplifies handling significantly.

Costs of parallelization

However, as already mentioned in the introduction, we want to point out once again that parallelization comes with costs. In addition to the implementation effort, you should consider the following:

- The parallel processes don't inherit any locks, which would be the case with update processes. Depending on the scenario, lock conflicts can arise that make parallelization impossible.
- There's no cross-process logical unit of work (LUW). Each process can only act transactionally on the database for itself.
- Parallelization can have an impact on system utilization. If a user in a highly parallelized application suddenly starts using 10 work processes simultaneously, this can be to the detriment of other users or the response speed of the system.
- The execution sequence and processing speed of the parallel processes can't be predicted. You should make sure that these processes aren't inadvertently based on partial results of other processes, which would introduce race conditions.
- Error analysis and debugging are made more difficult.

You should therefore only use the concept of parallelization if all other approaches to performance optimization have already been exhausted.

Chapter 13
File Upload

You probably know that most users like to save their data and then upload it to the SAP system so that it can be processed there. We'll cover file uploading in this chapter.

As already described in Chapter 1, it's no longer possible to use the `OPEN DATASET`/`CLOSE DATASET` statements in ABAP Cloud. However, uploading large files—such as images or Microsoft Excel files—is a common business requirement. For this purpose, the ABAP RESTful application programming model supports *OData streams*, which makes it possible to activate the management of large files (Large Objects [LOB]). On this basis, end users can upload external files in various file formats such as PDF, Excel, or various image formats. This enables media management. 13

This chapter describes how these files can be uploaded into ABAP RESTful application programming model applications and how they can then be displayed within the applications. In Chapter 14, we'll take a look at the handling of files based on the example of Excel files.

Upload images to reviews

In Section 13.1, we extend our recipe portal so that images can be added to reviews. To do this, we'll extend the Review entity to include the ability to save files. An image of a home-cooked dish can be uploaded and displayed along with a review. We also explain the basic options for displaying images in ABAP RESTful application programming model applications.

Using the Sample Application

The objects presented in this chapter are located in the `DATAMODEL` subpackage. For this chapter, the existing objects are extended and no new ones are created.

13.1 Extending the Sample Application to Include a File Upload Option

In this section, we explain how the Review business object can be extended to include the option of uploading files. To do this, the active database table and then the core data services (CDS) entities based on it must be extended first. We then expand the mapping within the behavior definition. Finally, we add the metadata extension accordingly.

Extending the database table

In the first step, we complete the definition of our database table ZACB_REVIEW for the review data, as shown in Listing 13.1. We add the attachment, mimetype, and filename fields.

```
define table zacb_review {
  key client              : abap.clnt not null;
  key review_id           : sysuuid_x16 not null;
...
  attachment              : abap.rawstring(0);
  mimetype                : abap.char(128);
  filename                : abap.char(128);
}
```

Listing 13.1 Extension of Database Table ZACB_REVIEW for File Uploads

The length of the added fields is fixed.

- **attachment**
  This field contains the binary data of the file in a RAWSTRING format.
- **filename**
  The name of the file is saved in this field.
- **mimetype**
  This field represents the content type of the uploaded file. No file can exist without the MIME type and vice versa.

[»]

MIME Type

A *MIME type* consists of the media type and the subtype. These specifications are separated by a slash. The media type specifies the group of the file, such as text for a text file or image for an image.

Table 13.1 lists some of the most common MIME types.

MIME Type	File Type
text/plain/html	HTML files
application/pdf	PDF files
image/jpeg	JPEG files
text/plain	Plain text files
video/mpeg	MPEG files

Table 13.1 Common MIME Types

Extending the draft table

We add the same fields to draft table ZACB_REVIEW_D that belongs to the database table (see Listing 13.2).

```
define table zacb_review_d {

  key client          : abap.clnt not null;
  key reviewid        : sysuuid_x16 not null;
...
  attachment          : abap.rawstring(0);
  mimetype            : abap.char(128);
  filename            : abap.char(128);
  "%admin"            : include sych_bdl_draft_admin_inc;
}
```

Listing 13.2 Extension of Draft Table ZACB_REVIEW_D for File Uploads

Extending the base entity

Now, we need to add these fields once in all relevant CDS entities of our data model. First, we add the fields in base entity ZACB_R_REVIEW (see Listing 13.3).


```
define view entity ZACB_R_Review
  as select from zacb_review
  association to parent ZACB_R_Recipe as _Recipe
    on $projection.RecipeId = _Recipe.RecipeId
{
  key review_id             as ReviewId,
      recipe_id             as RecipeId,
...
      created_at            as CreatedAt,
      attachment            as Attachment,
      mimetype              as Mimetype,
      filename              as Filename,
      _Recipe
}
```

Listing 13.3 Extension of the ZACB_R_REVIEW Base Entity

Extending the projection entity

We then extend the ZACB_C_Review projection entity accordingly to complete the extension of the fields (see Listing 13.4).

```
@EndUserText.label: 'Review projection entity'
@AccessControl.authorizationCheck: #NOT_REQUIRED
@Metadata.allowExtensions: true
define view entity ZACB_C_Review
  as projection on ZACB_R_Review as Review
{
```

```
  key ReviewId,
      RecipeId,
      Reviewtext,
...
      Attachment,
      Filename,
      Mimetype,
      _Recipe : redirected to parent ZACB_C_Recipe
}
```

Listing 13.4 Extension of the Projection Entity

Extending the behavior definition

We also need to add the mapping to the ZACB_R_REVIEW behavior definition so that the fields will also be saved in the database table (see Listing 13.5).

```
define behavior for ZACB_R_Review alias Review
...
{
...
  mapping for zacb_review
    {
      ReviewId          = review_id;
...
      CreatedAt         = created_at;
      CreatedBy         = created_by;
      Attachment        = attachment;
      Filename          = filename;
      Mimetype          = mimetype;
    }
}
```

Listing 13.5 Extension of the Mapping in the Behavior Definition

Supplementing the metadata extension

The fields must also be entered in the ZACB_C_REVIEW metadata extension to be able to map them in our SAP Fiori app (see Listing 13.6). The mimetype and filename fields are annotated as HIDDEN. Only the attachment field gets displayed in the user interface (UI). The recipe image test is added using the label annotation.

```
@Metadata.layer: #CORE
@UI: {
  headerInfo: {
    typeName: 'Review',
    typeNamePlural: 'Reviews'
  }
}
```

```
annotate view ZACB_C_Review with
{
...
  @UI.lineItem: [ {
    position: 40 ,
    importance: #MEDIUM,
    label: 'Recipe image'
  } ]
  @UI.identification: [ {
    position: 40 ,
    label: 'Recipe image'
  } ]
  Attachment;

  @UI.hidden: true
  Mimetype;

  @UI.hidden: true
  Filename;

}
```

Listing 13.6 Extension of the Metadata Definition

Testing the application

Now, we can start our SAP Fiori app to view the result. You can see it in Figure 13.1.

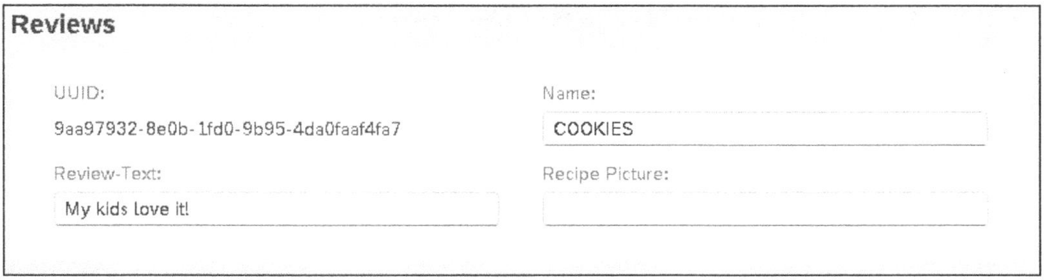

Reviews

UUID:
9aa97932-8e0b-1fd0-9b95-4da0faaf4fa7

Name:
COOKIES

Review-Text:
My kids love it!

Recipe Picture:

Figure 13.1 File Upload

The **Recipe Picture** field is displayed here in the UI and is available as a normal text field. To make it available as an upload dialog, additional annotations must be added to the fields.

Display as file upload field

The `@Semantic.mimeType: true` annotation must be added to the `Mimetype` field. This annotation tells the SAP Fiori app that the field can now be used for file types. In addition, the `Attachment` field in the `@Semantics.largeObject` annotation must be extended to indicate that it's a file upload. In Listing 13.7, these annotations are highlighted in bold.

```
define view entity ZACB_R_Review
  as select from zacb_review
  association to parent ZACB_R_Recipe as _Recipe
      on $projection.RecipeId = _Recipe.RecipeId
{
  key review_id              as ReviewId,
      recipe_id              as RecipeId,
...
      created_at             as CreatedAt,
      @Semantics.largeObject: {
       mimeType : 'Mimetype',
       fileName : 'Filename',
       contentDispositionPreference: #INLINE
      }
      attachment             as Attachment,
      @Semantics.mimeType: true
      mimetype               as Mimetype,
      filename               as Filename, _Recipe
}
```

Listing 13.7 Extension of Base Entity ZACB_R_REVIEW with Semantic Annotations

Testing the application again

When you now open the SAP Fiori app again, two icons are displayed below the **Recipe Picture** label (see Figure 13.2).

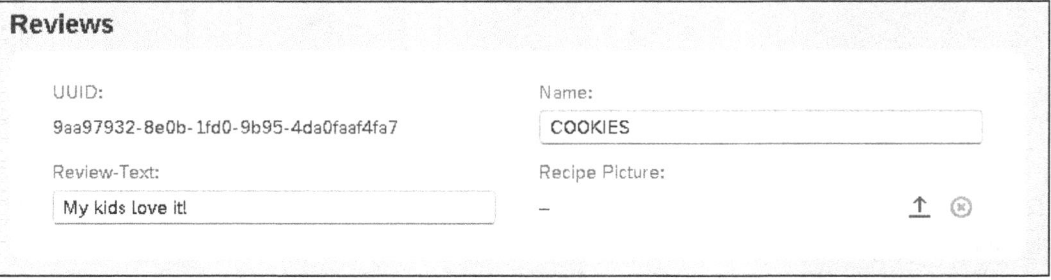

Figure 13.2 SAP Fiori App with Icon for Uploading Files

The following operations are possible via these icons in edit mode:

- **Upload File**
  Uploads the file using the [↑] icon.
- **Delete File**
  Deletes the file using the ⊗ icon.

Setting Up a Virus Scanner

When using the file upload for the first time, you may need to set up the virus scan profile /IWBEP/V4/ODATA_UPLOAD. Further information on setting up the virus scan profile can be found in the SAP documentation at *https://help.sap.com/doc/saphelp_gbt10/1.0/en-US/4e/0ba5960499001be10000000a42189c/content.htm?no_cache=true.*

Once you've uploaded a file, the filename gets displayed in the application (see Figure 13.3).

Reviews

UUID: 9aa97932-8e0b-1fd0-9b95-4da0faaf4fa7

Name: COOKIES

Review-Text: My kids love it!

Recipe Picture: cookies-in-basket.jpg

Figure 13.3 Display of the Filename After the Upload

Click on the **Apply** button to save the data record in the database. If you now check the data record in database table ZACB_RECIPE (see Chapter 4, Section 4.2, for information on how to do this using the ABAP development tools), you'll see that the filename, the MIME type, and the binary file content are saved there (see Figure 13.4).

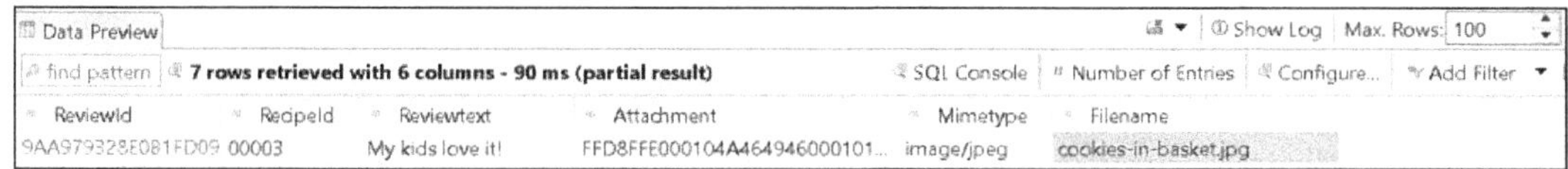

Data Preview — 7 rows retrieved with 6 columns - 90 ms (partial result)

ReviewId	RecipeId	Reviewtext	Attachment	Mimetype	Filename
9AA97932BE0B1FD09	00003	My kids love it!	FFD8FFE000104A464946000101...	image/jpeg	cookies-in-basket.jpg

Figure 13.4 Data Storage in Database Table ZACB_RECIPE

Displaying the images

In Listing 13.7, we've already added the @Semantics.largeObject.contentDispositionPreference annotation in the base entity. However, we haven't yet gone into what this annotation actually means. It tells the SAP Fiori app what happens to the file when the filename gets clicked. The following values can be specified for this annotation:

- `#INLINE`
  The file gets displayed in a new tab in the browser.
- `#ATTACHMENT`
  The file will be downloaded and can be opened in an image processing program of your choice.

In addition to these two options, you can also display the images directly. In the following sections, we'll show you an option for direct display within a table.

Images in tables

To display images in tables, for example, in our review overview table, it's necessary to add the `@Semantics.imageUrl: true` annotation to the `attachment` field that is supposed to be displayed as an image (see Listing 13.8).

```
define view entity ZACB_R_Review
  as select from zacb_review
  association to parent ZACB_R_Recipe as _Recipe
    on $projection.RecipeId = _Recipe.RecipeId
{
  key review_id              as ReviewId,
      recipe_id              as RecipeId,
…
      created_at             as CreatedAt,
      @Semantics.largeObject: {
       mimeType : 'Mimetype',
       fileName : 'Filename',
       contentDispositionPreference: #INLINE
      }
      @Semantics.imageUrl: true
      attachment             as Attachment,
      @Semantics.mimeType: true
      mimetype               as Mimetype,
      filename               as Filename, _Recipe
}
```

Listing 13.8 Extension of the Behavior Definition with the @Semantics.imageUrl Annotation to Display the File as an Image

Figure 13.5 shows the corresponding output in the list report for the reviews.

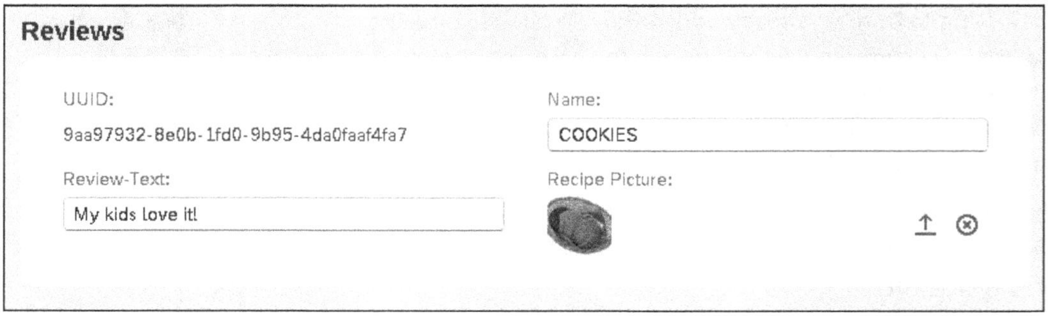

Figure 13.5 Direct Display of an Image

[+]

Annotations in ABAP RESTful Application Programming Model Applications

SAP provides a *showcase app* to present the various annotations that can be used in ABAP RESTful application programming model applications. It can be found in GitHub at *http://s-prs.de/v1064804*.

Restricting file types

Another setting you can make is to restrict the permitted files. This restriction can be defined using the `@Semantics.largeObject.acceptableMimeTypes` annotation. We only allow image formats for our file upload (see Listing 13.9).

```
define view entity ZACB_R_Review
...
      @Semantics.largeObject: {
       mimeType : 'Mimetype',
       fileName : 'Filename',
       contentDispositionPreference: #INLINE,
       acceptableMimeTypes: [ 'image/*' ]
      }
```

Listing 13.9 Annotation to Restrict the File Type

All image formats are permitted by means of the `'image/*'` value. If you want to further restrict the selection to a specific image format, you can specify `image/png`, for example. If you now want to upload a different file format in the app, a corresponding error message will display (see Figure 13.6).

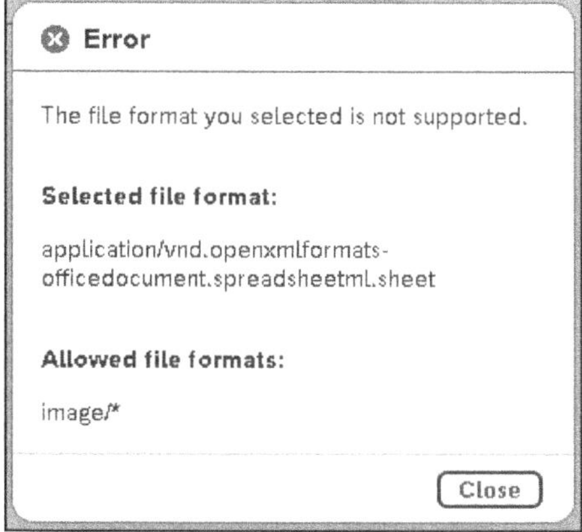

Figure 13.6 Error Message: Incorrect File Format

13.2 Summary

In this chapter, you've learned how to provide a way to upload files and store these files in database tables in ABAP RESTful application programming model applications. To demonstrate this using the example of the recipe portal, the database tables and the entities were first supplemented with the three fields for the filename, the MIME type, and the binary content of the file.

We then discussed the options for displaying the upload option as well as for displaying the files. You've learned how the files can be displayed in different ways by using annotations.

In the following chapter, we'll show you how you can continue using the uploaded files if they are Excel files.

Chapter 14
Using Excel Files

Many options are available for creating and reading Excel files in classic ABAP development. In ABAP Cloud, another one has been added: the XCO library. This chapter describes how you can use this library in both reading and writing scenarios.

Regardless of how ABAP development, the SAP product portfolio, or the global economic situation develop, one constant remains: The question "Can I also have this as an Excel file?" will probably never go away.

The integration of Microsoft Office documents has a long history in the ABAP platform. The export of data as Excel files in particular is an integral part of business applications. Accordingly, there are many different solutions for this task with different technological foundations. A running joke in the SAP community is the "first Excel post of the week", in which one of these techniques is picked up again in a blog post.

Different APIs

ABAP provides a large number of options for using Excel files. Each of those options has its advantages and disadvantages. Here are just a few:

- You can use *Object Linking and Embedding* (OLE) to programmatically address the Office applications installed by users with your own ABAP statements.
- Using classes of the SAP List Viewer (ALV) object model , you can easily convert internal tables into Excel files.
- You can use extensive Excel functions via the abap2xlsx open-source project.

There are also options available at the user interface (UI) level via the SAPUI5 libraries, such as the *UI5 Spreadsheet Importer*. Even OData provides some options: Excel can consume OData services directly as an external data source. Using the `$format=xlsx` parameter or a suitable HTTP header, Excel files can be generated directly via SAP Gateway for the supported OData services.

Evaluation criteria

When choosing the right technology, you should pay attention to a few criteria:

- Should it be possible to create the Excel file in a background job?
- Should the content be freely configurable, or is it just an internal table that is supposed to be output?
- Should data be output, read, or changed?

Excel files in ABAP Cloud

In this chapter, you'll learn about another option for connecting to Excel. The XCO library (extension components) contains a module that deals specifically with XLSX files. In contrast to many other options, it can be used in ABAP Cloud because the associated objects have been released for this use.

The range of functions and restrictions differ for the various technologies. You should therefore weigh up which option is best for your project based on the technologies you're using. If you've finally decided on developing using ABAP Cloud, there's currently no alternative to the XCO library, aside from developments in the SAPUI5 frontend or Tier 3 wrappers around classic APIs in SAP S/4HANA and SAP S/4HANA Cloud Private Edition.

Structure of this chapter

Section 14.1 deals with the program-based creation of new Excel files. In Section 14.2, existing Excel files are read and processed.

Excel Formats and Applications

This chapter deals exclusively with the Excel format with file extension *.xlsx* for spreadsheets. Files in this format can also be edited in applications other than Excel. For the sake of simplicity, we use the common term *Excel file*. If you need to work with XLS files (Excel 97–2003 workbooks), you can't use the techniques presented here. The handling of CSV files is also not discussed in this chapter. Although these can be opened and edited in Excel, they are a pure text format that has nothing to do with Excel itself.

Mass change of recipes

With regard to our recipe portal, this chapter uses the mass change of recipes and their ingredients as an application example. In preparation for this, we've already developed an SAP Fiori app using SAP Fiori elements and the ABAP RESTful application programming model, which you can see in Figure 14.1 and Figure 14.2.

Mass change process

This app allows you to create mass changes. In the first step, users should be able to download an Excel file as a template that contains the current data on recipes and ingredients in two worksheets. Users can then edit the file and upload and process the edited file again in the app. As part of this processing, the system updates the data records in the database tables based on the uploaded file. The technical requirements for uploading and downloading files have already been created. Details can be found in Chapter 13. This chapter is about implementing the actions for generating the template and processing the changed data using the API for Excel files provided in the XCO library.

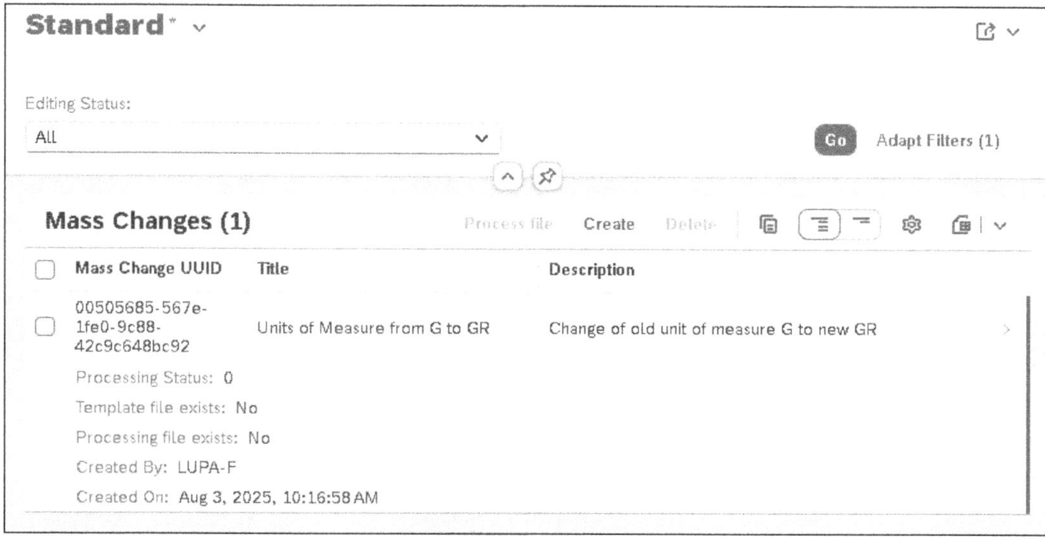

Figure 14.1 List Report of the Mass Changes App

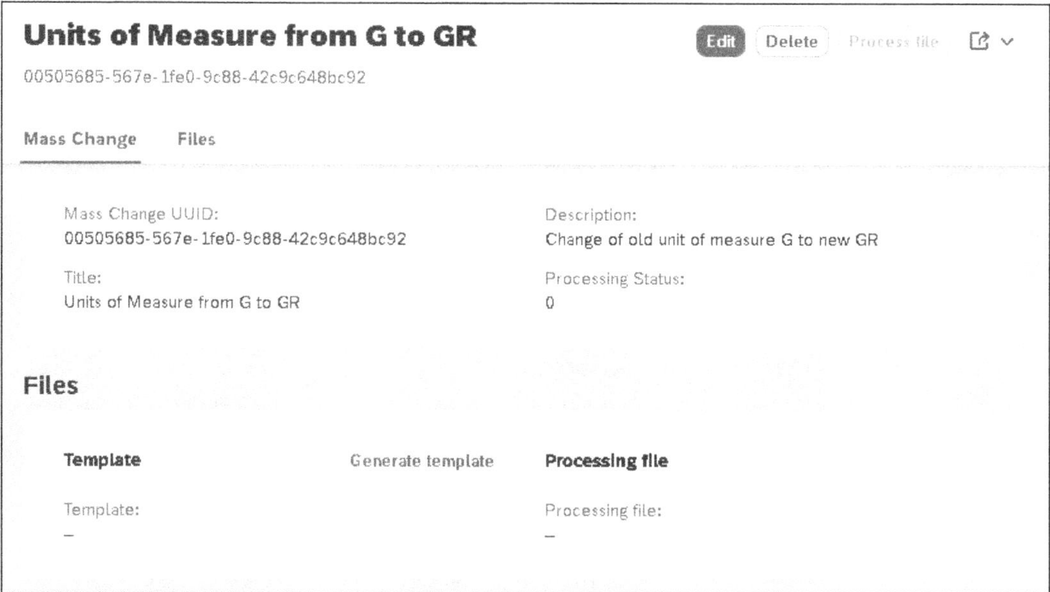

Figure 14.2 Object Page of the Mass Changes App

Architecture of the app

The Mass Changes app uses the ABAP RESTful application programming model and SAP Fiori elements. It therefore has a very similar architecture to the primary components of the recipe portal, which we described in Chapter 2. The basis of the Mass Changes app is the ZACB_MASSC database table, which contains mass changes. A mass change consists of a description, the template file, and the processing file as well as a processing status. In conjunction with the implementing ZBP_ACB_R_MASS_CHANGE class, the ZACB_R_MassChange root entity contains the business logic for carrying out mass changes to recipes and ingredients. An overview of the relevant development objects is shown in Figure 14.3.

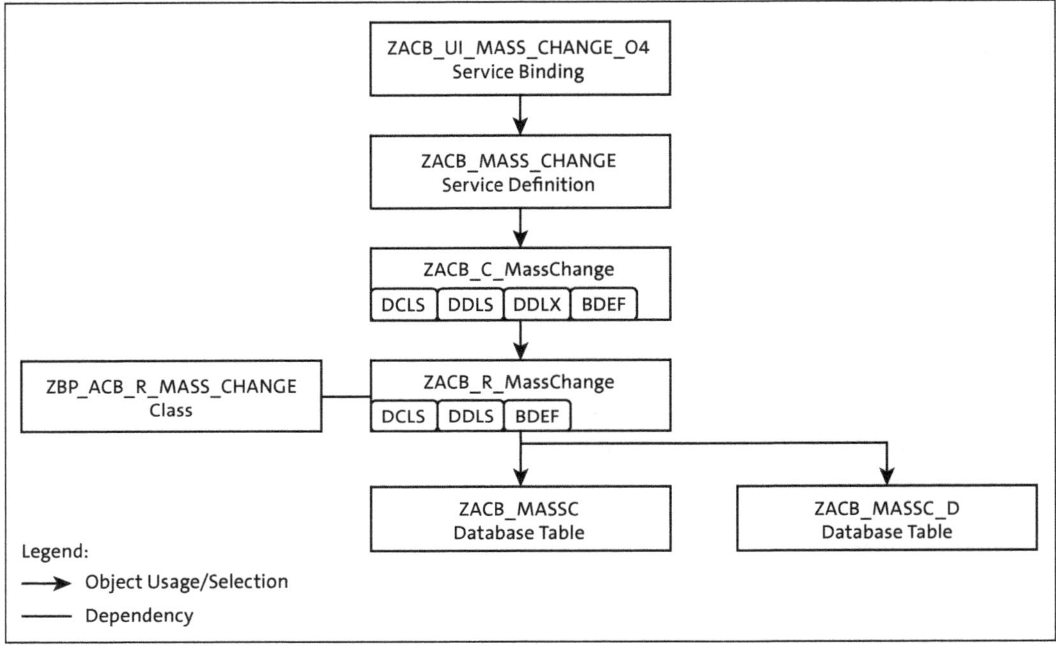

Figure 14.3 Architecture of the Mass Changes App in the Recipe Portal

[»]

Using the Sample Application

You can call the Mass Changes app (based on the ABAP RESTful application programming model) for mass changes to recipes and ingredients as a preview app via the ZACB_UI_MASS_CHANGE_04 service binding. To do this, open the **Open ABAP Development Object** dialog box in the ABAP development tools for Eclipse using the keyboard shortcut [Ctrl] + [Shift] + [A], and enter the name of the service binding: "ZACB_UI_MASS_CHANGE_04". Select the one hit that is displayed in the list. Make sure that the binding is published in your system. Then, select the **MassChange** entity set on the right-hand side, and click on the **Preview** button. All objects from this chapter can be found in the download material in the EXCEL subpackage.

14.1 Creating an Excel File

In the Mass Changes app, the user must first create a new mass change. The data records to be changed must be specified and described in this object before the data records can actually be changed.

It should be possible to download a template for easier handling. This should provide recipes and their ingredients in an easy-to-use Excel format. The user should then be able to adapt this template and upload it again. To generate this file, an action was defined in the Mass Change business object of the ABAP RESTful application programming model and positioned as the **Generate template** button in the SAP Fiori frontend next to the download option for the file (see Figure 14.4).

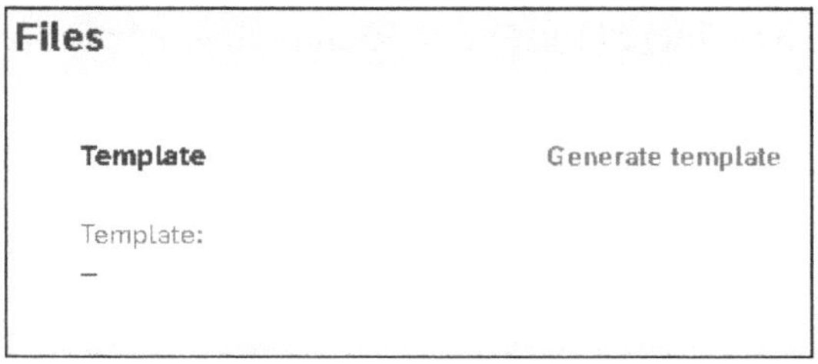

Figure 14.4 Generate Template Action in the Mass Changes App

14.1.1 Creating the Action

Action for template generation

In the ABAP RESTful application programming model, *actions* allow you to provide operations on the business object that aren't covered by the standard create, read, update, and delete (CRUD) actions. In our example, the action should generate an Excel file and make it available for download at the current instance of the Mass Change business object. This action, which is named generateTemplate, was defined as follows in the ZACB_R_MASSCHANGE behavior definition:

```
action ( features : instance )
  generateTemplate result [1] $self;
```

Using the quick fix **Add method for action in local class** function, which can be accessed via the keyboard shortcut Ctrl+1, we then started to implement the action in the local handler class LHC_MASSCHANGE (see Listing 14.1).

```
CLASS lhc_masschange DEFINITION
  INHERITING FROM cl_abap_behavior_handler.
  PRIVATE SECTION.
    ...
    METHODS generateTemplate FOR MODIFY
```

```
      IMPORTING keys
      FOR ACTION MassChange~generateTemplate
      RESULT result.
ENDCLASS.
CLASS lhc_masschange IMPLEMENTATION.
  METHOD generateTemplate.
    " TODO
  ENDMETHOD.
ENDCLASS.
```

Listing 14.1 Implementation of the generateTemplate Action

We want to create an Excel file in the implementation of this action and save it in the Mass Change object. It should contain all recipes and ingredients. We'll first focus on the handling of the Excel file and integrate it into the implementation of the action at the end of this section.

14.1.2 Creating the Excel Document

API class XCO_CP_XLSX

The point of entry to the XLSX module of the XCO library is API class XCO_CP_XLSX. In the module, all relevant class and interface identifiers are provided with the XLSX name component. Details on handling API classes and the concepts of the XCO library can be found in Chapter 3.

The API class provides two static attributes: coordinate and document. We want to create a new Excel file and therefore navigate deeper into the API for documents via the document attribute. We'll take a look at the coordinate attribute later in this section.

The document attribute

The instance behind the document attribute provides two methods for obtaining a document object:

- **empty**
  This method returns a reference to an empty document.
- **for_file_content**
  This method expects the content of an Excel file as a binary string. It creates a document object based on the transferred data and returns a reference to it.

First, we decide to use the empty method. We then describe how to read files using the for_file_content method in Section 14.2. The returned object expects the selection of an access type. Only write access is offered, which makes sense for an empty document. We therefore directly call the next method called write_access:

```
DATA(document) = xco_cp_xlsx=>document->empty(
  )->write_access( ).
```

As is usual in the XCO library, the XLSX module also works a lot with the concatenation of method calls and attribute accesses.

14.1.3 Creating the Workbook

Access to documents, workbooks, and worksheets

The write view of the document, which is represented via the IF_XCO_XLSX_WA_WORKBOOK interface, enables further navigation via two methods:

- **get_file_content**
  We can use this method to obtain the content of the Excel file we've assembled as a binary string. We'll call this method at the very end.
- **get_workbook**
  At this point, this method takes us further. It returns a representation of the *workbook*, which itself contains *worksheets*.

A document object created with the empty method automatically contains a worksheet. We can access it using the at_position method with parameter 1:

```
DATA(sheet) = document->get_workbook(
  )->worksheet->at_position( 1 ).
```


By default, the worksheet is called *Sheet1*. For us, the first worksheet should contain the recipes. So we're going to rename it. You can rename worksheets using the set_name method:

```
sheet->set_name( 'Recipes' ).
```

14.1.4 Writing the Recipe Data to the Worksheet

Write access to a worksheet

You can use the select method of the worksheet object to access the write access to the cells of a worksheet. It allows you to access multiple cells at the same time, which you must select using a *selection pattern*. You can imagine a selection pattern as a marking of columns and rows in the worksheet. In our example, we simply want to insert all recipe data records into the worksheet starting from cell A2. We want to add a header row to the first rows containing the column identifiers.

Selection pattern

For this purpose, we first create a selection pattern. The API class XCO_CP_XLSX_SELECTION provides an instance via the pattern_builder attribute that enables you to create a selection pattern by means of the builder design pattern. We create such a pattern for the cells from row 2 onward (see Listing 14.2).

```
DATA(from_second_row) =
  xco_cp_xlsx_selection=>pattern_builder
    ->simple_from_to(
```

```
)->from_row(
  xco_cp_xlsx=>coordinate->for_numeric_value( 2 )
)->get_pattern( ).
```

Listing 14.2 Creation of a Selection Pattern

Coordinates

As you can see, row 2 wasn't specified directly via an elementary data object, but via a *coordinate object*. Using the factory class provided with the xco_cp_xlsx=>coordinate attribute, you can obtain a coordinate object for a column (using the get_alphabetic_value method) or for a row (using the get_numeric_value method), which we need here for the selection pattern.

Row-based write access to a selection

You must pass the selection pattern to the select method and will then receive a *selection object*. Currently, this selection object provides only a method for filling cells with content row by row, based on the selection. The method is called row_stream. We want to call it directly too:

```
DATA(stream) = sheet->select( from_second_row
  )->row_stream( ).
```

Depending on the use case, you can largely save intermediate variables by concatenating calls.

The *row stream* obtained in this way provides write access via the operation attribute, but so far, this only includes write access for transferring data from an internal table using the write_from method. We use this write access to transfer the data from our database table ZACB_RECIPE. We opt for indirect access via the ZACB_R_Recipe CDS entity and transfer the internal table to the write_from method. This method expects the internal table as a data reference, so we use the constructor expression with the REF referencing operator. Finally, we run the execute method on the object that we receive back and that describes the write operation to actually perform the operation. You can see the complete code for running the write operation in Listing 14.3.

```
DATA(stream) = sheet->select( from_second_row
  )->row_stream( ).

SELECT FROM ZACB_R_Recipe
  FIELDS RecipeId, RecipeName, RecipeText
  ORDER BY PRIMARY KEY
  INTO TABLE @DATA(recipes).

stream->operation->write_from( REF #( recipes )
  )->execute( ).
```

Listing 14.3 Execution of a Write Operation on a Row Stream

14.1.5 Addition to the Header

The document now contains our recipe data in the first worksheet from row 2 onwards. Before we attach it as a file to the Mass Change business object, we want to fill the header of the document. This write access is a different use case than a row-based write access to the worksheet. We know that only one row is to be written to and have no internal table at hand. For this reason, we use cell-by-cell access.

Cell-based access via the cursor

The cell-based access is carried out via a *cursor*. You can think of it as clicking on a cell or navigating through the worksheet using the arrow keys. The worksheet object provides the `cursor` method to create just such a cursor. We call it for cell A1 and use the coordinate objects you've already become familiar with (see Listing 14.4).

```
DATA(cursor) = sheet->cursor(
  io_column =
    xco_cp_xlsx=>coordinate->for_alphabetic_value( 'A' )
  io_row    =
    xco_cp_xlsx=>coordinate->for_numeric_value( 1 )
).
```

Listing 14.4 Creation of a Cursor

The cursor provides direct write access to the cell via the `get_cell` method and the underlying `value` attribute. We use the write access to assign a heading to the first column using a literal:

```
cursor->get_cell( )->value->write_from( 'Recipe ID' ).
```

Repositioning the cursor

You can then reposition the cursor using the `move_up`, `move_right`, `move_down`, and `move_left` methods. Each of these methods contains an optional parameter to specify the number of cells by which you want to move in the specified direction. If the parameter doesn't get transferred, the cursor will be repositioned by exactly one cell. We move the cursor one cell to the right and assign the heading for the second column. The `move_*` methods each return the instance of the cursor as a `RETURNING` parameter so that the next write access can be made directly by concatenation:

```
cursor->move_right( )->get_cell(
  )->value->write_from( 'Recipe name' ).
```

Dynamic determination of the header

A generic implementation is available for the header row, which fills the cells with their identifiers based on the internal table displayed in the row below. This implementation is particularly useful in view of the second worksheet with the ingredients, which is still outstanding.

The *Run Time Type Services* (RTTS) that you may be familiar with from classic ABAP programming as a way of obtaining type information at runtime or generating and using data types at runtime are also available in ABAP Cloud. This makes it easy to determine the technical identifiers of the components of the row type of the internal table (see Listing 14.5).

```
DATA(descriptor) = CAST cl_abap_structdescr(
  CAST cl_abap_tabledescr(
    cl_abap_typedescr=>describe_by_data( recipes )
  )->get_table_line_type( ) ).
LOOP AT descriptor->get_components( )
     ASSIGNING FIELD-SYMBOL(<component>).
  cursor->get_cell( )->value->write_from(
    <component>-name ).
  cursor->move_right( ).
ENDLOOP.
```

Listing 14.5 Dynamic Determination of the Header

Restrictions in ABAP Cloud and in temporary packages

Accessing the data element texts, on the other hand, is difficult. The `get_ddic_field` method of the `CL_ABAP_ELEMDESCR` class can't be used in ABAP Cloud because the structure type of the `DFIES` result hasn't been released. Specifically for our example, however, the ABAP Repository API of the XCO library can't be used either because we've decided to develop the recipe portal in temporary packages for delivery purposes. The XCO library doesn't allow access to such objects and filters on the basis of the software component. We'll therefore stick to the technical designations for our purposes.

14.1.6 Adding Another Worksheet

Worksheet for the ingredients

In a second worksheet, we want to display the ingredients. To add another worksheet, you need to call the `add_new_sheet` method of the document object:

```
sheet = document->get_workbook(
  )->add_new_sheet( 'Ingredients' ).
```

Generic filling of worksheets

You can directly provide the name of the worksheet in the `iv_name` parameter of the method. As the logic for filling the ingredient worksheet in our example is identical to that of the recipe worksheet, we extract this function to a method. You can see the logic for filling the two worksheets in Listing 14.6.

```
CLASS lhc_masschange DEFINITION
  INHERITING FROM cl_abap_behavior_handler
  PRIVATE SECTION.
    ...
    METHODS add_table_to_worksheet
      IMPORTING !table TYPE STANDARD TABLE
                sheet  TYPE REF TO if_xco_xlsx_wa_worksheet.
ENDCLASS.

CLASS lhc_masschange IMPLEMENTATION.
  ...
  METHOD add_table_to_worksheet.
    DATA(from_second_row) =
      xco_cp_xlsx_selection=>pattern_builder
        ->simple_from_to(
        )->from_row(
          xco_cp_xlsx=>coordinate->for_numeric_value( 2 )
        )->get_pattern( ).
    DATA(stream) = sheet->select( from_second_row
      )->row_stream( ).
    stream->operation->write_from( REF #( table )
      )->execute( ).

    DATA(cursor) = sheet->cursor(
      io_column =
        xco_cp_xlsx=>coordinate->for_alphabetic_value(
          'A' )
      io_row    =
        xco_cp_xlsx=>coordinate->for_numeric_value( 1 ) ).

    DATA(descriptor) = CAST cl_abap_structdescr(
      CAST cl_abap_tabledescr(
        cl_abap_typedescr=>describe_by_data( table )
      )->get_table_line_type( ) ).
    LOOP AT descriptor->get_components( )
         ASSIGNING FIELD-SYMBOL(<component>).
      cursor->get_cell( )->value->write_from(
        <component>-name ).
      cursor->move_right( ).
    ENDLOOP.
  ENDMETHOD.
ENDCLASS.
```

Listing 14.6 Generic Filling of a Worksheet Using a Custom Method

14.1.7 Saving the Excel File

Saving as a file

Our workbook is now complete. We still have to save it to make it available to users. The term *saving* is interpreted broadly at this point. The API only enables the generation of a binary string. For this purpose, you need to call the get_file_content method of your document object. It returns the entire Excel file as a binary string, which you can continue using depending on the context.

In the classic provision of Excel documents in SAP GUI, the next step is to download it to the frontend or store it in the application server's file system. In the context of ABAP Cloud and our ABAP RESTful application programming model application, we save the file content in a database table. More details on handling files in ABAP Cloud can be found in Chapter 13. There we saved the image data in a database table.

Completing the action implementation

We now use the Entity Manipulation Language (EML) to process the business object instances addressed in the implementation of the action and save the generated file in the TemplateFileContent field. We also set the TemplateFileMimetype and TemplateFilename fields, which are required by the ABAP RESTful application programming model framework and SAP Fiori elements. The MIME type for an Excel file in XLSX format is application/vnd.openxmlformats-officedocument.spreadsheetml.sheet.

In our case, all instances are allowed to receive the same file. This is due to the fact that the file doesn't depend on the parameters of the mass change instance. As a rule, you would carry out the processing here specifically for each instance in a loop. You can see an example of this in Section 14.2.

Finally, you must use READ ENTITIES to read out the updated status and return it to the caller. The missing part of the action implementation in the generateTemplate method can be found in Listing 14.7.

```
...
DATA(content) = document->get_file_content( ).

MODIFY ENTITIES OF ZACB_R_MassChange IN LOCAL MODE
       ENTITY MassChange
       UPDATE FIELDS ( TemplateFileContent
                       TemplateFileMimetype
                       TemplateFilename
                       ProcessingStatus ) WITH VALUE #(
           FOR k IN keys
           ( %tky                  = k-%tky
             TemplateFileContent  = content
             TemplateFileMimetype =
               'application/vnd.openxmlformats-' &&
```

```
            'officedocument.spreadsheetml.sheet'
          TemplateFilename     = 'template.xlsx'
          ProcessingStatus     = '1' ) )
      REPORTED reported
      FAILED failed
      MAPPED mapped.

READ ENTITIES OF ZACB_R_MassChange IN LOCAL MODE
     ENTITY MassChange
     ALL FIELDS WITH CORRESPONDING #( keys )
     RESULT DATA(mass_changes).
result = VALUE #( FOR mass_change IN mass_changes
                  ( %tky   = mass_change-%tky
                    %param = mass_change ) ).
```

Listing 14.7 Saving the File via EML

14.1.8 Testing the Application

Downloading the file in the Mass Changes app

This completes the implementation of the template generation. In the Mass Changes app, after clicking on the **Generate template** button, the action is triggered, and the Excel file is generated, stored in the database table, and then made available to users for download (see Figure 14.5). You can see the content of our generated Excel file in Figure 14.6.

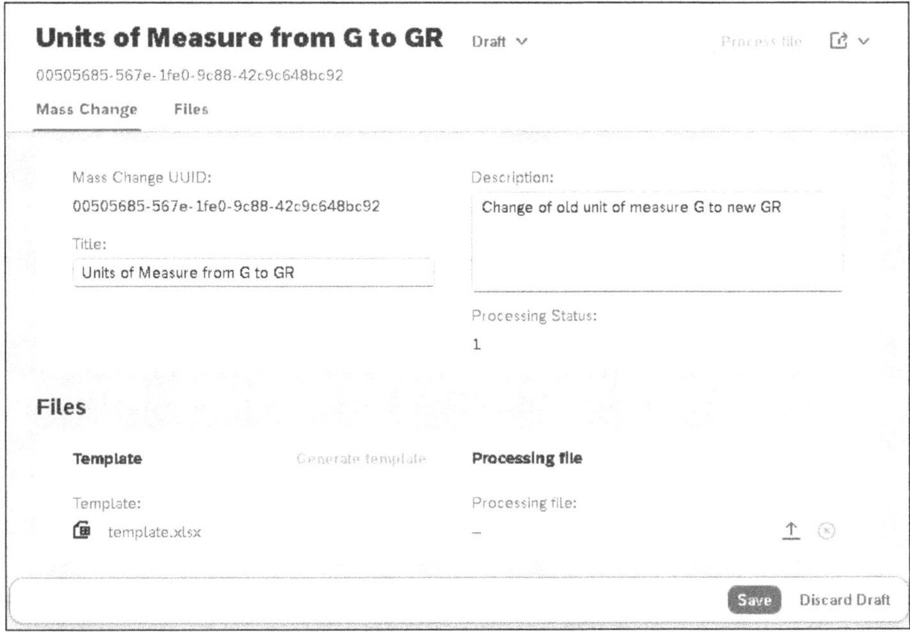

Figure 14.5 Object Page of the Mass Change App After Generating the Template

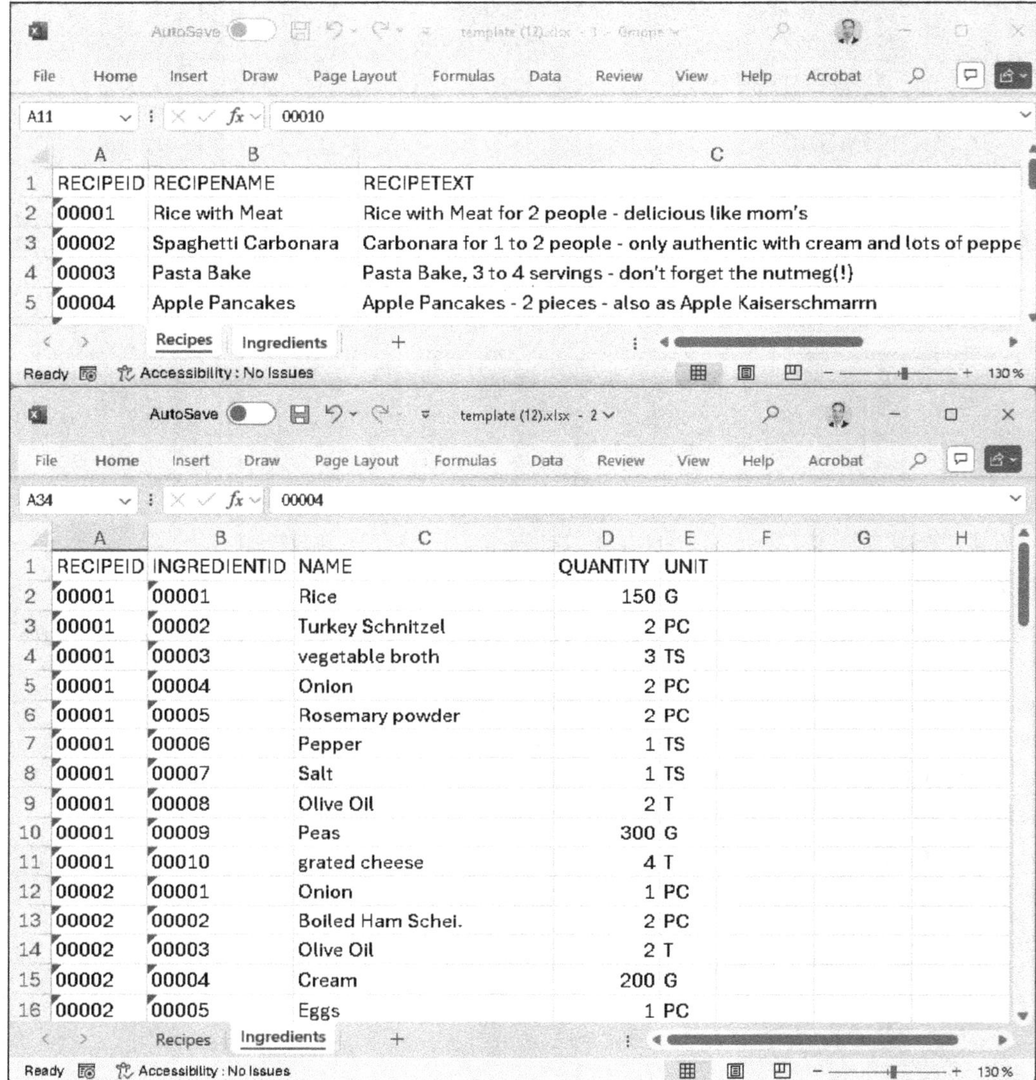

	A	B	C
1	RECIPEID	RECIPENAME	RECIPETEXT
2	00001	Rice with Meat	Rice with Meat for 2 people - delicious like mom's
3	00002	Spaghetti Carbonara	Carbonara for 1 to 2 people - only authentic with cream and lots of peppe
4	00003	Pasta Bake	Pasta Bake, 3 to 4 servings - don't forget the nutmeg(!)
5	00004	Apple Pancakes	Apple Pancakes - 2 pieces - also as Apple Kaiserschmarrn

	A	B	C	D	E
1	RECIPEID	INGREDIENTID	NAME	QUANTITY	UNIT
2	00001	00001	Rice	150	G
3	00001	00002	Turkey Schnitzel	2	PC
4	00001	00003	vegetable broth	3	TS
5	00001	00004	Onion	2	PC
6	00001	00005	Rosemary powder	2	PC
7	00001	00006	Pepper	1	TS
8	00001	00007	Salt	1	TS
9	00001	00008	Olive Oil	2	T
10	00001	00009	Peas	300	G
11	00001	00010	grated cheese	4	T
12	00002	00001	Onion	1	PC
13	00002	00002	Boiled Ham Schei.	2	PC
14	00002	00003	Olive Oil	2	T
15	00002	00004	Cream	200	G
16	00002	00005	Eggs	1	PC

Figure 14.6 Content of the Generated Excel Template

Converting values When you transfer the values from the ABAP variables to the cells of the worksheet, a conversion or *transformation* takes place. The transformation algorithm can be specified when writing via the cursor using the `set_transformation` method on the cell object, and when writing via an internal table using the `set_value_transformation` method before calling the `execute` method (see Listing 14.8).

```
stream->operation->write_from( REF #( table )
  )->set_value_transformation(
```

```
    xco_cp_xlsx_write_access=>value_transformation
      ->best_effort
  )->execute( ).
...
cursor->get_cell( )->value->set_transformation(
  xco_cp_xlsx_write_access=>value_transformation
    ->best_effort
  )->write_from( <component>-name ).
```

Listing 14.8 Specification of the Transformation Algorithm

The possible algorithms are provided via a factory in the value_transformation attribute of API class XCO_CP_XLSX_WRITE_ACCESS. For write accesses, there's currently only the *best effort transformation* available. This also applies implicitly if no algorithm has been specified. Details on the behavior of this transformation can be found in the documentation of the IF_XCO_XLSX_WA_VT_BEST_EFFORT interface. The transformation implementation takes care of, for example, calling conversion exits for language keys and units of measurement, converting Boolean values, and displaying date and time values in the numerical Excel-specific format.

Limitations of the best effort transformation

However, there are limitations to this type of transformation. The conversion of Boolean values from X to 1 and from <blank> to 0 only takes place for the abap_bool type from the ABAP type group, but not for the abap_boolean data element. In addition, some data types aren't supported at all. In our example, the **Quantity** column in the **Ingredients** worksheet would be empty. The quantity of an ingredient is described in the sample application using dictionary type INT2. This type isn't supported by the algorithm, so the column would be empty in the Excel file. To avoid this issue without subsequently changing the data type, we've used a CAST expression in the data selection (see Listing 14.9).

```
SELECT FROM ZACB_R_Ingredient
  FIELDS RecipeId,
         IngredientId,
         Name,
         CAST( Quantity AS INT4 ) AS Quantity,
         Unit
  ORDER BY PRIMARY KEY
  INTO TABLE @DATA(ingredients).
add_table_to_worksheet( table = ingredients
                        sheet = sheet ).
```

Listing 14.9 CAST Expression in the SELECT Command

14.2 Reading an Excel File

Action for processing a file

In the preceding section, an Excel file was created with the data of all recipes and ingredients. Users of the recipe portal can download and edit it as a template. In the next step in our sample application, someone uploads the file they have edited and clicks on the **Process file** button.

14.2.1 Creating the Action

The associated action in the Mass Change business object of the ABAP RESTful application programming model is called `processFile` and is defined as follows in the behavior definition:

```
action ( features : instance )
  processFile result [1] $self;
```

It calls the `processFile` method of the local `LHC_MASSCHANGE` handler class, which now needs to be implemented (see Listing 14.10). This handler class is supposed to read the file edited by the user and make the desired changes to the data via EML.

```
CLASS lhc_masschange DEFINITION
  INHERITING FROM cl_abap_behavior_handler.
  PRIVATE SECTION.
    ...
    METHODS processFile FOR MODIFY
      IMPORTING keys
      FOR ACTION MassChange~processFile
      RESULT result.
ENDCLASS.
CLASS lhc_masschange IMPLEMENTATION.
  ...
  METHOD processFile.
    " TODO
  ENDMETHOD.
ENDCLASS.
```

Listing 14.10 Implementation of the processFile Action

14.2.2 Creating the Document Object

Individual processing of files

We'll initially focus on the handling of Excel files with the XCO library again and take up the integration into the action of the ABAP RESTful application programming model business object later. To do this, we first create a separate method in the private section of the `LHC_MASSCHANGE` handler class, which takes care of processing a specific file. This method is called process_single_file and has the following signature:

```
METHODS process_single_file
  IMPORTING file_content TYPE xstring
  RAISING   lcx_processing_error.
```

It therefore simply accepts the file content as a binary string and processes it. In the event of an error, it triggers a locally defined class-based exception.

Read access

The initial situation for reading Excel files is that you have the file content as a binary string. You must take care of obtaining this binary string yourself in advance. In our case, the file content was read in advance by the caller of the method using READ ENTITIES. To read the file, you must use the XCO_CP_XLSX API class with the document attribute again. This time, you don't select the empty method as the point of entry, but the for_file_content method. You can pass the binary string to this method to obtain read access to the Excel document:

```
DATA(document) = xco_cp_xlsx=>document->for_file_content(
  file_content )->read_access( ).
```


Similar to creating a new Excel document, you specify an access type to get started. The read_access method returns a document object that is designed for read access.

[«]

Write Access to Existing Files

Write access to imported Excel files is located in the same place as read access. To change an existing file, you need to call the write_access method. However, this method isn't yet available in SAP S/4HANA 2023. For this reason, changing Excel files isn't covered in this chapter.

Further handling of the API for read access is very similar to that described for write access in Section 14.1. You'll therefore recognize many methods and access patterns. Technically, however, these are often other classes and interfaces, which are specific to the access type. With write access, for example, you received a document object of type IF_XCO_XLSX_WA_DOCUMENT. Now for read access, the type is IF_XCO_XLSX_RA_DOCUMENT. The name components WA and RA differentiate between write access and read access.

14.2.3 Reading a Worksheet

Read access to a worksheet

You can then navigate back to the workbook using the get_workbook method. In contrast to a new creation, the folder should now already contain worksheets. You can use the at_position and for_name methods to obtain an instance of the worksheet object.

We want to search specifically for a worksheet named *Recipes* because that's the name we also used in our template, and therefore select the for_name method (see Listing 14.11). We then use the exists method to ensure that the worksheet exists in the workbook.

```
DATA(sheet) = document->get_workbook(
  )->worksheet->for_name( 'Recipes' ).
IF NOT sheet->exists( ).
  cl_message_helper=>set_msg_vars_for_clike(
    |The "Recipes" worksheet does not exist| ).
  RAISE EXCEPTION TYPE lcx_processing_error USING MESSAGE.
ENDIF.
```

Listing 14.11 Receipt of an Existing Worksheet

[!]

Validating User Input

When reading an existing Excel file, you must actively take care of the input validation. Checking whether the worksheet in question even exists is the least of it. Macros or JavaScript components for an attack could also be hidden in the file, which are executed in another user's Excel client because you've transferred them to the database without checking them. For this reason, you should add suitable validation checks when processing external data.

14.2.4 Reading the Header

Interpreting the header via the cursor

To process the data, we first need to know the structure. One option is to assume that the user hasn't adjusted the column layout in the worksheet—but this involves risks. In real life, you should implement extensive checks for mass maintenance such as the one described here, so that inconsistencies don't arise inadvertently or even intentionally. For display reasons, we'll limit ourselves to the case where the user may have changed the order of the columns in the file compared to the template. We therefore want to read the header and interpret the columns using the technical identifiers shown there.

Because we only read the header row cell by cell for this interpretation, we use a cursor again (see Listing 14.12) and start in cell A1.

```
DATA(cursor) = sheet->cursor(
  io_column =
    xco_cp_xlsx=>coordinate->for_alphabetic_value( 'A' )
  io_row    =
    xco_cp_xlsx=>coordinate->for_numeric_value( 1 ) ).
```

Listing 14.12 Creation of a Cursor

We then use the cursor to move to the right through the header and analyze the cell content. We interpret this content as a component of a view entity and dynamically create a structure type. This allows us to later assign the remaining lines in bulk and type-safe to a suitable internal table. You can use the has_cell method to query whether the cursor is pointing to a valid cell. You can use the get_cell( )->has_value( ) method chain to query whether the cell contains a value. You can use get_value( )->write_to( ) to write the value to a target data object by reference. We navigate to the right until we reach an empty cell. The entire interpretation logic is shown in Listing 14.13.


```
DATA components     TYPE cl_abap_structdescr
                          =>component_table.
DATA column_header TYPE string.

WHILE cursor->has_cell( ) AND
      cursor->get_cell( )->has_value( ).
  cursor->get_cell( )->get_value( )->write_to(
    REF #( column_header ) ).
  IF column_header IS INITIAL.
    EXIT.
  ENDIF.

  cl_abap_typedescr=>describe_by_name(
    EXPORTING  p_name         =
      |ZACB_R_Recipe-{ column_header }|
    RECEIVING  p_descr_ref    = DATA(column_descriptor)
    EXCEPTIONS type_not_found = 1
               OTHERS         = 2 ).
  IF sy-subrc <> 0.
    cl_message_helper=>set_msg_vars_for_clike(
      |Field { column_header } unknown| ).
    RAISE EXCEPTION TYPE lcx_processing_error USING MESSAGE.
  ENDIF.
```

```
  APPEND VALUE #( name = column_header
                  type = CAST #( column_descriptor ) )
         TO components.

  cursor->move_right( ).
ENDWHILE.
```

Listing 14.13 Interpretation of the Header via the Cursor

Finally, we use RTTC to generate an internal table of our analyzed type and have thus created a target data object into which we can read the other rows of the worksheet in the next step (see Listing 14.14).

```
DATA recipes TYPE REF TO data.
DATA(table_descriptor) = cl_abap_tabledescr=>get(
  cl_abap_structdescr=>get( components ) ).
CREATE DATA recipes TYPE HANDLE table_descriptor.
```

Listing 14.14 Dynamic Generation of the Internal Table for the Recipe Data via RTTC

14.2.5 Reading Data into an Internal Table

Row-based read access to a selection

To read rows into an internal table, you can use row-based read access to a selection in the same way as write access. This again requires a selection pattern. To transfer data to the target data object—in our case, the internal table created at runtime behind the recipe data reference—you want to use the write_to method (see Listing 14.15).

```
DATA(from_second_row) =
  xco_cp_xlsx_selection=>pattern_builder->simple_from_to(
    )->from_row(
      xco_cp_xlsx=>coordinate->for_numeric_value( 2 )
    )->get_pattern( ).
sheet->select( from_second_row )->row_stream(
  )->operation->write_to( recipes )->execute( ).
```

Listing 14.15 Execution of a Read Operation on a Row Stream

The internal table then contains all rows starting from worksheet coordinate A2 as specified by the structured row type of the internal table.

Reading ingredients

We repeat the code for the second worksheet with the ingredients (see Listing 14.16). It makes sense to extract parts of the logic, so we move the header analysis and the creation of the target data object to a separate create_table_for_sheet method.

```
sheet = document->get_workbook( )->worksheet->for_name(
  'Ingredients' ).
IF NOT sheet->exists( ).
  cl_message_helper=>set_msg_vars_for_clike(
    |The "Ingredients" worksheet does not exist| ).
  RAISE EXCEPTION TYPE lcx_processing_error USING MESSAGE.
ENDIF.

DATA(ingredients) = create_table_for_sheet(
  view_entity = 'ZACB_R_Ingredient'
  sheet       = sheet ).

sheet->select( from_second_row )->row_stream(
  )->operation->write_to( ingredients )->execute( ).
```

Listing 14.16 Reading the Ingredients Data

Converting values

Similar to write access, the cell values are also converted in read access. In contrast to write access, however, there are other options available for read access in addition to the best effort transformation. In API class XCO_CP_XLSX_READ_ACCESS, the following implementations are provided via the factory behind the value_transformation attribute:

- **Identity transformation**
  No conversion takes place. The cell value is accepted unchanged and written to the data object.
- **String value transformation**
  All rows are interpreted as a string. The string type should therefore also be selected as the target data object. This implementation is suitable if the analysis is to be implemented for the specific type in ABAP.
- **Best effort transformation**
  This implementation is already known from write access. It converts language keys and units of measurement using the corresponding conversion exits as well as date and time values using Excel-specific numerical formatting. It provides special handling for the abap_bool type. Target data objects with unsupported data types are handled in the same way as the identity transformation.

You can select the transformation implementation via the set_value_transformation method for the cell value object or the write operation object of the row stream. If you don't call the method, the best effort transformation will be used.

14.2.6 Executing the Mass Change

Executing the mass change

As a result, we have two variables: `recipes` and `ingredients`. Both of them contain a reference to an internal table with the imported data. The components of the row types of these internal tables have identifiers that correspond to those of our associated ABAP RESTful application programming model objects. We can therefore send an EML command directly, which carries out the desired changes to recipes and ingredients together with all other necessary activities, such as updating the change document (see Listing 14.17).

```
MODIFY ENTITIES OF ZACB_R_Recipe
       ENTITY Recipe
       UPDATE FIELDS ( RecipeName RecipeText )
       WITH CORRESPONDING #( recipes->* )
       ENTITY Ingredient
       UPDATE FIELDS ( Name Quantity Unit )
       WITH CORRESPONDING #( ingredients->* )
       FAILED DATA(failed).
IF failed IS NOT INITIAL.
  cl_message_helper=>set_msg_vars_for_clike(
    |Error on mass change| ).
  RAISE EXCEPTION TYPE lcx_processing_error USING MESSAGE.
ENDIF.
```

Listing 14.17 Mass Change via MODIFY ENTITIES

At this point, we assume that the **Recipe name**, **Recipe text**, **Ingredient name**, **Ingredient quantity**, and **Ingredient unit** fields were present in the Excel spreadsheet with the correct column identifier. Otherwise, these fields would be interpreted as initial and emptied via the EML statement. We've only taken into account the case that the user re-sorts columns. In a production application, other precautions would have to be taken as well.

Integration into the action implementation

The processing now only needs to be integrated into the `processFile` action of our ABAP RESTful application programming model business object. For this purpose, we supplement the call of our `process_single_file` method and the surrounding handling of the action implementation. For multi-instance capability, all addressed instances must first be read and then processed in a loop.

If successful, we set the processing status to the value `3` (processing completed), issue a success message, and set the updated status of the instances in the result variable. If an error occurs, we set an error message for the instance of the mass change and mark it as incorrectly processed. The entire implementation of the method is shown in Listing 14.18.

```
METHOD processFile.
  READ ENTITIES OF ZACB_R_MassChange IN LOCAL MODE
       ENTITY MassChange
       FIELDS ( ProcessingFileContent )
       WITH CORRESPONDING #( keys )
       RESULT DATA(mass_changes).

  LOOP AT mass_changes
       ASSIGNING FIELD-SYMBOL(<mass_change>).
    TRY.
        process_single_file(
          <mass_change>-ProcessingFileContent ).
        INSERT VALUE #(
          %tky = <mass_change>-%tky
          %msg = new_message_with_text(
            severity = if_abap_behv_message
                       =>severity-success
            text     = |Mass change was | &&
                       |successfully completed| ) )
               INTO TABLE reported-masschange.
        MODIFY ENTITIES OF ZACB_R_MassChange IN LOCAL MODE
               ENTITY MassChange
               UPDATE FIELDS ( ProcessingStatus )
               WITH VALUE #(
               ( %tky             = <mass_change>-%tky
                 ProcessingStatus = '3' ) ).
      CATCH lcx_processing_error INTO DATA(error).
        INSERT VALUE #(
          %tky = <mass_change>-%tky
          %msg = new_message_with_text(
            severity = if_abap_behv_message
                       =>severity-error
            text     = error->get_text( ) ) )
               INTO TABLE reported-MassChange.
        INSERT VALUE #(
          %tky        = <mass_change>-%tky
          %fail-cause = if_abap_behv=>cause-unspecific )
               INTO TABLE failed-MassChange.
    ENDTRY.
  ENDLOOP.
```

```
    READ ENTITIES OF ZACB_R_MassChange IN LOCAL MODE
         ENTITY MassChange
         ALL FIELDS
         WITH CORRESPONDING #( keys )
         RESULT mass_changes.
    result = VALUE #( FOR mass_change IN mass_changes
                      ( %tky    = mass_change-%tky
                        %param = mass_change ) ).
ENDMETHOD.
```

Listing 14.18 Completed Implementation of the processFile Action

14.2.7 Testing the Application

Display in the Mass Changes app

Our implementation is now complete. A success message gets displayed in the Mass Changes app once processing is complete (see Figure 14.7).

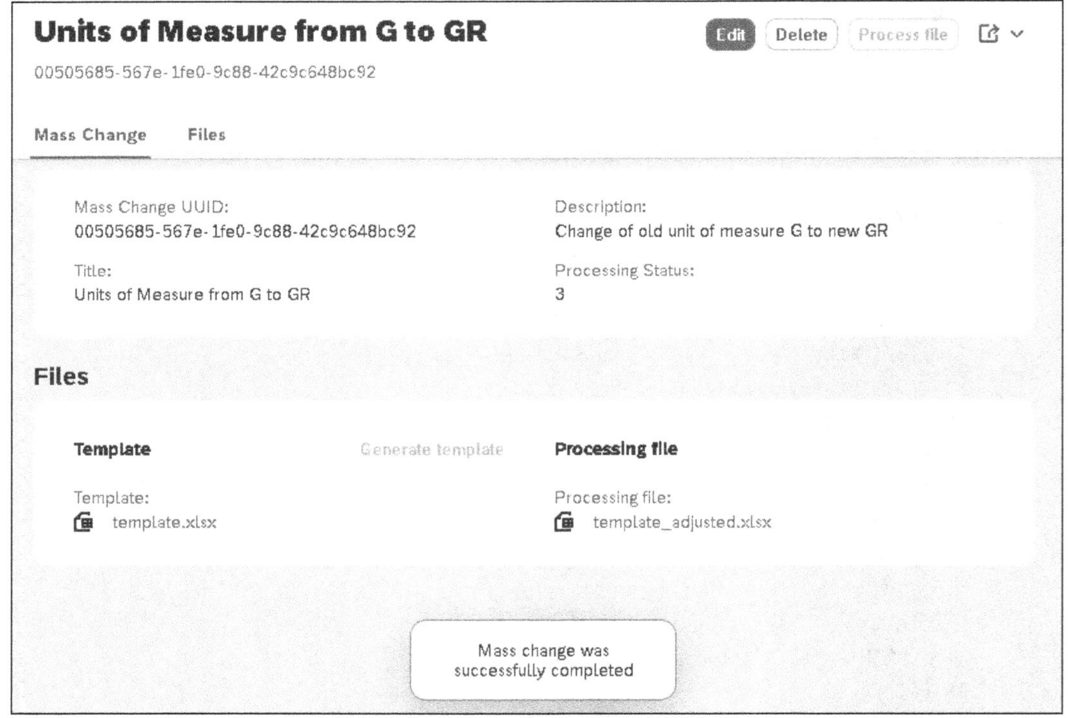

Figure 14.7 Successful Mass Change in the Mass Changes App

The recipes and ingredients have been changed on the basis of the modified file. In the event of an error, an error message will be displayed (see Figure 14.8).

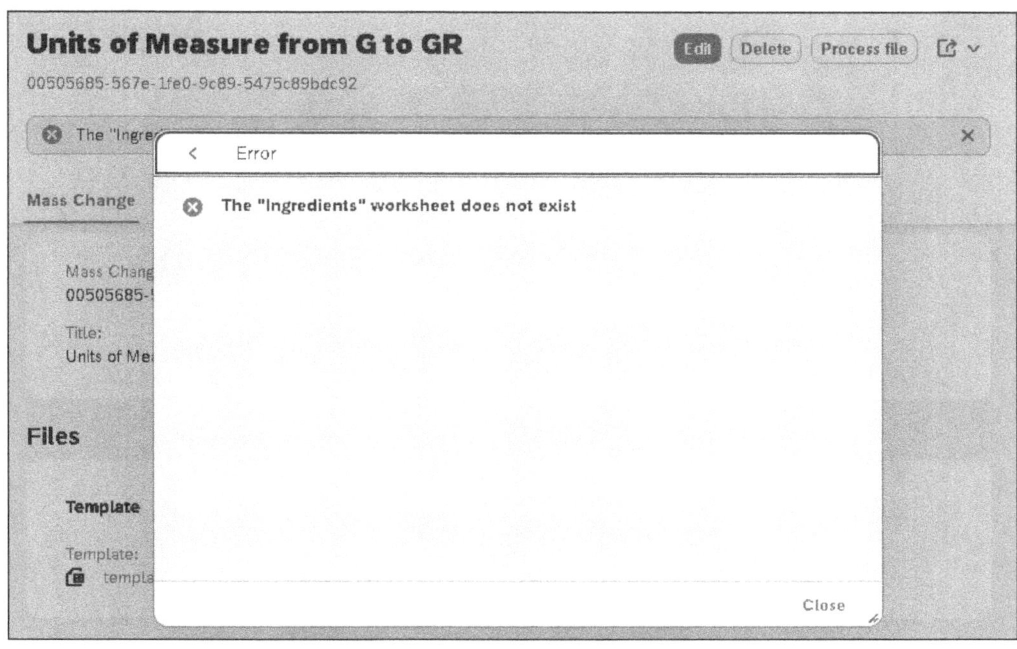

Figure 14.8 Display of an Error Message in the Mass Changes App

14

14.3 Summary

In this chapter, you've learned about the new ABAP Cloud–enabled API for creating and reading Excel files. This component of the XCO library requires some familiarization with the various classes, interfaces, and methods. However, once this has been done, Excel files can be created with comparatively simple means. The integration into the concept of file uploads and downloads with SAP Fiori elements and the ABAP RESTful application programming model is seamless because the API is only based on or generates binary strings. The app developed in the example already provides an extensive range of functions and could be extended very easily. Few of the functions shown are specifically designed for recipes or ingredients. Most of the implementation parts could be adapted to any technical background without major adjustments.

However, some functions are still missing. In SAP S/4HANA 2023, for example, it's not yet possible to edit Excel files, but this function is already available in the SAP BTP ABAP environment and in SAP S/4HANA Cloud Public Edition. It's also currently only possible to format cells, for example, their background color or font, in the public cloud environments. The API for Excel documents is still under active development, which is why its range of functions will continue to grow.

Chapter 15
Documenting Development Objects

You're probably familiar with the following scenario: You're working on a bug in a program you developed 15 years ago, but you don't remember anything about it. You decided not to provide any developer documentation at the time because you thought the program code was self-explanatory. This chapter explains how you can take care of the tedious documentation of development objects in ABAP Cloud.

Many projects lack a budget, which is why seemingly unnecessary time wasters are avoided. This often includes the documentation of developments.

Documentation in the comments

There are two different types of comment in the code to document the developments:

- **Documenting comments**
  These comments adhere to specific formatting conventions so that documentation can be created automatically from these comments.
- **Explanatory comments**
  These comments explain which special features, hacks, or problems exist.

As a general rule, you should comment as little as possible, but as much as necessary. In addition to the comments on the documentation, the names of the development objects, such as the classes or methods, should already provide information about their purpose. For example, if a method has the task of loading user data from a database, it should be called `loadUser` and not just `selectData`.

This chapter focuses on the documenting comments that can be used in the ABAP world. In Section 15.1, we first explain the concept of ABAP Doc comments, which are primarily used for object-oriented development. We then present the knowledge transfer document in Section 15.2, which is primarily used for the new development objects in the ABAP RESTful application programming model.

Documenting the recipe portal application

In this chapter, we create a new example class named ZCL_ACB_RECIPE, which we extend with ABAP Doc comments. We use this class to demonstrate how you can handle ABAP Doc comments. We also create a new core data services (CDS) view named ZACB_I_RECIPE_DOCU. For this CDS view, we're going to create a knowledge transfer document.

[»]

> **Using the Associated Sample Application**
>
> The objects presented in this chapter are located in the DOCUMENTATION subpackage.

15.1 ABAP Doc

ABAP Doc enables the documentation of ABAP objects on the basis of specific comments. It can be used to document global objects such as classes and interfaces as well as their methods and attributes. ABAP Doc documentation can also exist for local objects such as local variables. In a development environment such as the ABAP development tools for Eclipse, the content of ABAP Doc comments is evaluated, converted to HTML, and then displayed in a suitable manner.

ABAP Doc comment

An ABAP Doc comment is introduced by the "! string:

```
"! Describing the ingredients
```

The ABAP Element Info view

The documentation created can be displayed in the ABAP development tools in the **ABAP Element Info** view. You can call this view by clicking on an object documented with ABAP Doc and pressing the F2 key. A popup window appears in which all the information that has been documented gets displayed (see Figure 15.1).

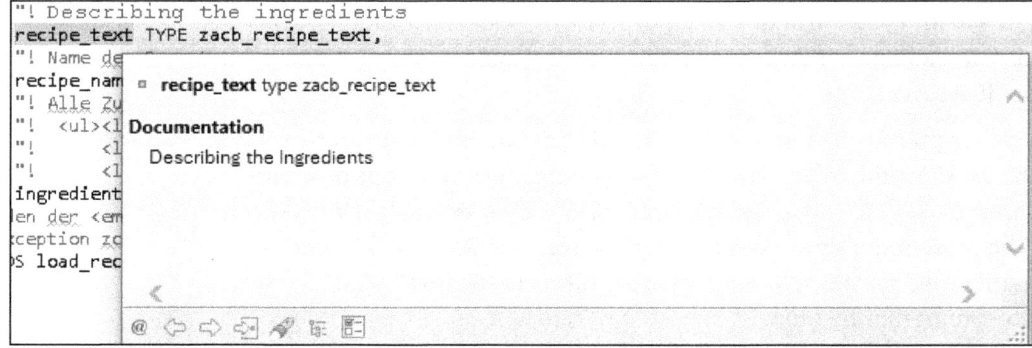

Figure 15.1 The ABAP Element Info View

Multiline comment

To use multiple lines for the documentation of source code, each line must start with "!:

```
"! Load the recipe information using
"! the recipe ID in table ZACB_RECIPE
```

Formatting for output

Table 15.1 lists the various tags that can be used to format the output of ABAP Doc comments.

Formatting option	Tag
Line break	` ` or ` </br>`
Paragraph	`<p>...</p>`
Highlighted text	`<em>...</em>`
Strongly highlighted text	`<strong>...</strong>`
Unsorted lists	`<ul><li>...</li></ul>`
Sorted lists	`<ol><li>...</li></ol>`

Headings	`<h1>...</h1>`
	`<h2>...</h2>`
	`<h3>...</h3>`

Table 15.1 Formatting Options for ABAP Doc Comments

The use of these formatting options is shown as an example in Listing 15.1.

```
"! All ingredients for the sample recipe
"!  <ul><li>Quantity: 100</li>
"!      <li>Unit: g</li>
"!      <li>Name: Flour</li></ul>
  ingredients TYPE TABLE OF zacb_ingredient.
"! Load the <em>recipe information</em> using the
"! recipe ID in table <strong>ZACB_RECIPE</strong>
METHODS load_recipe RAISING zcx_acb_recipe_not_found.
```

Listing 15.1 Formatting Options for ABAP Doc Comments

These formatting options are implemented visually in the **ABAP Element Info** view (see Figure 15.2).

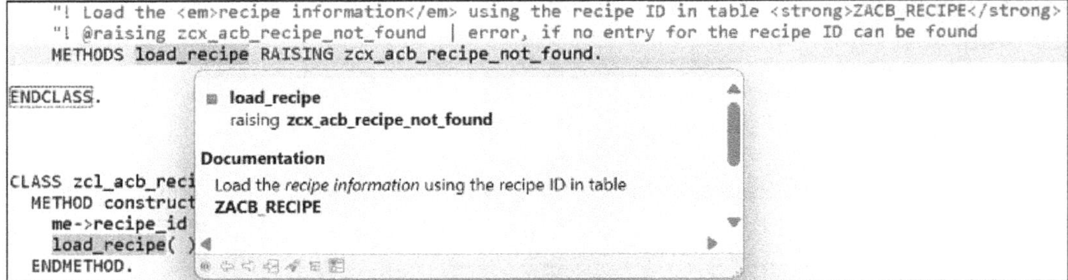

"! Load the <em>recipe information</em> using the recipe ID in table <strong>ZACB_RECIPE</strong>
"! @raising zcx_acb_recipe_not_found | error, if no entry for the recipe ID can be found
METHODS load_recipe RAISING zcx_acb_recipe_not_found.
ENDCLASS.
load_recipe
raising zcx_acb_recipe_not_found
Documentation
CLASS zcl_acb_reci
METHOD construct
me->recipe_id
load_recipe( )
ENDMETHOD.
Load the recipe information using the recipe ID in table
ZACB_RECIPE

Figure 15.2 Display of Formatting Options in the ABAP Element Info View

Parameters and exceptions

The parameters and exceptions can be documented for methods, events, function modules, and subroutines. Table 15.2 shows the syntax used for parameters and exceptions.

Documentation For	Syntax
Parameter	@parameter <parameter name> | <parameter documentation>
Class-based exceptions	@raising <exception name> | <exception documentation>
Classic exceptions	@exception <exception name> | <exception documentation>

Table 15.2 Documentation for Parameters and Exceptions

Listing 15.2 shows an example of the documentation of parameters and exceptions.

```
"! @parameter recipe_id | Transfer of a recipe ID
"! @raising zcx_acb_recipe_not_found | error, if no
"! entry for the recipe ID can be found.
METHODS constructor IMPORTING recipe_id
  TYPE zacb_recipe_id RAISING zcx_acb_recipe_not_found.
```

Listing 15.2 ABAP Doc Comment for a Parameter and an Exception

This ABAP Doc comment results in the display in the **ABAP Element Info** view shown in Figure 15.3.

Integrating short text descriptions

ABAP Doc and the classic short text descriptioncan also be combined. This is made possible by the "shorttext" attribute. A short text description looks as follows in ABAP Doc:

```
"! <p class="shorttext">Class for recipes</p>
```

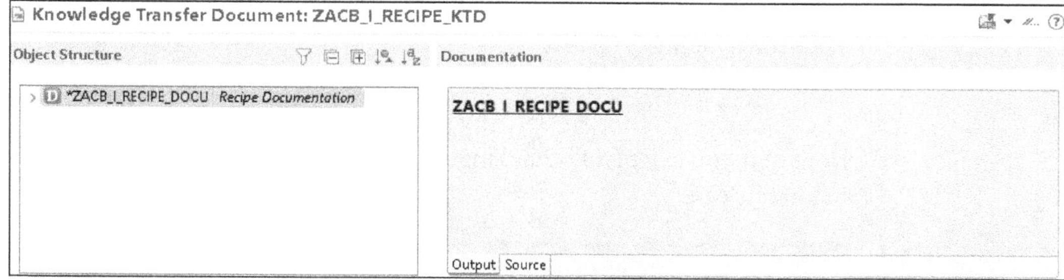

Figure 15.3 Output of ABAP Doc Comments for Parameters and Exceptions

If the "synchronized" attribute is now also added, short text descriptions and ABAP Doc comments can be synchronized:

```
"! <p class="shorttext synchronized">Class for recipes</p>
```

Figure 15.4 shows the connection between the ABAP Doc comment and the description for the ZCL_ACB_RECIPE class in the editor and in the project structure.

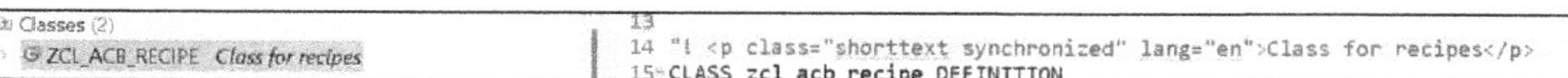

Figure 15.4 Linking the Texts

[«]

Length of Short Text Descriptions

The length of the description is limited to the length for short text descriptions of the respective repository object.

Importing existing descriptions

You can generate ABAP Doc from existing descriptions of global classes and interfaces. To import descriptions into classes or interfaces, simply open the development object, and select the **Source Code • Import ABAP Doc from Descriptions** item in the menu bar (see Figure 15.5). The imported texts are automatically synchronized.

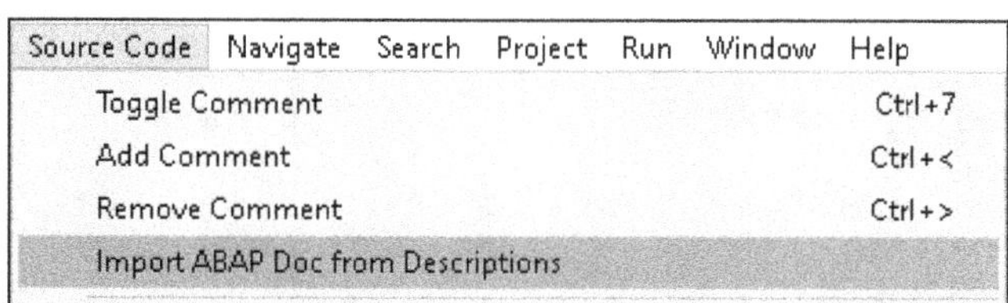

Figure 15.5 Import of Descriptions to ABAP Doc

The lang attribute

The imported texts also have the lang attribute, shows the main language of the development object:

```
"! <p class="shorttext synchronized lang="en">Class for recipes</p>
```

Note that you can't add other languages. The editor would display a warning in this case.

[»]

The Synchronized Attribute

The synchronized attribute should be used if not only the ABAP development tools but also the ABAP Workbench is used in the company.

Reference to other repository objects

An ABAP Doc comment can also contain a reference to other repository objects:

```
... {@link [[[kind:]name.]...][kind:]name} ...
```

A path to a repository object can be specified in curly brackets after @link to reference its documentation. This is followed by the following attributes:

- Name
  The name of a repository object is specified here.
- Kind
  The type of repository object is specified here.

The following is an example of a reference to the documentation of the ZCX_ACB_RECIPE_NOT_FOUND exception class:

```
"! Error if no entry can be found for recipe ID
"! {@link ZCX_ACB_RECIPE_NOT_FOUND}
```

If you now call the **ABAP Element Info** view via the F2 function key, this link is displayed (see Figure 15.6).

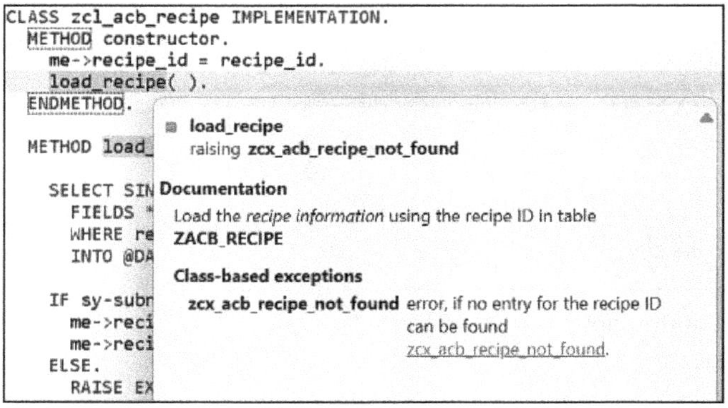

```
CLASS zcl_acb_recipe IMPLEMENTATION.
  METHOD constructor.
    me->recipe_id = recipe_id.
    load_recipe( ).
  ENDMETHOD.

  METHOD load_

    SELECT SIN
      FIELDS *
      WHERE re
      INTO @DA

    IF sy-subr
      me->reci
      me->reci
    ELSE.
      RAISE EX
```

load_recipe
raising zcx_acb_recipe_not_found

Documentation
Load the *recipe information* using the recipe ID in table
ZACB_RECIPE

Class-based exceptions
zcx_acb_recipe_not_found error, if no entry for the recipe ID can be found
zcx_acb_recipe_not_found.

Figure 15.6 Representation of the Reference

Exporting the documentation

In the ABAP development tools, you can export ABAP Doc documentation for classes and interfaces in HTML files. To do this, follow these steps:

1. In the project explorer, select a package, class, or interface. Select the **Export** item from the context menu (see Figure 15.7). Alternatively, you can also select **File • Export** in the menu bar.

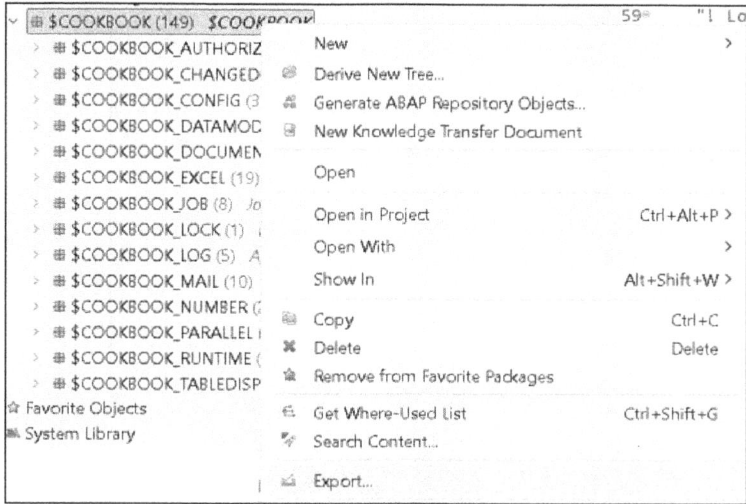

Figure 15.7 Using the Context Menu to Select Export

2. In the wizard that opens, select the **ABAP Doc** item under **ABAP** (see Figure 15.8).

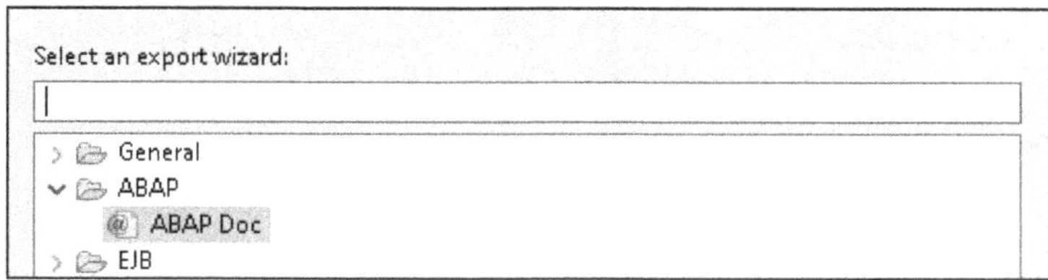

Figure 15.8 Selecting ABAP Doc

3. Confirm your selection by clicking the **Next** button.
4. In the next step, select classes, interfaces, and entire packages (see Figure 15.9). In addition, choose the visibility (**Public**, **Public and protected**, or **Public, protected and private**) to define which type of methods should be documented. For example, only public methods should be exported.
5. Specify the storage location in the **Directory** field.

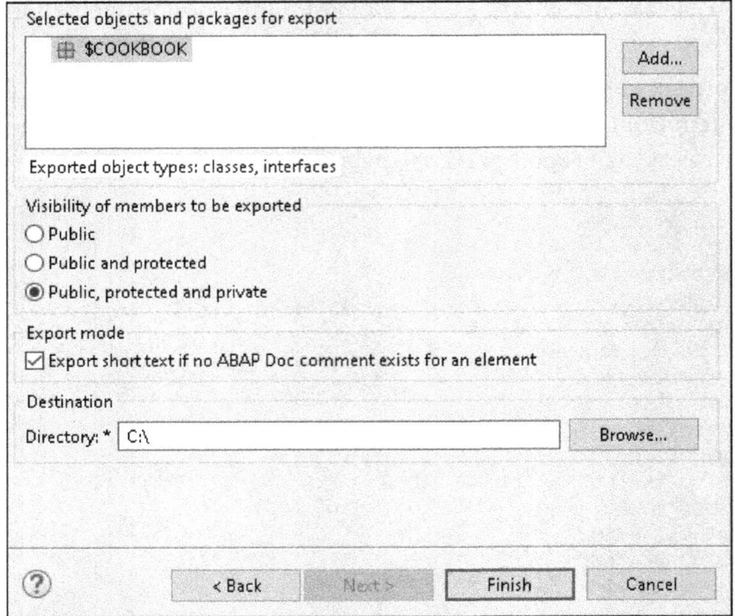

Figure 15.9 Export Details in the Export Wizard

6. Once you've entered all the details, start the export by clicking the **Finish** button.

You'll be notified of the successful export (see Figure 15.10) and can go directly to the corresponding folder from there.

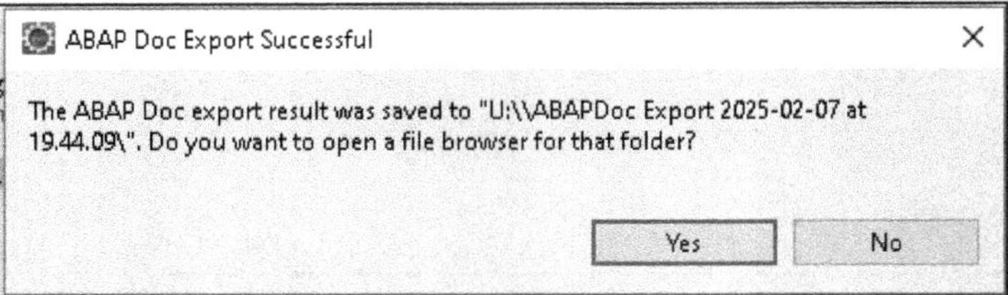

Figure 15.10 Export Success Message

HTML documentation For each documented class, an HTML page containing the ABAP Doc documentation was generated. In Figure 15.11, you can see this page for our `ZCL_ACB_RECIPE` class.

Class ZCL_ACB_RECIPE

public final create public

Documentation

Class for recipes

Attributes

Visibility and Level	Name	Documentation
private instance	**ingredients** type table of zacb_ingredient	All ingredients for the sample recipe • Quantity: 100 • Unit: g • Name: Flour
private instance	**recipe_id** type zacb_recipe_id	Unique identifier for a recipe
private instance	**recipe_name** type zacb_recipe_name	Name des Rezepts
private instance	**recipe_text** type zacb_recipe_text	Describing the ingredients

Methods

Visibility and Level	Name	Documentation
public instance	**constructor** importing **recipe_id** type zacb_recipe_id raising **zcx_acb_recipe_not_found**	**Parameters** **recipe_id** Transfer of a recipe ID **Class-based Exceptions** **zcx_acb_recipe_not_found** error, if no entry for the recipe ID can be found *zcx_acb_recipe_not_found.*
private instance	**load_recipe** raising **zcx_acb_recipe_not_found**	Load the *recipe information* using the recipe ID in table **ZACB_RECIPE** **Class-based Exceptions** **zcx_acb_recipe_not_found** error, if no entry for the recipe ID can be found *zcx_acb_recipe_not_found.*

Figure 15.11 Display of the Exported ABAP Doc Documentation in a HTML Document

15.2 Knowledge Transfer Document

The *knowledge transfer document* is the next stage in the documentation of development objects. It focuses primarily on ABAP RESTful application programming model development objects such as CDS views, behavior definitions, or service bindings. However, it can also be used for many other repository objects such as packages or data elements.

15.2.1 Creating a Knowledge Transfer Document

To create a knowledge transfer document, follow these steps:

1. In the ABAP development tools project explorer, select the object to be documented, and then right-click and select **New Knowledge Transfer Document** from the context menu.
2. In the creation wizard that opens, enter a name for the knowledge transfer document (see Figure 15.12). If necessary, you can still change the referenced object using the **Browse** button. Continue by clicking **Next**.

15

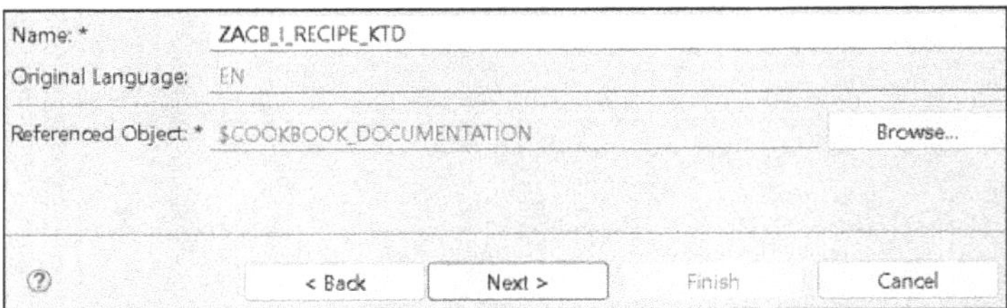

Figure 15.12 Wizard for Creating a Knowledge Transfer Document

3. Select a transport request.
4. Complete the installation by clicking the **Finish** button. Then, the knowledge transfer document gets created in the same package as the referenced object.

Editing the knowledge transfer document

After the creation, the knowledge transfer document is automatically opened in the knowledge transfer document editor (see Figure 15.13). Now, you can start documenting.

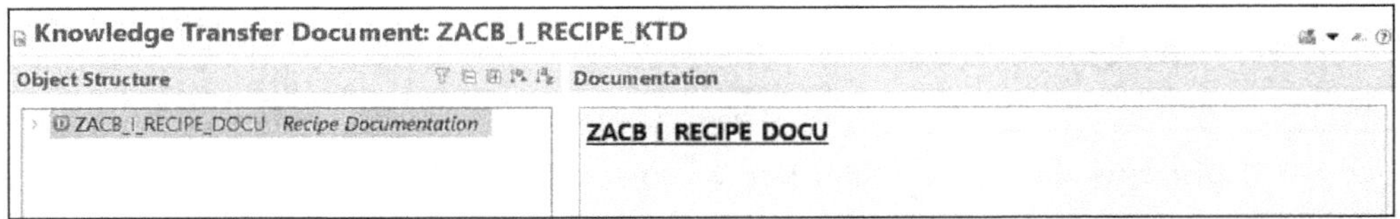

Figure 15.13 Knowledge Transfer Document Editor

In the knowledge transfer document editor, you can select the object or subobject to be documented on the left. On the right, you can switch back and forth between the **Source** (for entering the documentation) and **Output** (for checking the output of the entered data) tabs. *Markdown* is used as the syntax for knowledge transfer documents.

[»]

Markdown

Markdown is a very simple markup language that was designed with the primary aim of being easy to read in its raw state, without any conversion.

Formatting options

You can now write the documentation in Markdown on the **Source** tab. To display all available formatting options (see Figure 15.14), you can press the keyboard shortcut `Ctrl`+`space bar` in the editor.

In addition to the classic formatting options such as strong (bold) or italics, an additional special formatting option called **Link with title** enables a link to another development object. The corresponding syntax looks as follows:

```
[Displayed text](development object name)
```

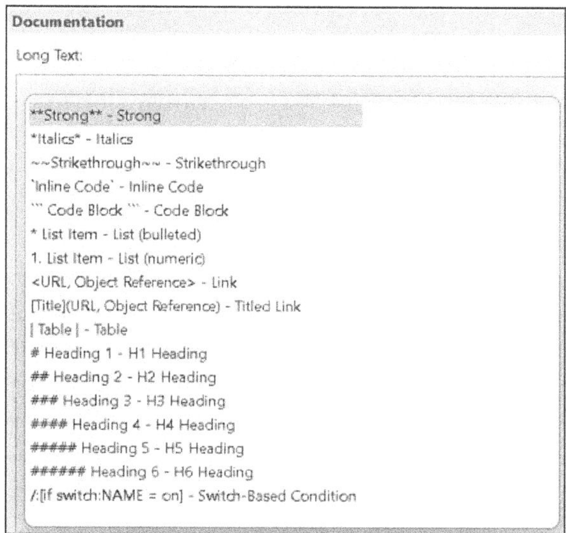

Figure 15.14 Display of the Available Formatting Options

You can use the keyboard shortcut `Ctrl`+`space bar` to search for the development object names.

Output of the documentation

Once you've created the documentation in the Markdown syntax, the result will be displayed on the **Output** tab. Figure 15.15 shows an input on the **Source** tab in the upper area and the output on the **Output** tab in the lower area.

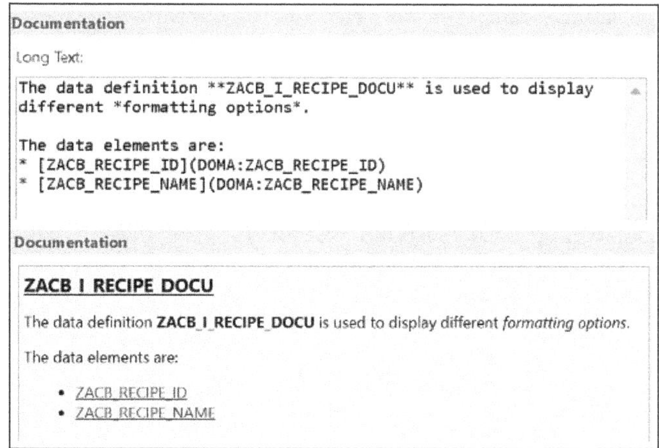

Figure 15.15 Documentation on the Source and Output Tabs

After the entry, the knowledge transfer document must still be activated. It's then displayed in the development package under **Texts • Knowledge Transfer Documents** (see Figure 15.16).

Core Data Services (1)
Data Definitions (1)
ZACB_I_RECIPE_DOCU Recipe Documentation
Documentation
ZACB_I_RECIPE_KTD
ZACB_I_RECIPE_DOCU
Source Code Library (2)
Classes (2)
Texts (1)
Knowledge Transfer Documents (1)
ZACB_I_RECIPE_KTD Documentation for ZACB_I_RECIPE_DOCU

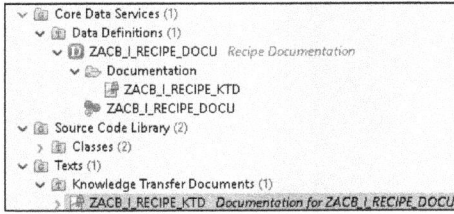

Figure 15.16 Knowledge Transfer Document for a Development Object

Querying the documentation status

The ABAP development tools also provide a way to check which documentation exists or is pending for a package. To do this, you need to open the context menu of the package and select **Expand <Folder> By**. In the dialog box that opens, you must then select the **Documentation Status** option (see Figure 15.17).

Figure 15.17 Querying the Expand-By Option

This changes the tree view in the project explorer, as shown in Figure 15.18. Now, it displays for which objects which type of documentation is possible (e.g., SAPSCRIPT_POSSIBLE), for which one exists (e.g., KTD) or for which there is no possibility at all (e.g., DOCUMENTATION NOT POSSIBLE).

$COOKBOOK_DOCUMENTATION (4) Documentation
Documentation not possible (1)
Texts (1)
Knowledge Transfer Documents (1)
ZACB_I_RECIPE_KTD Documentation for ZACB_I_RECIPE_DOCU
KTD documentation exists (1)
Core Data Services (1)
Data Definitions (1)
ZACB_I_RECIPE_DOCU Recipe Documentation
SAP Script documentation possible (2)
Source Code Library (2)
Classes (2)
ZCL_ACB_RECIPE Class for recipes
ZCX_ACB_RECIPE_NOT_FOUND Error class for recipes not found

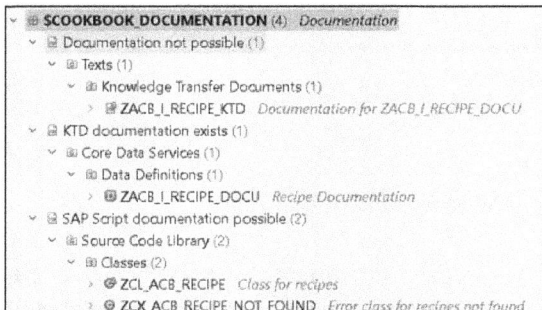

Figure 15.18 Tree View in the Project Explorer

15.2.2 Linking a Knowledge Transfer Document to a Development Object

Because the knowledge transfer document and the development object are directly related, various operations on the development object can have an effect on the knowledge transfer document. The knowledge transfer document is directly linked to the development object. Each element of an object can be documented separately, and separate documentation for each element isn't required. Figure 15.19 illustrates these relationships.

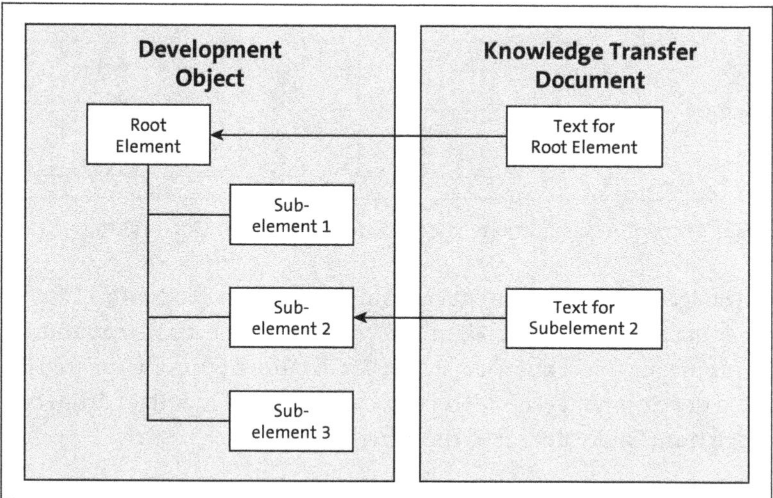

Figure 15.19 Connection Between the Development Object and Knowledge Transfer Document

Package assignment

If another package is assigned to a development object, a corresponding package assignment is automatically made for the knowledge transfer document as well (see Figure 15.20).

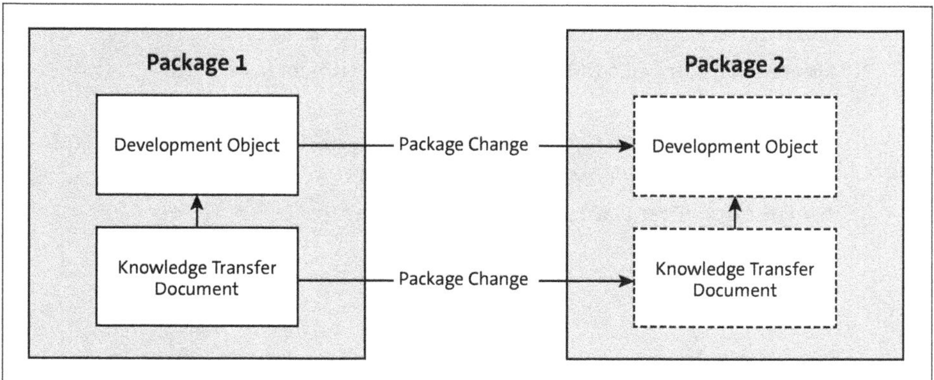

Figure 15.20 Automatic Package Change of the Knowledge Transfer Document

Transport request assignment

If the development object is assigned to a transport request, the knowledge transfer document isn't automatically assigned to the same transport request. However, it can be manually assigned to the same or a different transport request and transported separately from the development object (see Figure 15.21).

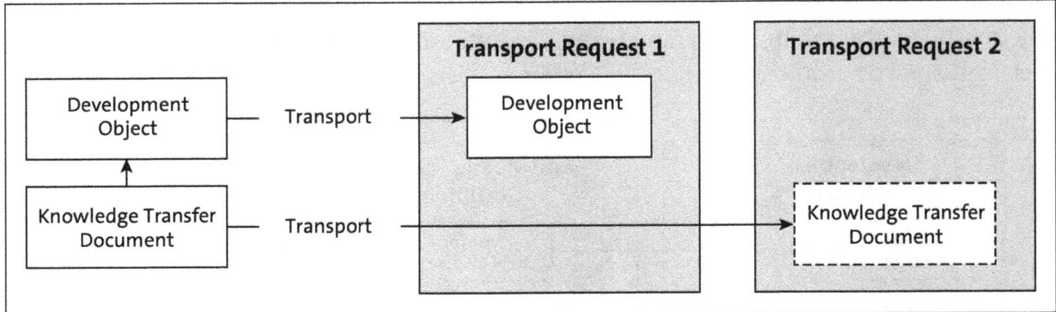

Figure 15.21 Transport Assignment of a Knowledge Transfer Document

Copy of a development object

The knowledge transfer document isn't automatically copied with a copy of the development object (see Figure 15.22). It can't be copied manually either. This means a new knowledge transfer document must be created for the copied development object. The text can be copied manually from one knowledge transfer document to the other.

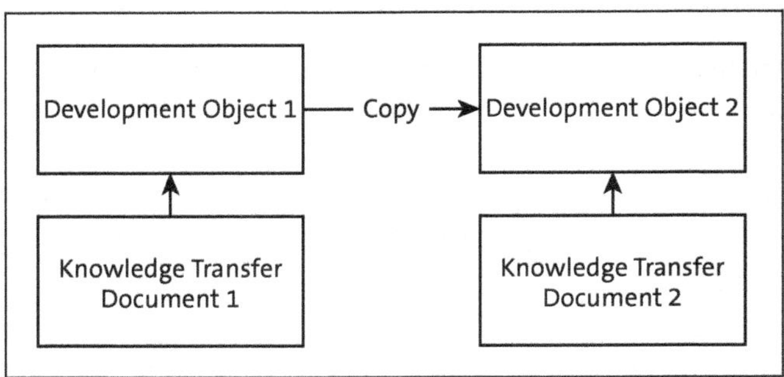

Figure 15.22 Copy of a Knowledge Transfer Document

Deleting a development object

If the development object is deleted, the associated knowledge transfer document gets deleted too (see Figure 15.23). However, the knowledge transfer document can also be deleted separately.

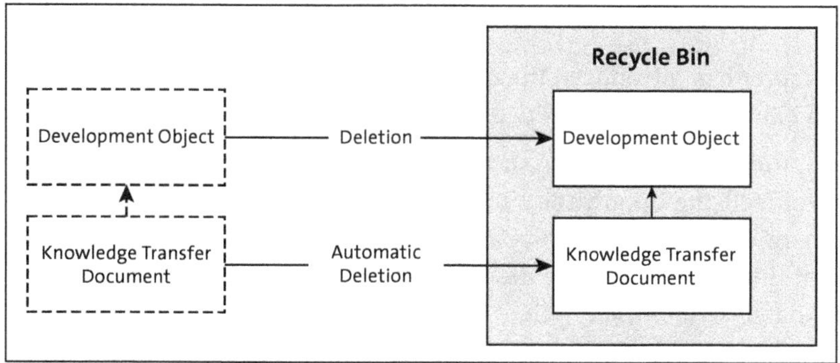

Figure 15.23 Deletion of a Knowledge Transfer Document

Multilingual documentation

Texts for a development object in different languages can also be added to a knowledge transfer document (see Figure 15.24).

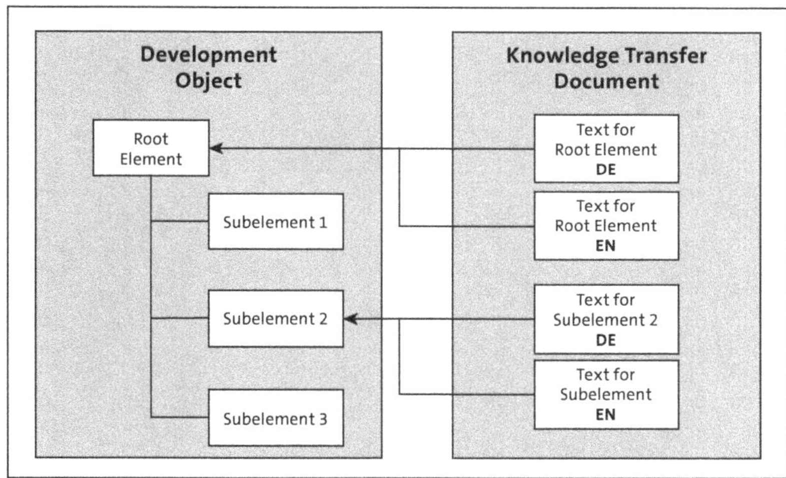

Figure 15.24 Different Languages of a Knowledge Transfer Document

The language is entered in the **Original Language** field when the knowledge transfer document gets created (see Figure 15.25). If another language needs to be added, the project must be added again in another language in the ABAP development tools. A new knowledge transfer document must then be created for the referenced object.

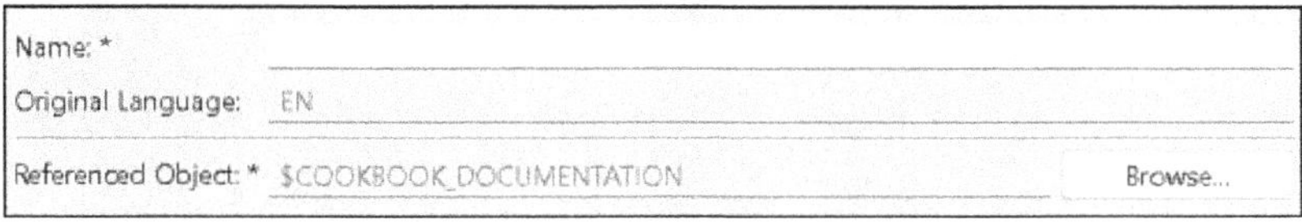

Figure 15.25 Original Language Field During the Creation of a Knowledge Transfer Document

15.3 Summary

This chapter has introduced the development documentation types ABAP Doc and knowledge transfer document in ABAP Cloud.

Development documentation should be created to give other developers or even yourself the opportunity to understand the code. It's important to document not the what, but the why. In addition to maintaining the development documentation, the development objects should already have a name that describes them well.

The first type of documentation we introduced is ABAP Doc. ABAP Doc is used to document object-oriented developments. In addition to classes and interfaces, this also includes their methods with parameters. The other type of documentation we've introduced are knowledge transfer documents. Knowledge transfer documents were developed by SAP for ABAP RESTful application programming model development objects. They are written in Markdown syntax.

Chapter 16
Authorizations

Imagine a customer calls saying that they can see data from another company code. This is a data privacy incident, and you must implement an authorization check as quickly as possible.

Authorization checks are also necessary in ABAP Cloud to check which roles can access an authorization object. Only by checking the authorizations in the code can you guarantee data security and data privacy, as well as ensure that only authorized SAP users can create, change, display, or delete data. In this chapter, you'll learn what you need to know to protect your ABAP Cloud applications by using authorizations.

Section 16.1 deals with authorizations for read operations. There, we'll present core data services (CDS) access controls. In Section 16.2, we describe the different authorizations for change operations in the ABAP RESTful application programming model environment.

16

Read and write operations in the recipe portal

In this chapter, we extend the recipe portal to include authorizations for read and change operations. To demonstrate the authorizations for read operations, we've added the **Published** (PUBLISHED) field to our app (see Figure 16.1).

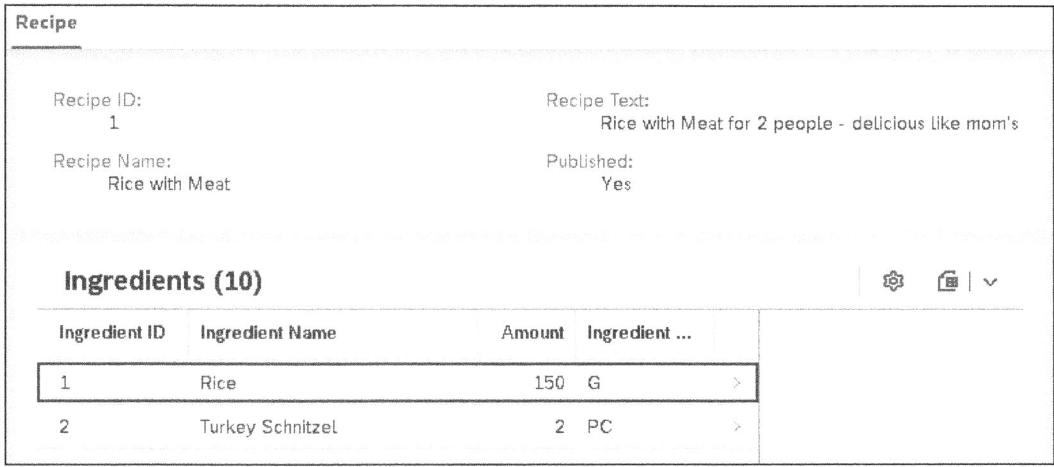

Figure 16.1 SAP Fiori App with the New "Published" Field

This addition must range from the database table to the metadata extension. The idea behind this extension is that users can only see published recipes, while the author of a recipe should also be able to see their own unpublished recipes.

In addition, only specific users may carry out certain change operations. Only the author of a recipe may delete and change it. Another user may create reviews for the recipe.

[»]

> **Using the Sample Application**
>
> The objects created in this chapter are located in the AUTHORIZATION subpackage.

16.1 Authorization Checks for Read Operations

Data Control Language

To protect data from unauthorized read access, ABAP CDS provides its own authorization concept based on a Data Control Language (DCL). To restrict read access to ABAP RESTful programming application model business objects, it's sufficient to model a *CDS access control* for the CDS entities used in these business objects by means of DCL. The access control allows you to restrict the results returned by a CDS entity to those that you want to authorize a user to view.

Creating a CDS access control

Follow these steps to create an access control:

1. Right-click on the CDS entity, and select **New Access Control** from the context menu.
2. In the window that opens, you must maintain the following values (Figure 16.2):
   - **Name**: Enter a suitable name for the access control; for our example, we entered "ZACB_R_RECIPE".
   - **Description**: Enter a description.
   - **Protected Entity**: Enter the CDS entity to be protected in this field.

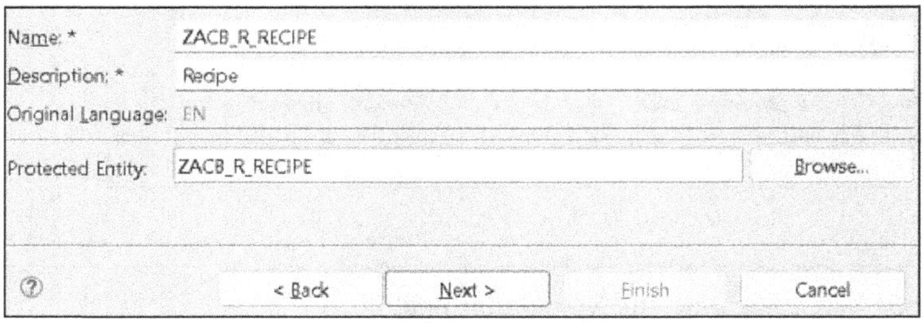

Figure 16.2 Creating an Access Control

3. Confirm your input by clicking the **Next** button.
4. Select a transport request in the subsequent window. Click the **Next** button to go to the template selection in Figure 16.3.

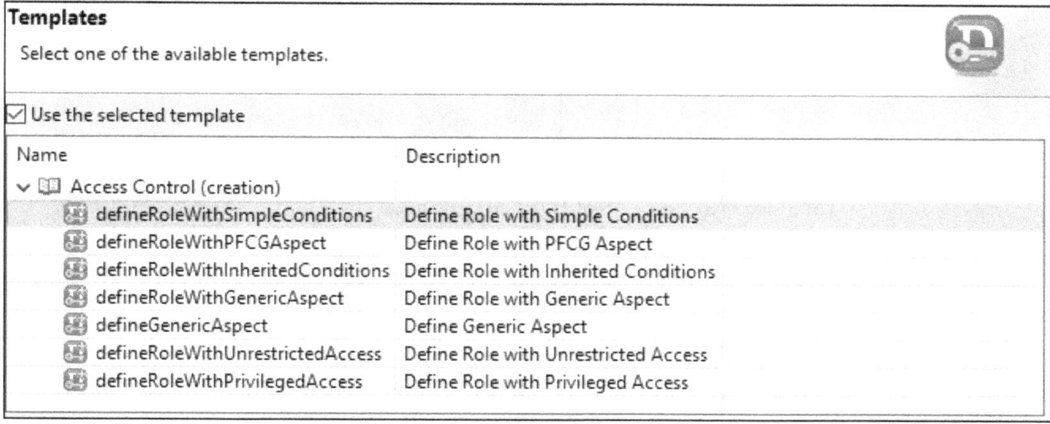

Figure 16.3 Templates for Access Control

These templates are used to provide code modules with keywords and placeholders for various types of access controls. Table 16.1 describes the various templates. When creating an access control, the placeholders in the source code are replaced by the actual values, if possible.

16

Access Control Template	Description
`DefineRoleWith-SimpleConditions`	Defines basic access control with static conditions. This is used if access is restricted due to fixed conditions such as specific field values.
`DefineRoleWith-PFCGAspect`	Defines roles created in Transaction PFCG. Ideal for scenarios that require seamless integration with SAP authorization roles.
`DefineRoleWith-InheritedConditions`	Adopts conditions from another role. Useful for reusing existing access logic to implement a consistent and hierarchical control.
`DefineRoleWith-GenericAspect`	Implements generic access control using user-specific attributes or parameters. Applicable for dynamic access restrictions based on user context or organizational data.
`DefineGenericAspect`	Defines reusable generic aspects for multiple roles. Used to create a common logic to establish consistency across multiple DCL roles.

Table 16.1 Access Control Templates

Access Control Template	Description
DefineRoleWith-UnrestrictedAccess	Allows unrestricted access to the CDS views. Suitable for scenarios in which no restrictions are required, such as for public or open data access.
DefineRoleWith-PrivilegedAccess	Enables privileged access based on special conditions or roles. Used for comprehensive access control, such as for data visibility at the administration level or for audit purposes.

Table 16.1 Access Control Templates (Cont.)

5. For our example, select the DefineRoleWithSimpleConditions template to restrict access to a specific CDS entity. After selecting the template, the access control shown in Listing 16.1 is generated.

```
@EndUserText.label: 'Recipe'
@MappingRole: true
define role ZACB_R_RECIPE {
    grant
        select
            on
                ZACB_R_RECIPE
                    where
                        entity_element_1 =
'literal_value'
                        or entity_element_2 =
aspect user;

}
```

Listing 16.1 Automatically Generated Access Control

The ZACB_R_RECIPE value was automatically added to the code template.

6. Add the implementation logic to the access control, as shown in Listing 16.2.

```
@EndUserText.label: 'Recipe'
@MappingRole: true
define role ZACB_R_RECIPE {
    grant
        select
            on
                ZACB_R_RECIPE
                where
```

```
                        Published = 'X' or
                        CreatedBy = aspect user;
}
```

Listing 16.2 Custom Implementation Logic for Access Control

We've thus implemented an authorization check that takes effect depending on whether the recipe (ZACB_R_RECIPE) has already been published (Published = 'X') or whether the current user has created the recipe (CreatedBy = aspect user).

Activating the access control

However, the access control created in this way won't automatically be used by the associated CDS entity. We still need to adapt the CDS entity accordingly. By default, it contains an annotation called @AccessControl.authorizationCheck. This annotation can have the following values:

- **#NOT_REQUIRED**
  No access control is required. All users have full access.
- **#MANDATORY**
  An access control object must be present.
- **#NOT_ALLOWED**
  There must be no access control object. If present, it will be ignored.
- **#CHECK**
  A warning will be issued if there is no access control object.

We now change the value of the @AccessControl.authorizationCheck annotation of the entity to #CHECK (see Listing 16.3).

```
@AccessControl.authorizationCheck: #CHECK
@EndUserText.label: 'CDS entity Recipe'
define root view entity ZACB_R_Recipe
  as select from zacb_recipe
  composition [0..*] of ZACB_R_Ingredient as _Ingredient
  composition [0..*] of ZACB_R_Review as _Review

{
  key recipe_id as RecipeId,
  recipe_name as RecipeName,
...
```

Listing 16.3 Changing the Value of the @AccessControl.authorizationCheck Annotation to #CHECK

Data preview

If you now call CDS entity ZACB_R_RECIPE via the data preview in the ABAP development tools, you'll see all the data that corresponds to our authorization check (see Figure 16.4). This means that only all published recipes are

16

displayed, which you can recognize by an **x** in the **Published** column. If you've created a recipe yourself, it will also be displayed in unpublished status.

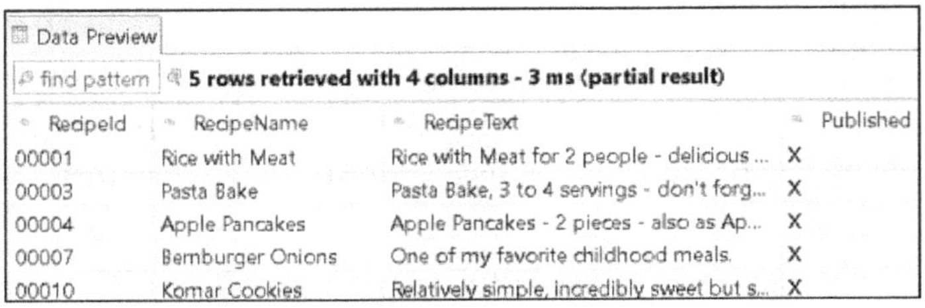

Data Preview

find pattern | **5 rows retrieved with 4 columns - 3 ms (partial result)**

RecipeId	RecipeName	RecipeText	Published
00001	Rice with Meat	Rice with Meat for 2 people - delicious ...	X
00003	Pasta Bake	Pasta Bake, 3 to 4 servings - don't forg...	X
00004	Apple Pancakes	Apple Pancakes - 2 pieces - also as Ap...	X
00007	Bernburger Onions	One of my favorite childhood meals.	X
00010	Komar Cookies	Relatively simple, incredibly sweet but s...	X

Figure 16.4 Display of Published Recipes in the Data Preview

Testing the SAP Fiori app

When you open our SAP Fiori app, however, you'll see that it still displays all recipes (see Figure 16.5).

Recipes (10) Create Delete

Recipe ID	Recipe Name	Recipe Text
1	Rice with Meat	Rice with Meat for 2 people - delicious like mom's
2	Spaghetti Carbonara	Carbonara for 1 to 2 people - only authentic with cream and lots of pepper
3	Pasta Bake	Pasta Bake, 3 to 4 servings - don't forget the nutmeg(!)
4	Apple Pancakes	Apple Pancakes - 2 pieces - also as Apple Kaiserschmarrn
5	Bahmi Goreng	Baked Noodles with Meat
6	Farmer's Breakfast	A Breakfast Classic
7	Bernburger Onions	One of my favorite childhood meals.
8	Simple Cookies	Quick to make - but quite hard
9	Meatballs w/ Potatoe	A dish for when you need something quick - tastes quite good.
10	Komar Cookies	Relatively simple, incredibly sweet but soft

Figure 16.5 Display of All Recipes in the Recipe Portal

Access control at the projection level

For the authorization check to also take effect in the SAP Fiori app, an access control must be created for the projection entity. This access control must contain the logic shown in Listing 16.4.

```
@EndUserText.label: 'Recipe'
@MappingRole: true
define role ZACB_C_RECIPE {
    grant
```

```
        select on ZACB_C_RECIPE
         where inheriting conditions from entity
               ZACB_R_Recipe;
}
```

Listing 16.4 Inheritance of the Access Control

The `inheriting conditions from entity` command is used to inherit the authorization check from the specified root entity. When you now open the app again, the data display will be limited to the published recipes (see Figure 16.6).

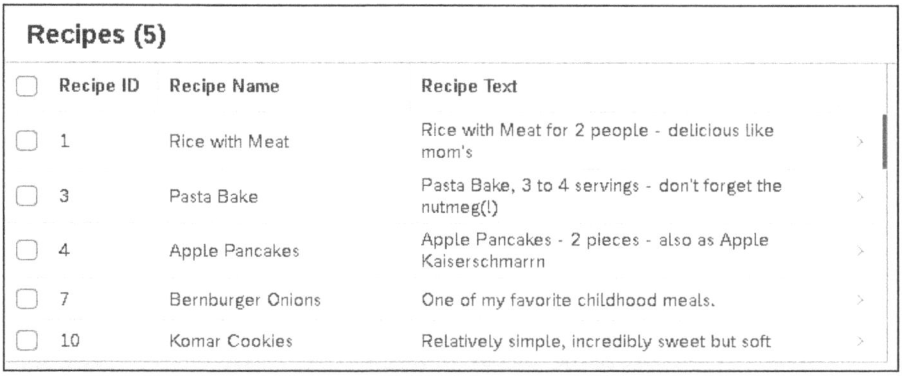

Recipes (5)

	Recipe ID	Recipe Name	Recipe Text
☐	1	Rice with Meat	Rice with Meat for 2 people - delicious like mom's
☐	3	Pasta Bake	Pasta Bake, 3 to 4 servings - don't forget the nutmeg(!)
☐	4	Apple Pancakes	Apple Pancakes - 2 pieces - also as Apple Kaiserschmarrn
☐	7	Bernburger Onions	One of my favorite childhood meals.
☐	10	Komar Cookies	Relatively simple, incredibly sweet but soft

Figure 16.6 Display of Published Recipes in the Recipe Portal

16.2 Authorization Checks for Change Operations

Authorization methods

In ABAP RESTful programming application model business objects, change processes can be checked for unauthorized access at runtime. For this purpose, specific authorization implementation options are available. The authorizations are defined in the behavior definition and implemented in the corresponding *authorization methods*. These special authorization methods are called at runtime at a specific time prior to the actual change operation. This allows you to prevent change operations via access control before the actual change operation gets executed in the transaction buffer.

The following authorization methods can be used:

- `GLOBAL AUTHORIZATION`
- `INSTANCE AUTHORIZATION`
- `PRECHECK`

In this section, you'll get to know each of these authorization methods. We explain their use in each case using our example.

16.2.1 Global Authorizations

The GLOBAL AUTHORIZATION authorization method

The GLOBALAUTHORIZATION authorization method is used to check whether a user is generally permitted to perform an operation. To enable global authorization checks in a ABAP RESTful programming application model application, you need to add the keyword global in the behavior definition. Listing 16.5 shows this for our ZBP_ACB_R_RECIPE behavior definition.

```
managed implementation in class zbp_acb_r_recipe unique;
strict ( 2 );
with draft;

define behavior for ZACB_R_Recipe alias Recipe
persistent table zacb_recipe
...
authorization master ( global )
early numbering
{

  field ( readoOnly )
  CreatedAt,
...
```

Listing 16.5 Addition of the "global" Keyword

Method in behavior implementation

Next, you need to add the get_global_authorization method in the behavior implementation class. You can use the quick fix function to automatically generate the method declaration in the behavior implementation class. Then, the method declaration will be generated, as shown in Listing 16.6.

```
METHODS:
      get_global_authorizations FOR GLOBAL AUTHORIZATION
        IMPORTING
        REQUEST requested_authorizations FOR Recipe
        RESULT result.
```

Listing 16.6 Automatic Generation of the get_global_authorization Method Declaration

The importing requested_authorizations parameter describes the operation for which an authorization check is to be carried out. Table 16.2 lists the possible values for this parameter.

Operation	Description
%create	Create
%update	Update
%delete	Delete
%action-Edit	Edit

Table 16.2 Possible Operations to Be Authorized

As shown in Listing 16.7, you must implement an authorization logic in the get_global_authorization method.

Implementing the authorization logic

```
METHOD get_global_authorizations.
  DATA(myself) =
    cl_abap_context_info=>get_user_technical_name( ).

  SELECT FROM zacb_user
  FIELDS COUNT( * )
  WHERE username = @myself.

  IF sy-dbcnt <> 0.
    result-%create = if_abap_behv=>auth-allowed.
    result-%update = if_abap_behv=>auth-allowed.
    result-%action-Edit = if_abap_behv=>auth-allowed.
    result-%delete = if_abap_behv=>auth-allowed.
  ELSE.
    result-%create = if_abap_behv=>auth-unauthorized.
    result-%update = if_abap_behv=>auth-unauthorized.
    result-%action-Edit = if_abap_behv=>auth-unauthorized.
    result-%delete = if_abap_behv=>auth-unauthorized.
  ENDIF.
ENDMETHOD.
```

Listing 16.7 Authorization Logic in the get_global_authorization Method

For our sample application, we check whether the current user (cl_abap_context_info=>get_user_technical_name( )) is registered as such in our recipe portal. If that isn't the case, the user must not perform any operation. This is ensured by the assignment in the return value, such as result-%create = if_abap_behv=> auth-unauthorized. The operations in the result return value correspond to those in the requested_authorization import parameter (see Table 16.2). Table 16.3 lists the possible constant values for the authorization method.

Constant	Description
if_abap_behv=>auth-unauthorized	The operation must not be performed.
if_abap_behv=>auth-allowed	The operation may be performed.

Table 16.3 Constants for the Authorization Method

Passing on authorization checks to child entities

You should also add the authorization dependent by command to all associated entities (see Listing 16.8). This command is used to delegate the authorization checks from the root entity.

```
define behavior for ZACB_R_Ingredient alias Ingredient
persistent table zacb_ingredient
with additional save
lock dependent by _Recipe
authorization dependent by _Recipe
draft table zacb_ingredien_d

//etag master <field_name>
{
```

Listing 16.8 Addition of the "authorization dependent by" Keyword to the CDS Entities

Testing the application

If you now start our SAP Fiori app, you can see what access looks like if there's no authorization to create a recipe or if there is such authorization. To do this, click on the **Create** button (shown earlier in Figure 16.5). If there is no authorization, a popup window appears with a corresponding message (see Figure 16.7). You'll be informed that you're not authorized to perform this operation.

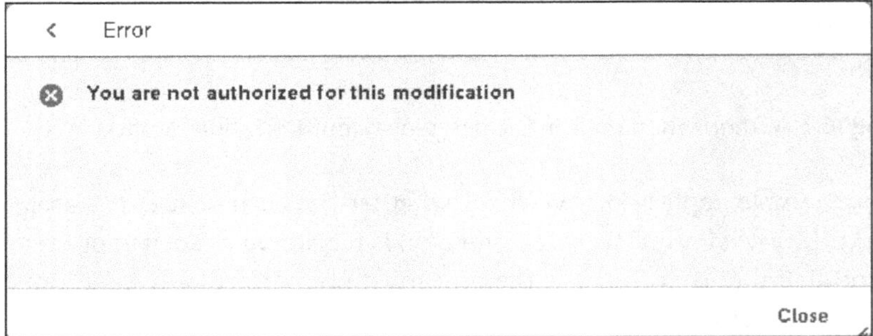

Figure 16.7 Popup Window If Authorization for the Create Operation Is Missing

If you have the authorization required, you can create a recipe for the recipe portal as described in Chapter 2, Section 2.5.

16.2.2 Instance-Dependent Authorizations

The INSTANCE AUTHORIZATION method

The INSTANCE AUTHORIZATION method is used for authorization checks that determine which operation is possible depending on the respective instance. In our example, we implement the INSTANCE AUTHORIZATION method in such a way that it checks whether the delete operation is possible for the current user, depending on the recipe in question. To do so, you must perform the steps described here.

Behavior definition

In the first step, add the instance addition to the authorization master command in the behavior definition (see Listing 16.9).

```
managed implementation in class zbp_acb_r_recipe unique;
strict ( 2 );
with draft;

define behavior for ZACB_R_Recipe alias Recipe
persistent table zacb_recipe
...
authorization master( global , instance )
early numbering
{
```

Listing 16.9 Addition of the "instance" Keyword to the Behavior Definition

As with the GLOBAL AUTHORIZATION method, the editor then informs you that a method hasn't yet been implemented. In this case, it's the method for the INSTANCE_AUTHORIZATION operation (see Figure 16.8).

```
Operation INSTANCE_AUTHORIZATION for ZACB_R_RECIPE not implemented
persistent table zacb_recipe
draft table zacb_recipe_d
```

Figure 16.8 The INSTANCE_AUTHORIZATION Quick Fix Option

Using a quick fix, the get_instance_authorization method can then be generated automatically in the behavior implementation class. Listing 16.10 shows the result.

```
METHODS:
get_instance_authorizations FOR INSTANCE AUTHORIZATION
  IMPORTING keys REQUEST requested_authorizations
  FOR Recipe RESULT result.
```

Listing 16.10 Automatic Generation of the get_instance_authorizations Method Declaration

Implementing the authorization logic

This method has the same parameters as the get_global_authorization method (see Table 16.2). You can now add your individual authorization logic to the method. Listing 16.11 shows this for our example.

```
METHOD get_instance_authorizations.
  READ ENTITIES OF zacb_r_recipe IN LOCAL MODE
       ENTITY recipe
       FIELDS ( RecipeId CreatedBy )
       WITH CORRESPONDING #( keys )
       RESULT DATA(recipes)
       FAILED failed.

  DATA(myself) =
    cl_abap_context_info=>get_user_technical_name( ).

  LOOP AT recipes INTO DATA(recipe).
    IF myself <> recipe-CreatedBy.
      APPEND VALUE #(
              %tky    = recipe-%tky
              %delete = if_abap_behv=>auth-unauthorized
                    ) TO result.
    ENDIF.
  ENDLOOP.
ENDMETHOD.
```

Listing 16.11 Implemented Authorization Logic for INSTANCE AUTHORIZATION

We implement a logic that checks whether the logged-in user is assigned to the selected recipe as an author. If that isn't the case, the user isn't authorized for the delete operation.

Testing the application

Now, start the SAP Fiori app again to test this logic. Select a recipe. On the details page, the author can now delete the recipe using the **Delete** button. If you're not the author of the recipe, you won't be able to delete it, and the corresponding button will be grayed out (see Figure 16.9).

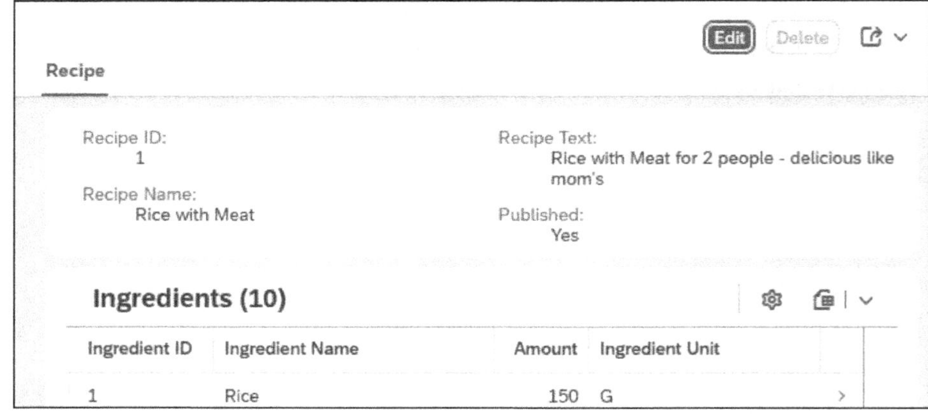

Figure 16.9 The User Can't Delete the Recipe

16.2.3 Authorization Precheck

The PRECHECK authorization method

Using the third authorization method—PRECHECK—we can check incoming values and reject incoming requests before the data reaches the transaction buffer. This preliminary check is called at runtime prior to the assigned change operation and removes all entries from the change request for which the condition in the precheck isn't fulfilled. The steps discussed next are necessary to implement such a precheck.

Behavior definition

If a precheck is to be implemented for the operation, the precheck command must first be added to the behavior implementation (see Listing 16.12).

```
managed implementation in class zbp_acb_r_recipe unique;
...
define behavior for ZACB_R_Recipe alias Recipe
persistent table zacb_recipe
...
{
...
  create;
  update(precheck);
  delete;
...
```

Listing 16.12 Addition of the PRECHECK Authorization Option to the Behavior Definition

As with the other checks, you're then informed that the corresponding UPDATE method hasn't yet been implemented (see Figure 16.10).

```
27   create;
28   Precheck for UPDATE ZACB_R_RECIPE is not implemented
29   delete;
```

Figure 16.10 Quick Fix Message for the PRECHECK Operation

The precheck_update method implementation is generated automatically using the quick fix function:

```
METHODS:
precheck_update FOR PRECHECK
 IMPORTING entities FOR UPDATE Recipe.
```

Implementing the authorization logic

It still needs to be supplemented by an implementation logic. This implementation includes a check as to whether the current user is the author of the recipe. If that isn't the case, the user will receive a message that the recipe can't be changed (see Listing 16.13).

```
METHOD precheck_update.
  DATA(myself) =
    cl_abap_context_info=>get_user_technical_name( ).
  LOOP AT entities INTO DATA(entity).

    READ ENTITIES OF zacb_r_recipe IN LOCAL MODE
    ENTITY Recipe
    FIELDS ( CreatedBy ) WITH VALUE #( (
                                        %key = entity-%key
                                      ) )
    RESULT DATA(recipes).
    LOOP AT recipes INTO DATA(recipe).

      IF myself <> recipe-CreatedBy.
        APPEND VALUE #( %tky =  entity-%tky )
         TO failed-ingredient.

        APPEND VALUE #(
                      %tky = entity-%tky
                      %msg = new_message_with_text(
           severity = if_abap_behv_message=>severity-error
           text = 'You cannot add any ingredients
                      here...'
             ) ) TO reported-ingredient.
      ENDIF.
    ENDLOOP.
  ENDLOOP.
ENDMETHOD.
```

Listing 16.13 Implementation of the precheck_update Method

In our example, the corresponding message is displayed in a popup window (see Figure 16.11).

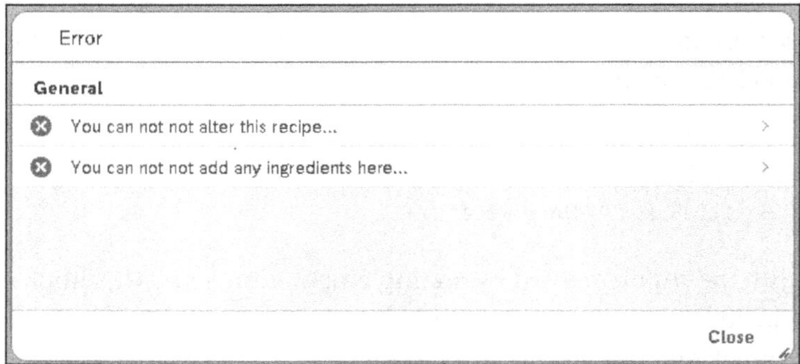

Figure 16.11 Popup Window with the Message That No Change Is Possible

16.3 Summary

In this chapter, you've learned how to handle authorizations in ABAP Cloud and how to implement authorizations for read operations. In this context, you've become familiar with the access controls for CDS entities. There are three different authorization options for change operations in ABAP RESTful programming application model applications:

- `GLOBAL`
  You can check for change operations that are to be made available globally.
- `INSTANCE`
  Depending on the respective instance, the system checks whether a change operation is possible.
- `PRECHECK`
  Prior to saving, you can check whether a change option is possible.

We've shown you these three authorization options using our recipe portal.

Chapter 17
Using APIs

In ABAP Cloud, the use of released application programming interfaces is an elementary component of application development. You must be able to find these APIs, evaluate them, and use them in a smart way. This chapter teaches methods for doing this.

Application programming interfaces (APIs) are defined points that allow an application to be used programmatically from outside. These APIs make it possible to develop stable software from the provider's point of view, and the implementation details can be subsequently changed without affecting the components used. On the user's side, external components can be integrated without the risk of having to invest a lot of time and effort for modifications in the case of updates. These are static components such as data types or constants as well as procedures that influence the program execution.

API state and release contracts

Traditionally, the ABAP platform has the release indicator for function modules or Business Application Programming Interfaces (BAPIs) that take on the role just described. Such a separation between private and public objects of a component can also be made possible via package interfaces. In ABAP Cloud, there's a new concept for this, which has already been briefly described in Chapter 1, Section 1.4: the *API state* and the *release contracts.* Both can be used to classify development objects as those that support external use.

APIs include executable code and procedures as well as type definitions and database access. The correct use of objects is checked syntactically on the user's side when using the ABAP language version ABAP for Cloud Development. On the provider's side, actions that would constitute an incompatible change are prevented.

APIs are therefore particularly relevant in ABAP Cloud because nonreleased objects from other software components are prevented from being used by the syntax check. So, in SAP S/4HANA Cloud Public Edition and in the SAP BTP ABAP environment, there's no way around APIs. In SAP S/4HANA and SAP S/4HANA Cloud Private Edition, the clean core concept accelerates their use. If you stick to using only released objects, the next upgrade will be much easier and you'll also minimize the dependencies of your software so that it can also be ported to cloud systems.

Structure of this chapter

In this chapter, you'll learn how to use APIs. In Section 17.1, you'll learn how to find APIs in the first place and which technical options can support you in your search. Section 17.2 deals specifically with calling APIs based on the ABAP RESTful application programming model. They are the successor to BAPIs for calling functions in the standard SAP system.

[»]

Sample Application

From this chapter onward, we move away from our recipe portal and refer to objects in the standard SAP system. Not all examples are directly reproducible in the SAP BTP ABAP environment, as no SAP ERP components are installed there.

17.1 Finding the Right APIs

Various search options are available to help you find the right API for your application, which we'll discuss in this section.

17.1.1 Successor Objects

Successor objects in syntax errors

The first method of finding a suitable API is the one that you might intuitively come into contact with anyway. You're used to using a familiar development object from the standard SAP system, which you're now prohibited from using in ABAP Cloud due to a syntax error. The object hasn't been released for use in ABAP Cloud. Depending on the specific object, the error message can directly suggest an alternative that you can use.

In Figure 17.1, an attempt was made to access database table VBAK using ABAP SQL.

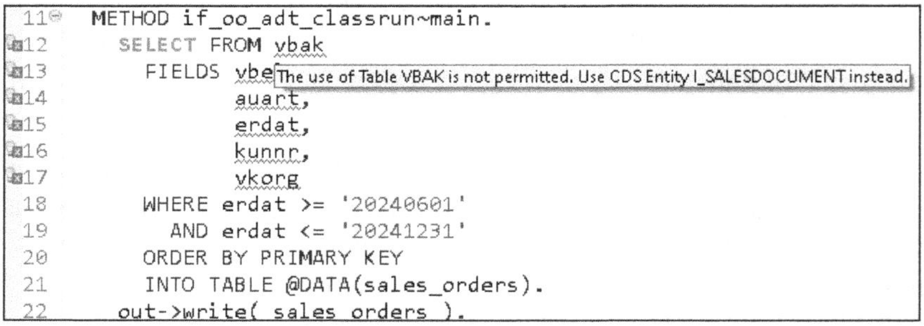

```
METHOD if_oo_adt_classrun~main.
  SELECT FROM vbak
    FIELDS vbe
           auart,
           erdat,
           kunnr,
           vkorg
    WHERE erdat >= '20240601'
      AND erdat <= '20241231'
    ORDER BY PRIMARY KEY
    INTO TABLE @DATA(sales_orders).
  out->write( sales_orders ).
```

The use of Table VBAK is not permitted. Use CDS Entity I_SALESDOCUMENT instead.

Figure 17.1 Syntax Error When Using Table VBAK

The code can't be activated because the ABAP language version option selected, **ABAP for Cloud Development**, forces the check for APIs and database table VBAK hasn't been released. In general, no standard SAP database tables have been released in ABAP Cloud because write access would automatically be possible by releasing the tables.

The I_SalesDocument CDS entity

The error message indicates that access to CDS entity I_SalesDocument is recommended instead of access to table VBAK. I_SalesDocument has been released and enables indirect read access to table VBAK. The code can be easily adapted. The CDS entity is part of the virtual data model (VDM) of SAP S/4HANA and therefore uses the standardized English identifiers for the fields. The corrected code is shown in Listing 17.1.

```
SELECT FROM I_SalesDocument
  FIELDS SalesDocument,
         SalesDocumentType,
         CreationDate,
         SoldToParty,
         SalesOrganization
  WHERE CreationDate >= '20240601'
    AND CreationDate <= '20241231'
  ORDER BY PRIMARY KEY
  INTO TABLE @DATA(sales_orders).
out->write( sales_orders ).
```

Listing 17.1 Indirect Access to Database Table VBAK via CDS Entity I_SalesDocument

Maintaining successor objects

For the system to propose the CDS entity, the relationship between table VBAK and CDS entity I_SalesDocument had to be maintained in advance by SAP. The CDS entity is the *successor object*. To display it, open the development object that hasn't been released, for example, using the keyboard shortcut [Ctrl]+[Shift]+[A]. Then, go to the **Properties** view, and open the **API State** area. If the relationship to the successor object has been maintained, you'll find a release contract for the object with release status **Not to Be Released**. Instead of the release, the successor object is maintained under **Successors** (see Figure 17.2).

Overview of successor objects in the system

Alternatively, you can also perform a table-based search through the successor objects. You can use the released I_APIsWithCloudDevSuccessor CDS entity directly in the system. This view entity selects all development objects with a successor object. You can open it using the keyboard shortcut [Ctrl]+[Shift]+[A] and then open the data preview using the [F8] function key. In Figure 17.3, we searched for objects that start with V and have a successor object. The link between table `VBAK` and CDS entity `I_SalesDocument` is included here alongside other hits.

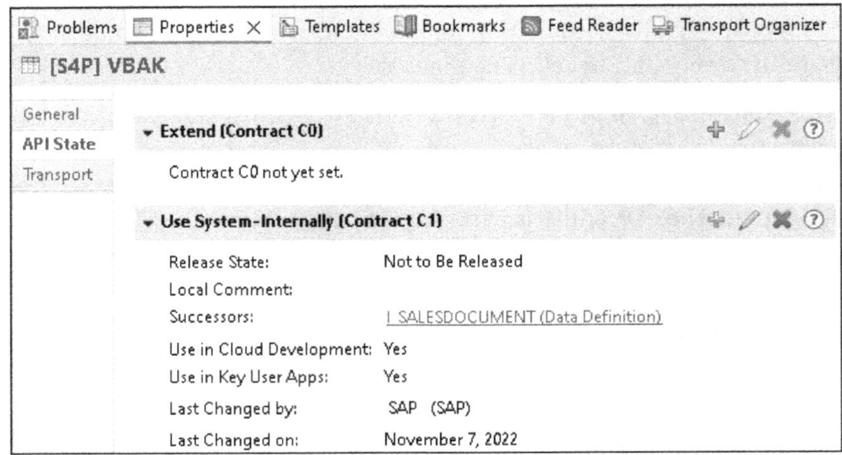

Figure 17.2 Maintaining the Successor Object

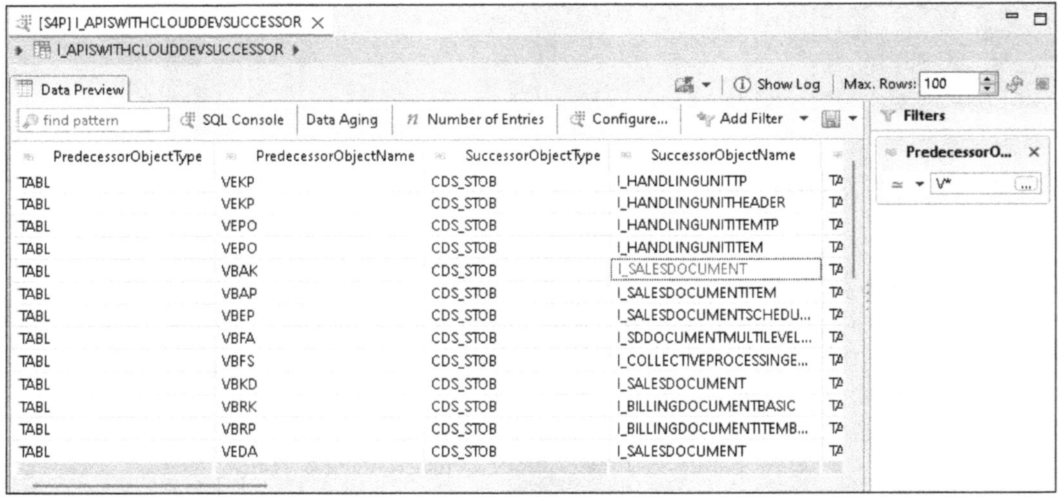

PredecessorObjectType	PredecessorObjectName	SuccessorObjectType	SuccessorObjectName
TABL	VEKP	CDS_STOB	I_HANDLINGUNITTP
TABL	VEKP	CDS_STOB	I_HANDLINGUNITHEADER
TABL	VEPO	CDS_STOB	I_HANDLINGUNITITEMTP
TABL	VEPO	CDS_STOB	I_HANDLINGUNITITEM
TABL	VBAK	CDS_STOB	I_SALESDOCUMENT
TABL	VBAP	CDS_STOB	I_SALESDOCUMENTITEM
TABL	VBEP	CDS_STOB	I_SALESDOCUMENTSCHEDU...
TABL	VBFA	CDS_STOB	I_SDDOCUMENTMULTILEVEL...
TABL	VBFS	CDS_STOB	I_COLLECTIVEPROCESSINGE...
TABL	VBKD	CDS_STOB	I_SALESDOCUMENT
TABL	VBRK	CDS_STOB	I_BILLINGDOCUMENTBASIC
TABL	VBRP	CDS_STOB	I_BILLINGDOCUMENTITEMB...
TABL	VEDA	CDS_STOB	I_SALESDOCUMENT

Figure 17.3 Table-Based Display of Successor Objects with CDS View I_APIsWithCloudDevSuccessor

17.1.2 Searching via CDS Entity I_APIsForCloudDevelopment

Overview of released APIs per CDS entity

Another use case is the search for APIs without a predecessor object that you already know. There are many different options here. One is the use of the released CDS entity `I_APIsForCloudDevelopment`. Not only does it show you successor objects but also all APIs in the system that can be used in

ABAP Cloud. In Figure 17.4, the data preview was searched for released global classes that use the ABAP string in the identifier.

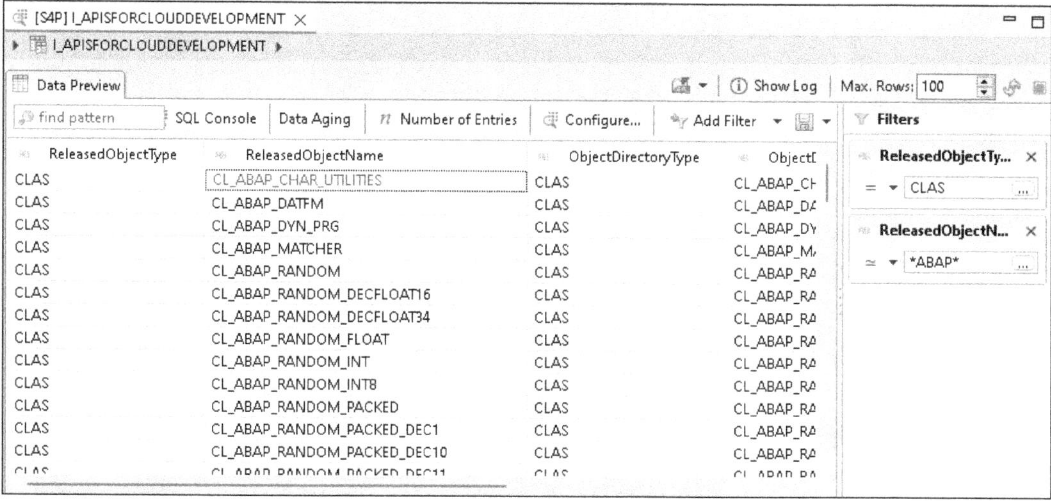

Figure 17.4 Table-Based Display of All APIs with CDS View I_APIsForCloud-Development

This CDS entity is very useful for an initial overview or program-based analysis of APIs. However, it's not suitable for a detailed search. More convenient options are available in the system for this purpose.

17.1.3 Searching via the Open ABAP Development Object Dialog Box

Open ABAP Development Object

In the **Open ABAP Development Object** dialog box, it's possible to search only for released objects. Open this dialog box using the keyboard shortcut `Ctrl`+`Shift`+`A`, and add the "api:" filter next to your usual search criteria. The autocomplete section, which you can call via the shortcut `Ctrl`+`space bar`, provides numerous options for filtering based on the API state (see Figure 17.5).

You could use the `HAS_CLOUD_DVLPMNT_SUCCESSOR` filter, for example, to filter for all objects for which a successor object has been maintained. In general, you'll probably search specifically for the objects you can use in ABAP Cloud. To do this, you should use the `USE_IN_CLOUD_DEVELOPMENT` filter. Alternatively, `USE_IN_CLOUD_DVLPMNT_ACTIVE` can also be used to filter based on the activation status of API catalogs. For extensible objects, you should use the `EXTEND_IN_CLOUD_DEVELOPMENT` filter. In Figure 17.6, we've searched for global classes that can be used in ABAP Cloud and have the name component `XML`.

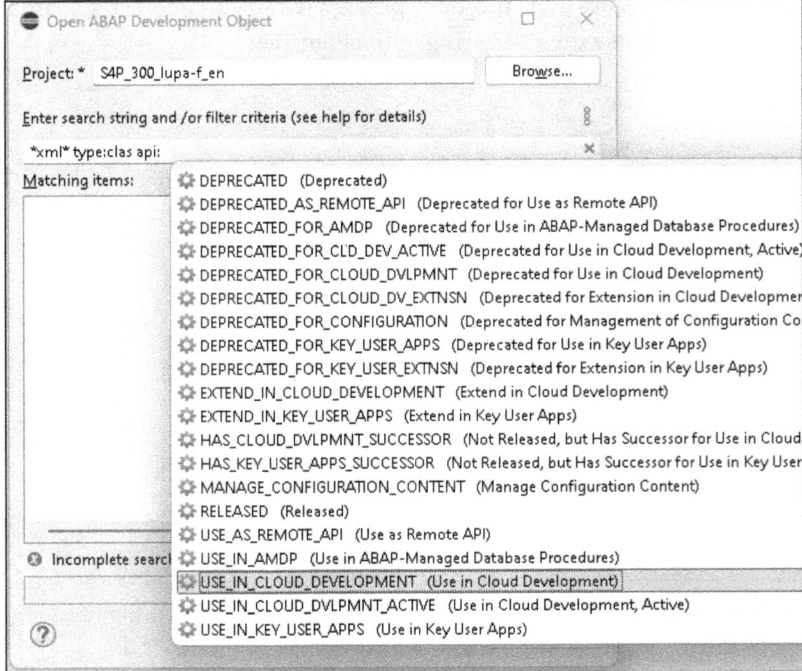

Figure 17.5 API Filter in the Open ABAP Development Object Dialog Box

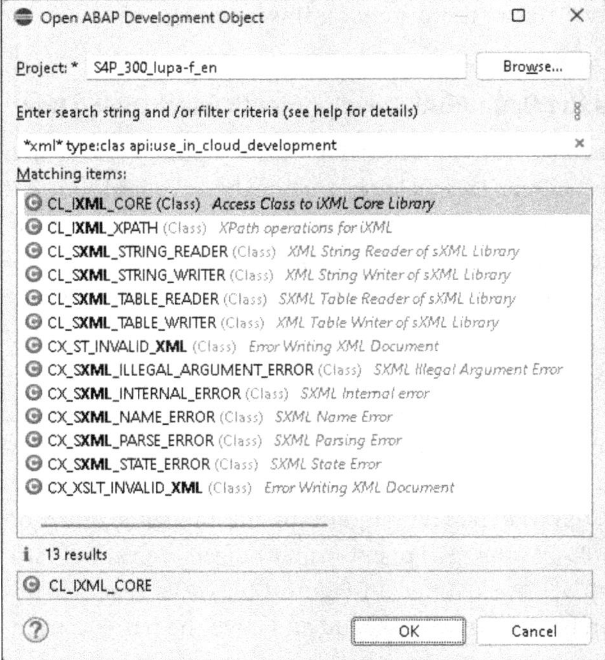

Figure 17.6 Hit List in the Open ABAP Development Object Dialog Box

One disadvantage of the **Open ABAP Development Object** dialog box is that you have to open it again if your selected entry wasn't the right one after all. You can solve this problem by activating the **Use Pattern from Previous Search** option, so that at least the search pattern doesn't have to be reentered each time (see Figure 17.7).

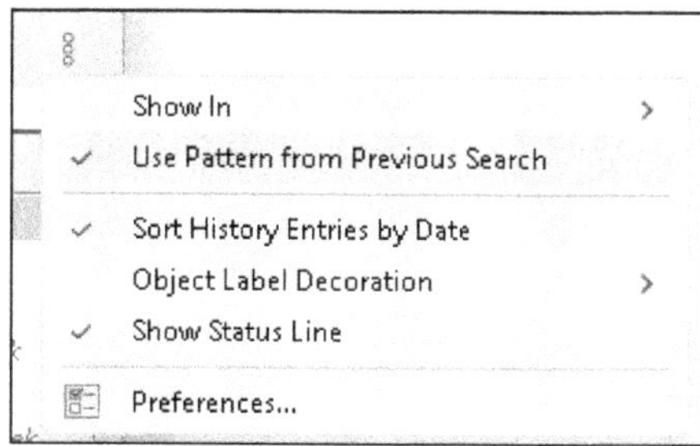

Figure 17.7 Reusing the Search Pattern

17.1.4 Searching via ABAP Object Search

ABAP object search

Another approach is the use of *ABAP object search*. You can use the shortcut [Ctrl]+[H] to access a search dialog similar to the one described in the preceding section (see Figure 17.8).

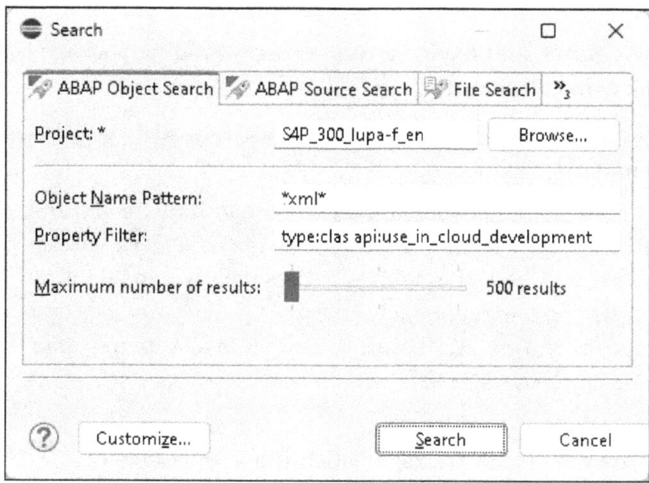

Figure 17.8 ABAP Object Search with API Filter

However, the results aren't displayed in a dialog box here, but as a separate view in the ABAP development tools, so that you can easily go through the hits one after the other (see Figure 17.9). You can display additional columns to help you with your search.

Problems | Properties | Templates | Bookmarks | Feed Reader | Transport ... | Search × | Console | Dictionary ...

[S4P] ABAP Object Search: '*xml* type:clas api:use_in_cloud_development' - 13 objects found

Object	Description	Package
CL_IXML_CORE	Access Class to iXML Core Library	SIXML_CORE
CL_IXML_XPATH	XPath operations for iXML	SIXML_XPATH
CL_SXML_STRING_READER	XML String Reader of sXML Library	SXML
CL_SXML_STRING_WRITER	XML String Writer of sXML Library	SXML
CL_SXML_TABLE_READER	SXML Table Reader of sXML Library	SXML
CL_SXML_TABLE_WRITER	XML Table Writer of sXML Library	SXML
CX_ST_INVALID_XML	Error Writing XML Document	S_ABAP_EXCEPTIONS
CX_SXML_ILLEGAL_ARGUMENT_ERROR	SXML Illegal Argument Error	SXML
CX_SXML_INTERNAL_ERROR	SXML Internal error	SXML
CX_SXML_NAME_ERROR	SXML Name Error	SXML
CX_SXML_PARSE_ERROR	SXML Parsing Error	SXML
CX_SXML_STATE_ERROR	SXML State Error	SXML
CX_XSLT_INVALID_XML	Error Writing XML Document	S_ABAP_EXCEPTIONS

Figure 17.9 Hit List of the ABAP Object Search

17.1.5 Grouping and Filtering in Project Explorer

Expand Folder By option in the project explorer

Another option is grouping and filtering in the project explorer. Using the techniques described previously, you can find CDS entity `I_SalesDocument`, for example. Now, you might be wondering what other CDS entities can be used in ABAP Cloud in sales, assuming that these entities are in the same package or in surrounding packages in the package hierarchy. In this case, follow these steps:

1. Open the known object, and navigate to the corresponding package by clicking the **Link with Editor** button.

System Library in Project Explorer

You first need to make sure that you have an ABAP repository tree for the **system library** in your ABAP project. It shows you all the packages available in the system in a hierarchy. If this subtree is missing, you can add it by right-clicking to open the context menu for the ABAP project and selecting **New • ABAP Repository Tree**. Make sure that the **Take this tree into account for linking** option is selected.

You're now in the `VDM_SD_SLS` package which is a subpackage of `VDM_SD` and `APPL` (see Figure 17.10).

2. You can load all the contents of the package by pressing the F5 function key. However, not all content can be used in ABAP Cloud. The subpackage alone contains 1,865 objects. To display the objects grouped by API stat, right-click on the desired package, and select the **Expand Folder by** option from the context menu (see Figure 17.11).

System Library <press F5 to load full content>
APPL <press F5 to load full content> *Structure Package: Logistics/Accounting*
VA *Application development R/3 Sales*
VDM_SD <press F5 to load full content> *Main Package Virtual Data Model for SD*
VDM_SD_ANALYTICS *VDM SD Analytical Views*
VDM_SD_BF *VDM SD Basic Functions*
VDM_SD_BIL *VDM SD Billing*
VDM_SD_MD *VDM SD Master Data*
VDM_SD_SLS <press F5 to load full content> *VDM SD Sales*
Core Data Services <press F5 to load full content>
Data Definitions <press F5 to load full content>
I_SALESDOCUMENT *Sales Document*

Figure 17.10 Project Explorer After Clicking Link with Editor

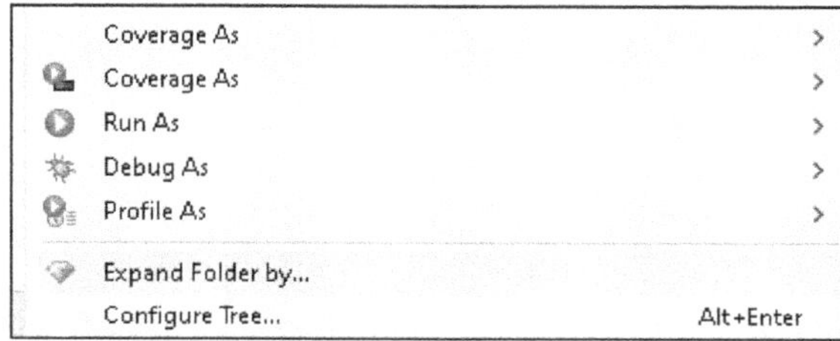

Figure 17.11 Project Explorer: Expand Folder By

3. Select the **API State** option (see Figure 17.12).

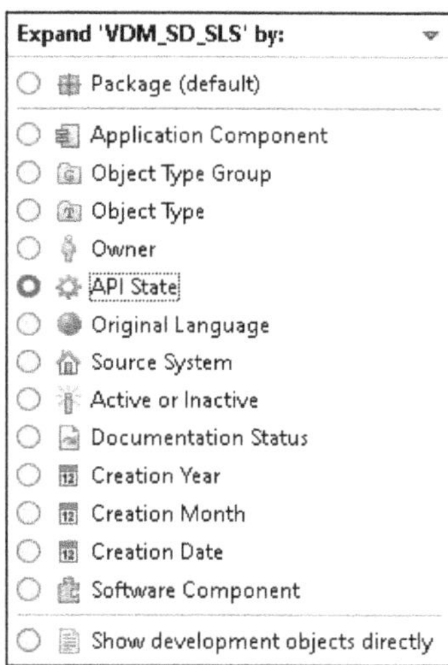

Figure 17.12 Expand Folder By: Grouping Options

4. All development objects in the package are displayed, grouped by API state. You can now find the objects that can only be used in ABAP Cloud directly by expanding the `USE_IN_CLOUD_DEVELOPMENT` subtree (see Figure 17.13). Of the 1,865 properties, 190 remain.

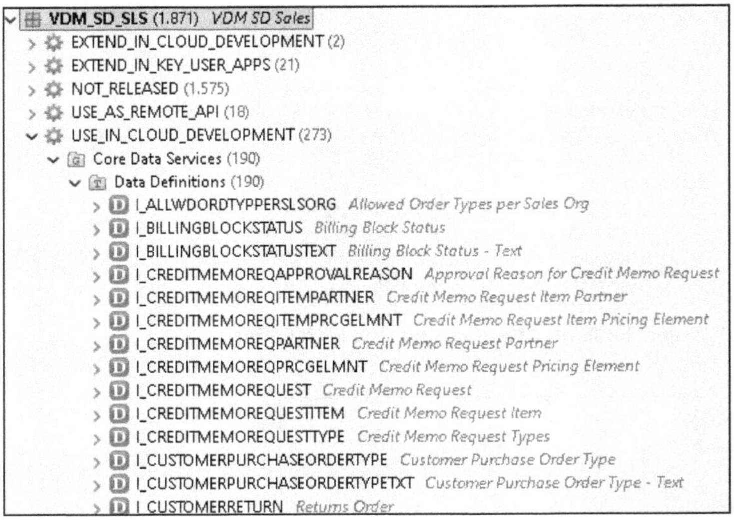

Figure 17.13 Package: Grouped by API State

ABAP repository trees

The **Expand Folder by** function is very useful for changing the display for a package on an ad hoc basis. If, on the other hand, you want to enter the project explorer directly with the aim of finding APIs, you can use *ABAP repository trees*. They allow you to define your own navigation structure in the project explorer. To find APIs, for example, it often makes less sense to take a technical approach from the perspective of packages and more from the perspective of the application components.

To create a new ABAP repository tree, follow these steps:

1. Right-click on your ABAP project in the project explorer, and select **New • ABAP Repository Tree** from the context menu.
2. To navigate by application components, select the **Application Component Hierarchy** template and confirm by clicking the **Next** button (see Figure 17.14).

Templates for Released and Obsolete Objects

Alternatively, by using the **Released Objects** and **Deprecated Objects** templates, you can directly use a template that sets filters for the API state. However, these objects aren't grouped by application component by default, so we choose the other template as a starting point.

3. To configure your new ABAP repository tree, you first need to enter "Released Objects by Application Component" in the **Tree Name** field and then adjust the **Property Filter** field so that only objects that can be used in ABAP Cloud are listed.

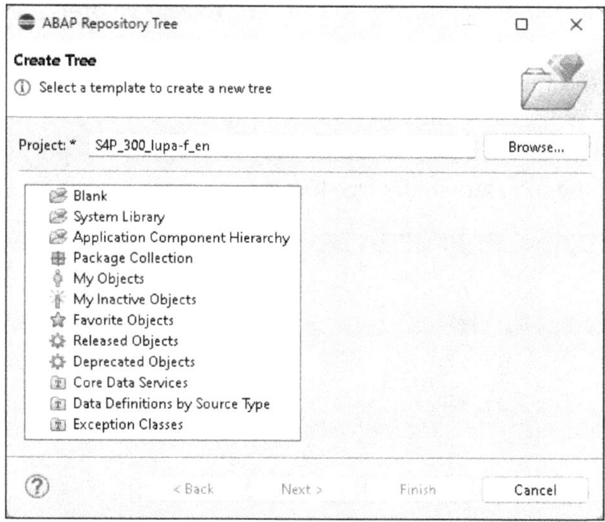

Figure 17.14 Selecting the Template for an ABAP Repository Tree

4. The default division by packages below the application component isn't relevant to us, so we remove the **Package** level from the **Selected Tree Levels**. The modification options are shown in Figure 17.15.

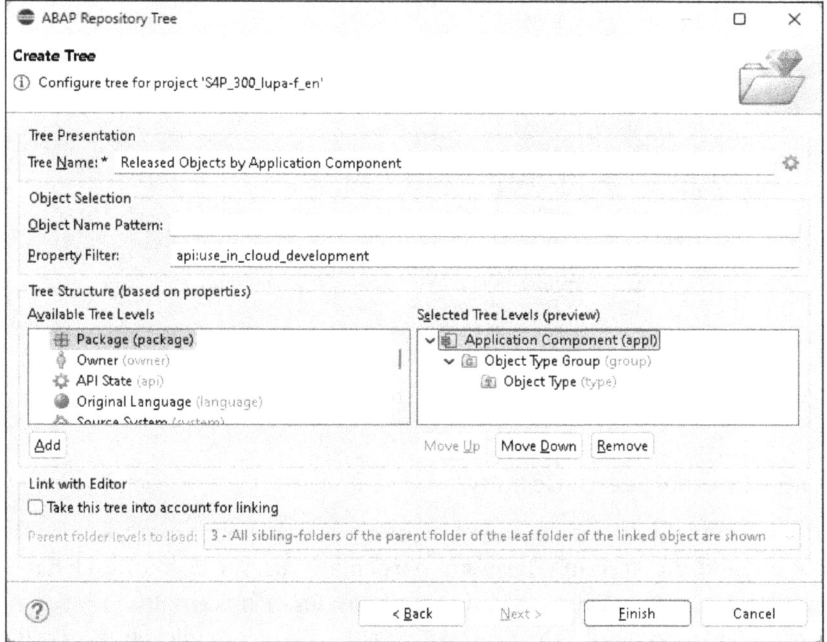

Figure 17.15 Options for Configuring an ABAP Repository Tree

5. Confirm your entries by clicking the **Finish** button.

Using the ABAP repository trees

You'll then find your new ABAP repository tree in the project explorer (see Figure 17.16). You can use it to find released objects using the functional point of entry, that is, the application component. Because ABAP repository trees are very flexible, they can be modified easily. If at this point you also want to see objects that haven't been released, you can remove the property filter and include the API state in the tree levels.

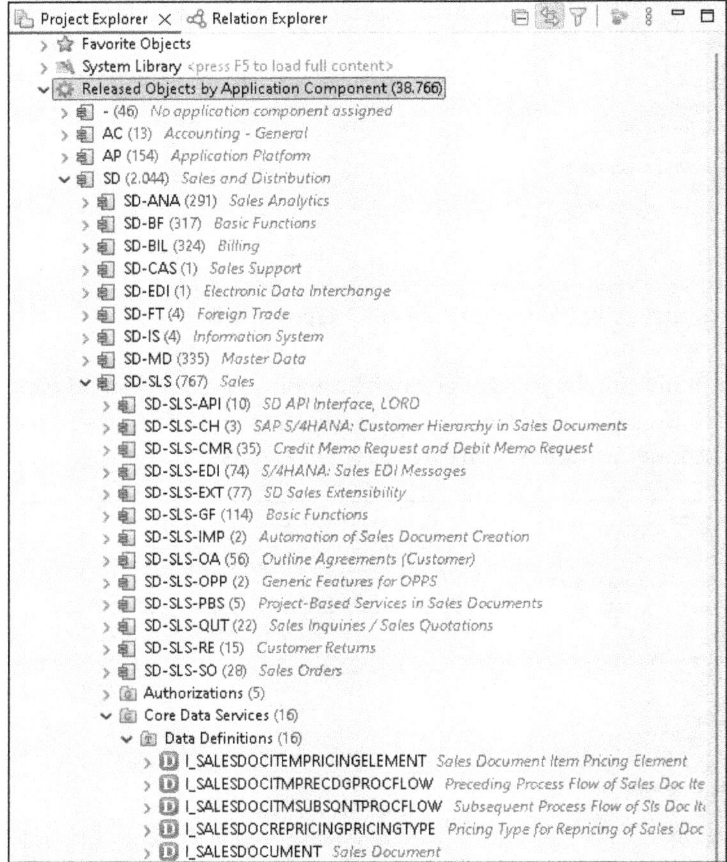

Figure 17.16 New ABAP Repository Tree in the Project Explorer

17.1.6 External Search Options

Cloudification repository

In addition to the options in the SAP system, search methods are also available outside the system. These are particularly useful if you don't have access to a specific release or want to consume an API externally. The *cloudification repository* published by SAP on GitHub (*https://github.com/SAP/abap-atc-cr-cv-s4hc*) contains data on released objects and successor objects in JavaScript Object Notation (JSON) and CSV format. This means you also have access to this information outside a system.

Cloudification repository viewer

There are several options available for displaying the data. With the *cloudification repository viewer*, SAP provides a web interface in which you can search for objects. For our example, we're looking for table `VBAK`. The hit list shows that the object of the `TABL` type hasn't been released, but that there's a successor object: `I_SalesDocument` (see Figure 17.17). This object gives us the same information as we already receive via other channels directly in the system—without having system access. You can access the cloudification repository viewer via the links in the README file of the cloudification repository or directly via the following URL: *https://sap.github.io/abap-atc-cr-cv-s4hc/*.

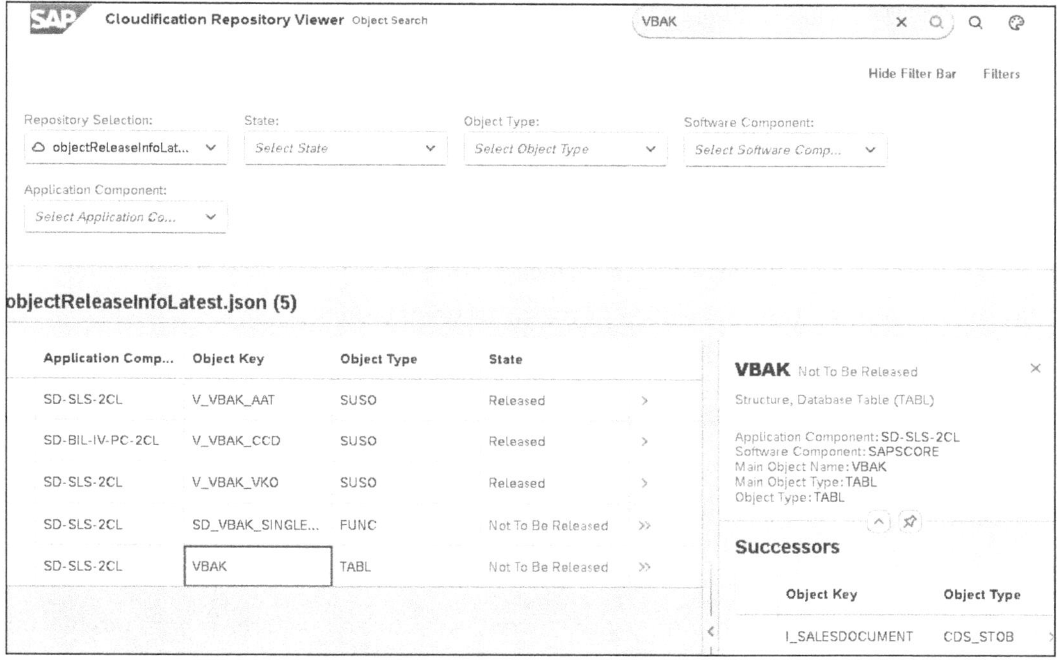

Figure 17.17 Searching for Table VBAK in the Cloudification Repository Viewer

Data sources in the cloudification repository

In the view, you can change the data source via the **Repository Selection** dropdown field:

- The *objectReleaseInfoLatest.json* data source is selected by default. It contains data from release contracts in SAP S/4HANA Cloud Public Edition. This includes all API states. In addition to released objects, this also includes nonreleased objects, such as database table `VBAK` with the successor relationship in our example.
- The *objectReleaseInfo_PCELatest.json* data source contains data on SAP S/4HANA Cloud Private Edition. In addition to **Latest**, you can also select a specific release here.

- The *objectClassifications.json* data source shows you data on *classic APIs*. These are objects that haven't been released for use in ABAP Cloud, but are classified as stable by SAP so that you can use these objects in SAP S/4HANA and SAP S/4HANA Cloud Private Edition if there's no alternative. However, this isn't technically possible with ABAP Cloud directly. You have to develop a wrapper with classic ABAP, encapsulate it using the three-tier model, and release it as an API yourself. You can't use classic APIs in SAP S/4HANA Cloud Public Edition and in the SAP BTP ABAP environment.

Other evaluation options

Other options for displaying APIs can also be found in the following places. These are very similar in terms of content, so they're not discussed in detail here:

- **Cloudification repository viewer from Software-Heroes**
  In the SAP Community, Software-Heroes offers an alternative cloudification repository viewer. Compared to the version provided by SAP, the display is somewhat clearer by object type, and you can directly see the successor objects in a table. You can access this viewer at *https://software-heroes.com/en/cloudification-repository-viewer.*
- **APIs in the ABAP keyword documentation**
  You can generate a list of released APIs in your system in the ABAP keyword documentation. To do this, follow the path **ABAP APIs • Released APIs • Released APIs of the Current System**.

SAP Business Accelerator Hub

Another option outside the system is *SAP Business Accelerator Hub* (*https://api.sap.com/*). It's primarily used to find APIs for the integration of external applications with an SAP product. However, it can also help to open up APIs in the system. To do this, follow these steps:

1. Go to the *https://api.sap.com/* website, and then select your product on the **Products** tab (here, **SAP S/4HANA**).
2. Use the **APIs** tab to go to the APIs that can be used externally. Switch to the **All** tab to search through all available APIs.
3. We're once again trying to find an API for sales documents. A search for "sales document" only leads indirectly to a hit. The APIs are more specific, so that we can, for example, specifically select the synchronous API **Sales Order (A2X)** for sales orders (see Figure 17.18).

   After branching to the detailed view, you'll receive all the necessary technical specifications for using this API.
4. Clicking on the **Documents** tab for **Business Documentation** takes you to the SAP Help Portal for the selected product, where you can find the technical name of the API in the system. In this specific case, **Sales Order**

(A2X) has been implemented as an OData V2 service with the SAP Gateway Service Builder project, API_SALES_ORDER_SRV.

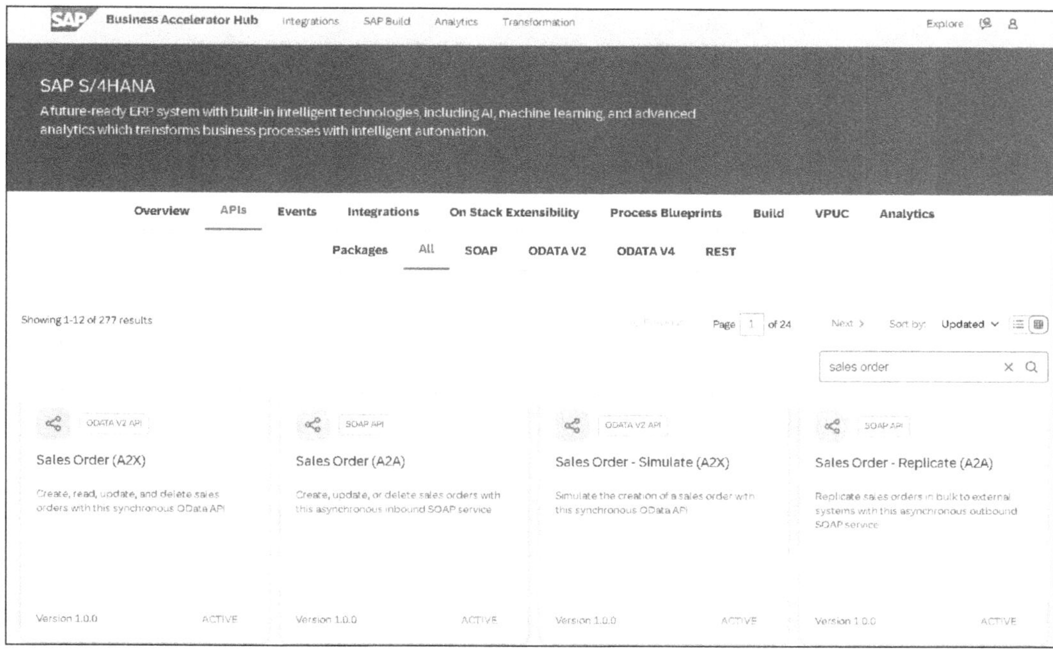

Figure 17.18 Searching for "Sales Order" in SAP Business Accelerator Hub

17.2 Calling APIs Based on the ABAP RESTful Application Programming Model via EML

In addition to data types, the definition of callable procedures is also a kind of API. For a long time, the BAPIs were the method of choice for interacting with business objects in the standard SAP system. With their function module–based interface and their remote capability via remote function call (RFC), BAPIs continue to be versatile—but not in ABAP Cloud. The ABAP RESTful application programming model also makes it possible to provide APIs for internal and external users. The current business objects of the standard modules are therefore gradually being covered with ABAP RESTful application programming model facades, and the BAPIs are therefore also being replaced.

Search for ABAP RESTful application programming model facade for customer orders

To call an API based on the ABAP RESTful application programming model in the system, you must first find this API. You've already learned about search tools in the previous section. As an example, let's say we want to change the **Customer Reference** field in the order header of a sales order. As a search tool, we select the **Open ABAP Development Object** dialog box and filter for usability in ABAP Cloud. We also enter a search pattern as an identifier and the BDEF object type for behavior definitions in the filter line (see Figure 17.19).

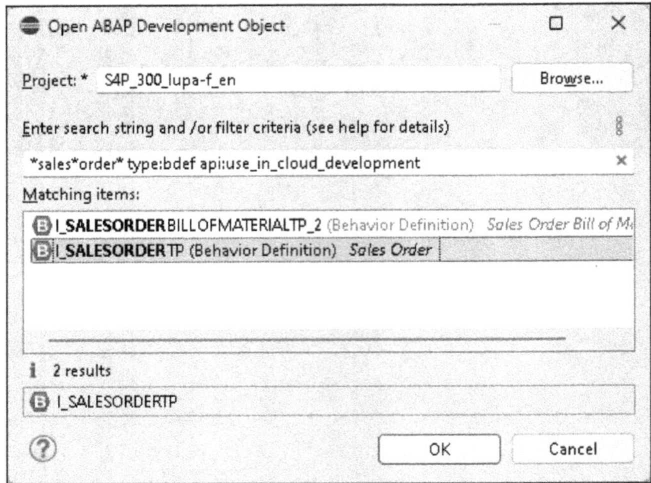

Figure 17.19 Searching for Behavior Definitions for Sales Orders

We select the one relevant hit, **I_SALESORDERTP**. This is a projection behavior definition with a transactional interface (see Figure 17.20).

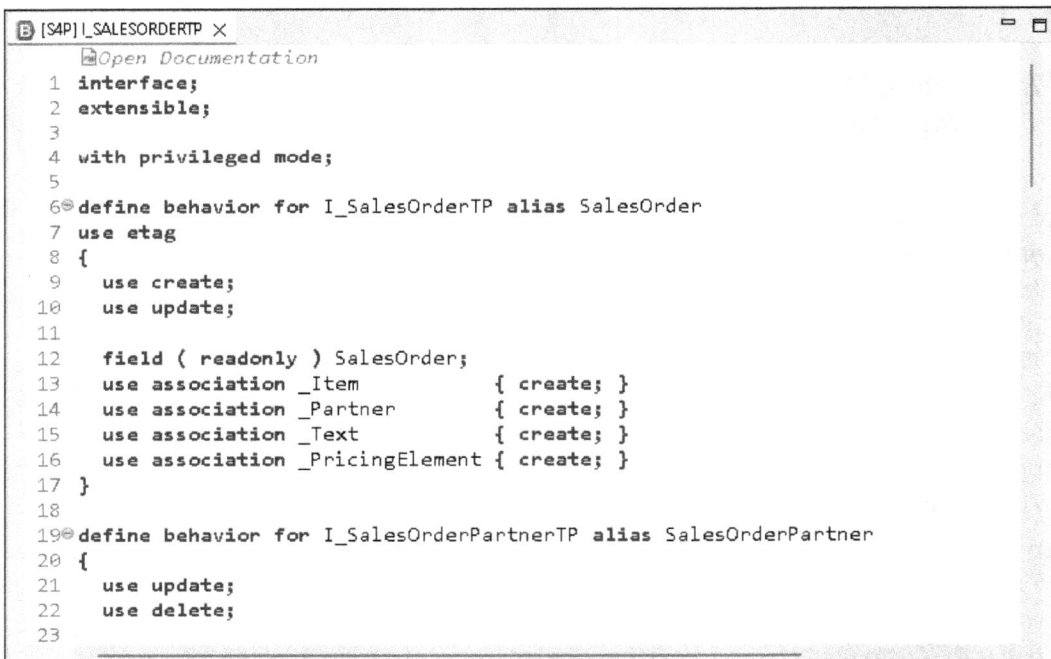

```
[S4P] I_SALESORDERTP
Open Documentation
interface;
extensible;

with privileged mode;

define behavior for I_SalesOrderTP alias SalesOrder
use etag
{
  use create;
  use update;

  field ( readonly ) SalesOrder;
  use association _Item           { create; }
  use association _Partner        { create; }
  use association _Text           { create; }
  use association _PricingElement { create; }
}

define behavior for I_SalesOrderPartnerTP alias SalesOrderPartner
{
  use update;
  use delete;

```

Figure 17.20 Projection Behavior Definition I_SalesOrderTP

Documentation of APIs based on the ABAP RESTful application programming model

The implementation isn't relevant to us for the time being. We want to find out how we can use the API based on the ABAP RESTful application programming model. SAP has kindly maintained documentation on this. You can open this documentation via the **Open Documentation** link at the top of the behavior definition. The relevant knowledge transfer document opens (see Figure 17.21). Details on handling this object type can be found in Chapter 15.

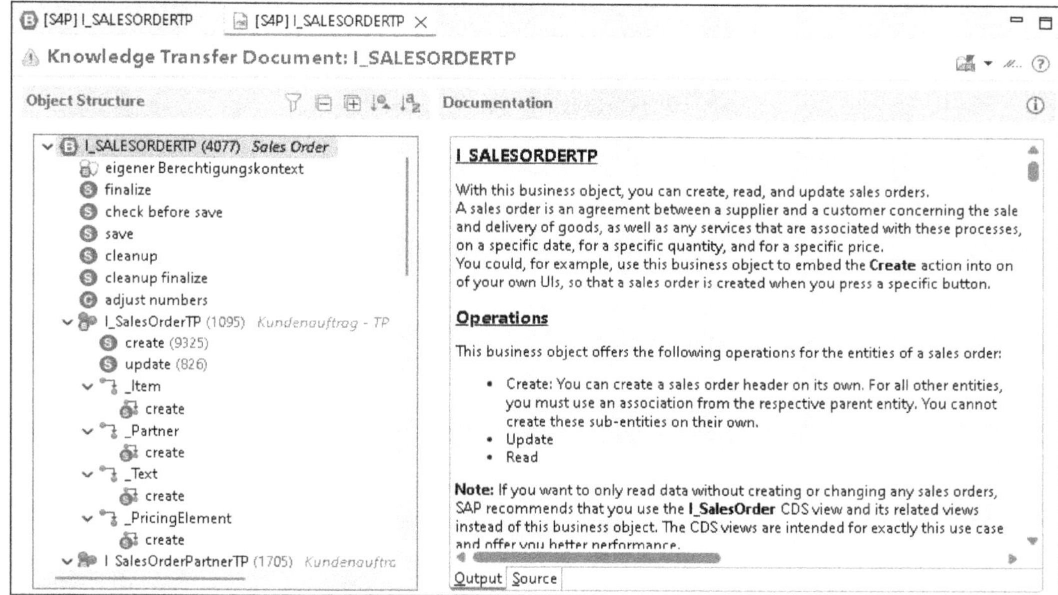

Figure 17.21 Documentation for Projection Behavior Definition I_SALESORDERTP

The document contains the specification of the API, including technical restrictions and sample calls. For some ABAP RESTful application programming model facades, there are currently technical restrictions based on the old implementation behind the facade. Specifically, in this case, it's not possible to process multiple sales orders within the same SAP LUW. This kind of information can be found in the documentation.

Changing a sales order

The documentation for changing an existing sales order can be found in the **Object Structure** area under **I_SalesOrderTP** in the **update** operation. This also contains a code example that we've adapted. The local interaction with the ABAP RESTful application programming model object in ABAP takes place via the Entity Manipulation Language (EML). We've already used this language in other chapters to interact with the transaction buffer when implementing an ABAP RESTful application programming model business object. In this case, we call an external business object via EML.

The implementation for modifying the customer reference is shown in Listing 17.2. We explain its details following the listing.

```
MODIFY ENTITIES OF I_SalesOrderTP
       ENTITY SalesOrder
       UPDATE
       FIELDS ( PurchaseOrderByCustomer )
       WITH VALUE #(
         ( PurchaseOrderByCustomer = '100005 (EML)'
           %key-SalesOrder         = '0000012246' ) )
       FAILED   DATA(failed)
       REPORTED DATA(reported).
IF failed IS NOT INITIAL.
  out->write( failed ).
  out->write( reported ).
  ROLLBACK ENTITIES.
ELSE.
  out->write( reported ).
  COMMIT ENTITIES.
ENDIF.
```

Listing 17.2 Changing a Sales Order via EML

EML syntax The EML syntax is complex and takes some getting used to. MODIFY ENTITIES is the statement for write accesses, such as a new creation or modification. We're referring here to the I_SalesOrderTP business object that we found beforehand. We select the SalesOrder entity because we want to change the order header data. We use the UPDATE operation to specify that an update is supposed to take place instead of a new creation. The FIELDS command is used to specify the fields that are to be updated. FIELDS is a way of providing a field list in EML. After the WITH VALUE addition, these fields are then supplied with their new values, and the keys are also specified.

Failed actions can be identified via the FAILED return parameter. REPORTED contains messages on the changes made or information on errors. If no actions have failed, we trigger the save process via COMMIT ENTITIES. In the event of errors, we roll back the transaction using ROLLBACK ENTITIES for security reasons.

The change worked in our case and can be seen in the sales order (see Figure 17.22 and Figure 17.23).

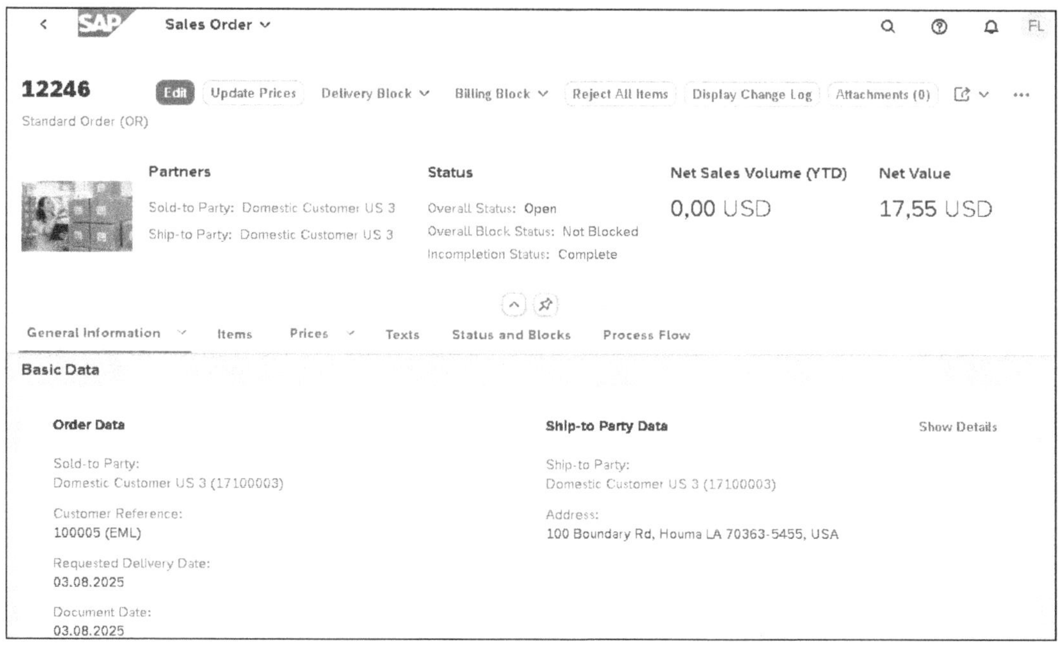

Figure 17.22 Changed Customer Reference in the Sales Order

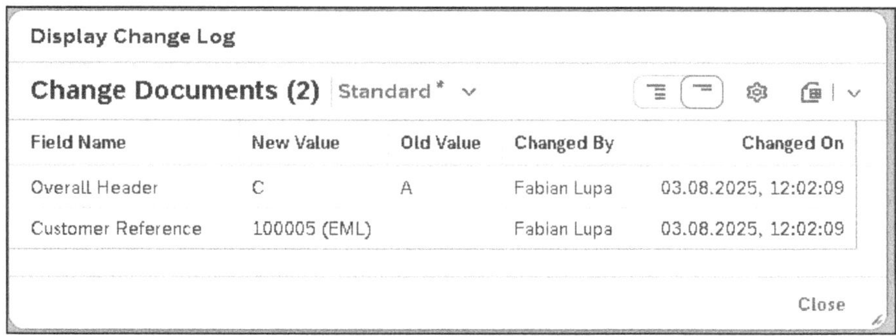

Figure 17.23 Change Documents for the Changed Customer Reference

17.3 Summary

In this chapter, you've learned how to use APIs. For this purpose, we've built on the basics from Chapter 1, and you've learned which tools you can use to find APIs. These include tools in the ABAP development tools but also those outside the system that are provided by SAP or the SAP community.

You've also learned how to call APIs based on the ABAP RESTful application programming model via EML. Up to this point, we had only used EML to call functions in the standard SAP system. The next chapter explains how you can make extensions that can also be addressed via EML, such as custom fields.

Custom APIs

What hasn't been addressed in this chapter is how you can create custom APIs. This can be relevant if you provide software to customers who expect stable APIs, or if you want to separate your ABAP cloud developments according to software components and have them communicate with each other via approved interfaces. APIs for clean core and the three-tier model are also relevant. Here, you use the release concept to be able to call objects in tier 3 from ABAP Cloud–compliant objects.

Chapter 18

Extensions in ABAP Cloud

In many customer projects, you have to deal with the requirement to extend a standard SAP Fiori app. Now, it's your turn to implement this requirement.

As standard software, SAP S/4HANA covers the core processes of a company—such as financial accounting or materials management. However, these core processes may need to be extended to adapt them to individual customer requirements, but extending the standard SAP system is a sensitive issue. The correct procedures must be selected so that extensions aren't overwritten during the next system upgrade, as would be the case with modifications, for example.

Extension options

SAP S/4HANA therefore provides a flexible platform for extending and integrating standard functions. ABAP Cloud provides the following options for extending existing processes:

- **Key user extensibility**
  No-code/low-code functions enable specialist users to make small adjustments themselves. Possible modifications include the addition of custom fields or simple logic. We take a look at key user extensibility in Section 18.1.
- **Developer extensibility**
  The options for developer extensibility are based on the traditional developer role. You can use released extension points to extend the standard SAP system. Developer extensibility is the subject of Section 18.2.

Moreover, in this chapter, we no longer refer to our recipe portal as a sample application. Because this is about extending the standard SAP system, we're going to add a checkbox for a voucher to a standard SAP sales order. We play through this extension with the two basic extension options, key user and developer extensibility.

18.1 Key User Extensibility

The key user extensibility allows end users with little technical know-how to make extensions to the system. There are a number of SAP Fiori apps that support such extensions.

18.1.1 Setting Up the Adaptation Transport Organizer

Adaptation Transport Organizer

The *Adaptation Transport Organizer* is a prerequisite for using the key user extensibility. The changes are usually made in a development system and must then be transported through the system landscape. As the changes within the scope of key user extensibility are made by users who have no development knowledge, the Adaptation Transport Organizer simplifies this process. The transport of the extensions is managed by the Adaptation Transport Organizer.

Configuration

By default, the use of key user extensibility is disabled for the standard SAP objects. To activate it, the Adaptation Transport Organizer must be set up as described here:

1. Call Transaction S_ATO_SETUP.

   You can see from the traffic light symbol next to the **Adaptation Transport Organizer configured** field whether or not the Adaptation Transport Organizer has already been configured for your system landscape (see Figure 18.1).

2. If that isn't the case, you can choose between a prefix and a namespace setup:

   - **Prefix setup**: This setup defines a customer prefix that can be used for all packages and transports, such as `ZZ1_<generated name of the object>`. You must enter this customer prefix in the **Prefix** field.
   - **Namespace setup**: This setup defines an SAP namespace that is used for all packages and transports, such as `/ABC/_<generated name of the object>`. You must enter the namespace in the **Namespace** field.

3. In the **Local package** field, you need to define a local package in which extensions are initially saved. Bear in mind that objects saved in this package can be transported later. The package is assigned to the `LOCAL` software component (see Figure 18.2). To display the software component of a package, switch to the package and its **Properties** in the ABAP development tools for Eclipse.

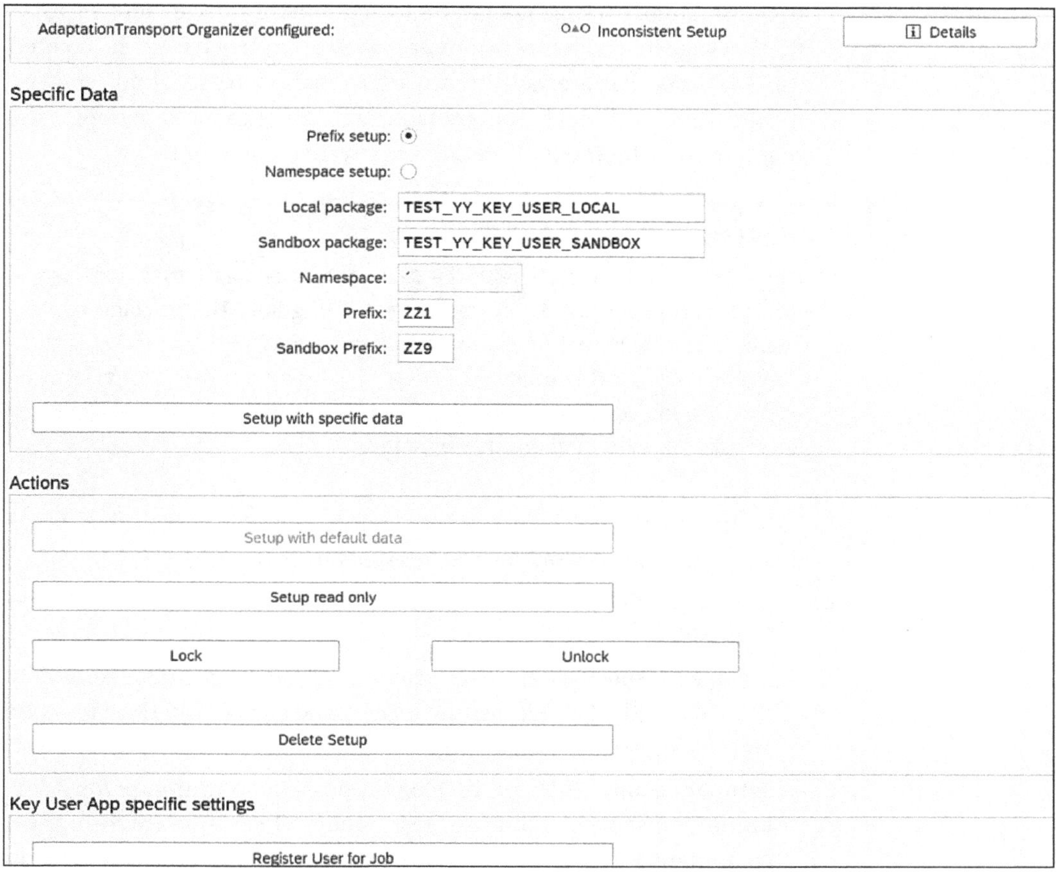

Figure 18.1 Transaction S_ATO_SETUP

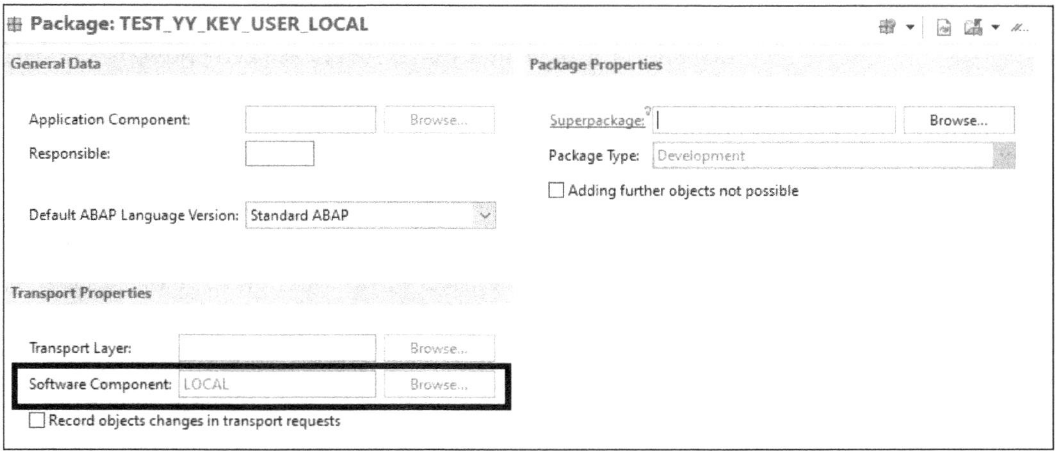

Figure 18.2 Local Package with "LOCAL" Software Component

4. In the **Sandbox package** field (refer to Figure 18.1), you can define a sandbox package in which extensions are saved after they have been created. ABAP objects that are saved there after creation are local objects. This means they will never be transported. The sandbox package also requires an assignment of the LOCAL software component.

[!]

The $ Character in the Package Name

If the package name starts with the $ character, such as $YATO_KEY_USER_LOCAL, this package will be deleted after an upgrade. The package name should rather begin with a character string such as TEST_, as in TEST_YY_KEY_USER_LOCAL, for example. The TEST_ package won't be lost after an upgrade. If you've already created extensibility elements in $ packages, you must assign them to another package.

5. You must also enter the prefix for sandbox objects. As described previously, these objects won't be transported any further.
6. You can now perform setup functions using the following buttons show in Figure 18.1:
   - **Setup with specific data**: The Adaptation Transport Organizer is configured and the SAP Fiori launchpad can be used to call the extension and transport apps for key users.
   - **Setup read only**: This option makes it possible to configure the Adaptation Transport Organizer in read mode, which is useful for nondevelopment systems to call the apps for extensions. This way you can check whether the extensions made in the development system are available in the successor system.
   - **Setup with default data**: This option configures the Adaptation Transport Organizer with default values that have been preset by SAP. The option is only active when the Adaptation Transport Organizer gets configured for the first time.
   - **Lock**: The Adaptation Transport Organizer and the SAP Fiori apps for extension and transport are locked.
   - **Unlock**: This option is used to unlock and activate the Adaptation Transport Organizer. Extension entries can now be changed, created, and transported.

[!]

Setting Up the Adaptation Transport Organizer in the Development System Only

SAP advises against allowing development within the scope of key user extensibility in multiple systems or clients of a system chain. The development of extensibility elements can lead to considerable problems, such as ABAP dumps like DBSQL_DUPLICATE_KEY_ERROR. This dump is triggered because entries are written twice to a client-independent table. SAP therefore recommends configuring the Adaptation Transport Organizer in only one development system and one client and then transporting the changes to other systems. In other environments, the Adaptation Transport Organizer should only be set up in read-only mode.

Once the Adaptation Transport Organizer has been set up, you can implement key user extensions. In the following sections, we'll first describe how to add custom fields and then how to add a custom logic.

18.1.2 Custom Fields

Custom Fields app

In the SAP Fiori launchpad, you'll now find the **Custom Fields** app tile (app ID F1481; see Figure 18.3). Depending on your system status, there's also a discontinued app tile called **Custom Fields and Logic (Deprecated)**. Make sure you use the newer app.

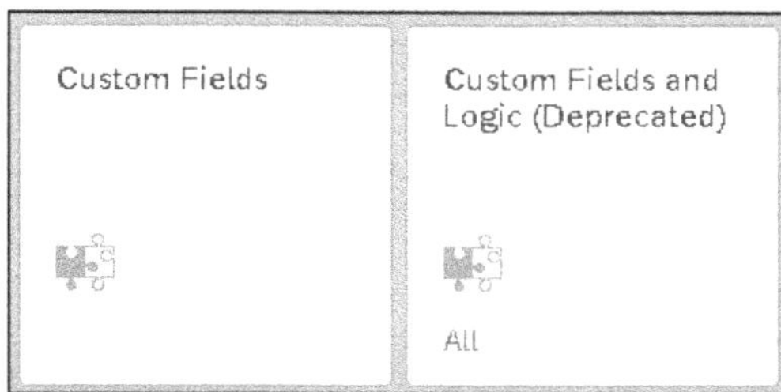

Figure 18.3 Tile for Opening SAP Fiori App Custom Fields

[«]

The Custom Fields App

Information on the Custom Fields app (app ID F1481) can be found in the SAP Fiori apps reference library at *http://s-prs.de/v1064810*. There you must first select your release. On the **IMPLEMENTATION INFORMATION** tab, you'll then find all the information you need to configure the app in your system, such as which standard roles are available and which Open Data Protocol (OData) services need to be activated.

Overview of custom fields

After opening the SAP Fiori app, you'll first see an overview of all custom fields in the system (see Figure 18.4). In our system, no custom fields have yet been maintained.

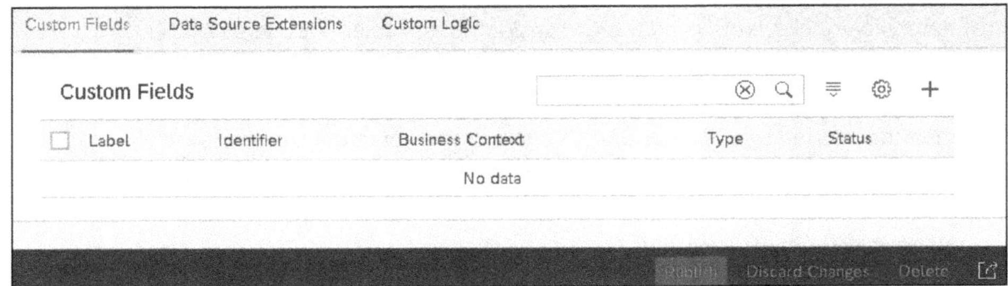

Figure 18.4 Overview of All Custom Fields

Adding a custom field

To add a new field, follow these steps:

1. Click on the plus icon (**Create**) on the right above the table.
2. A dialog box for creating a new field opens (see Figure 18.5). Enter the following details here:
   - **Business Context**: The business context combines all technical objects that form a logical unit, such as the SAP Fiori app to be extended and the associated CDS views and interfaces. In our example, the business context is the sales document (SD_SALESDOC).
   - **Label**: Insert a description of the field content here.
   - **Identifier**: Enter a freely selectable technical name for the field here.
   - **Tooltip**: Enter the information that should be displayed to users when they move the mouse pointer over the field.
   - **Type**: Select the field type here. Many of the defined types, such as checkbox, date or text, are self-explanatory. In our example, we want to insert a checkbox.

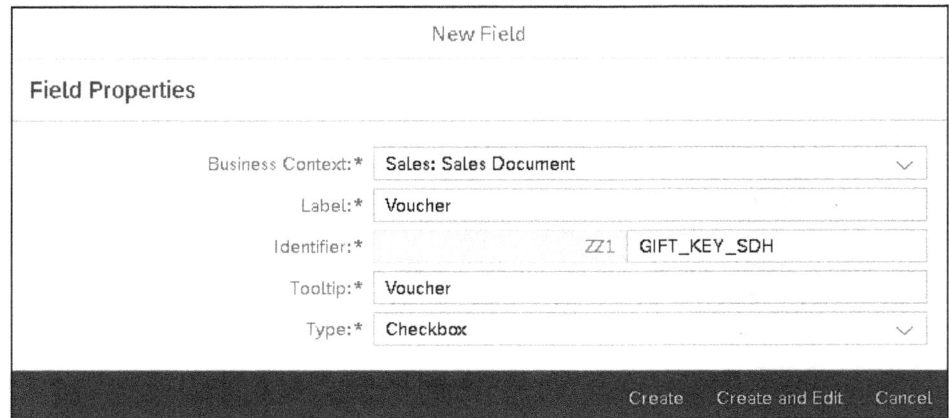

Figure 18.5 Dialog Box for Defining a New Field

3. Click on **Create and Edit**.

4. In the window that opens next (see Figure 18.6), you can add the field for various use cases such as APIs or business scenarios by clicking on **Enable Usage** in the line for the respective data source.

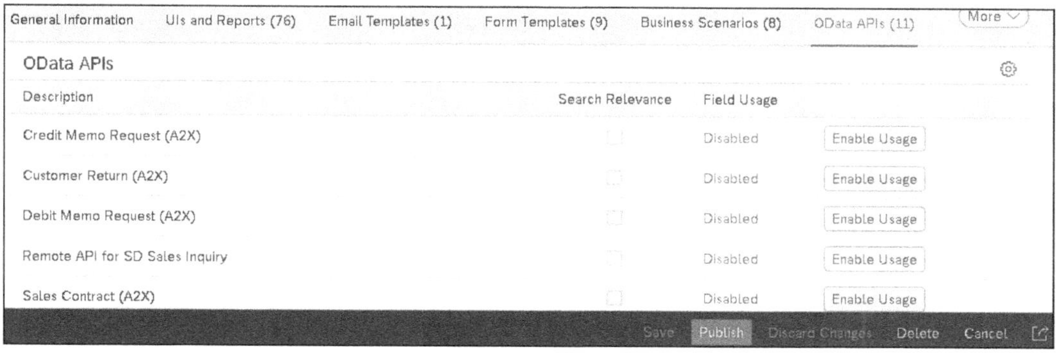

Figure 18.6 Use Cases for Activating the Field

5. Once you've selected your use cases, you can release the field by clicking the **Release** button. After that it will be available for the selected use cases.

Extension structure

Let's take a look at how the field extension was technically implemented. The custom fields are saved in a separate *extension structure* (see Listing 18.1). Extension structures end with the INCL_EEW_PS suffix.


```
@EndUserText.label : ''
@AbapCatalog.enhancement.category : #NOT_EXTENSIBLE
extend type sdsalesdoc_incl_eew_ps
  with zz1_p5aymcnlkxd7ubkr4fjfznvmra {
  zz1_gift_key_sdh  : zz1_gift_key not null;
}
```

Listing 18.1 Extension Structure

In our example, the ZZ1_GIFT_KEY_SDH field has been added to the SDSALES-DOC_INCL_EEW_PS structure using our extension. This extension structure can now be added to different repository objects, for example in database tables for persisting the data (see Listing 18.2).

```
@EndUserText.label : 'Sales Document: Header Data'
@AbapCatalog.enhancement.category : #EXTENSIBLE_CHARACTER_NUMERIC
@AbapCatalog.tableCategory : #TRANSPARENT
@AbapCatalog.deliveryClass : #A
@AbapCatalog.dataMaintenance : #RESTRICTED
define table vbak {
...
```

```
    vbak_status             : include vbak_status;
    tech_fields             : include vbak_tech_fields;
    ext                     : include sdsalesdoc_incl_eew_ps;
    include vbak_ext_glo;
}
```

Listing 18.2 Extension of a Database Table

Transferring the custom field into the app

The custom field isn't included by default in the SAP Fiori app to be extended. Figure 18.7 shows the Manage Sales Orders app screen (app ID F3893) with the default fields, but still without our custom checkbox for the voucher.

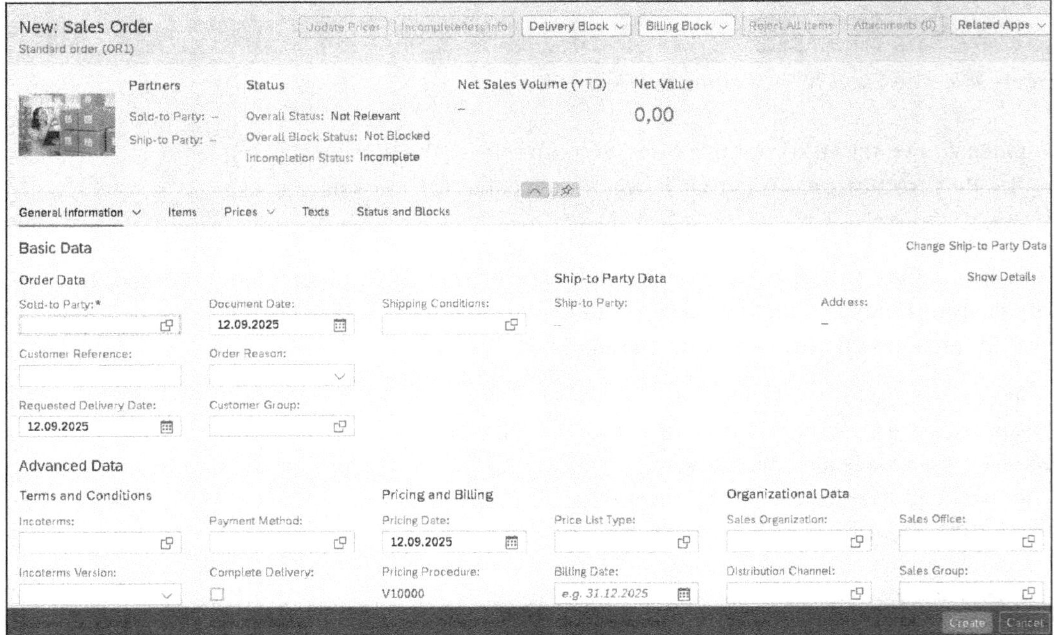

Figure 18.7 SAP Fiori App Without the Custom Field

User Interface Customizations by Application Experts

Setting up user interface (UI) customization by application experts is described in the SAP Help Portal at *http://s-prs.co/v619801*.

For the custom field to be adopted in the SAP Fiori app, it must be added to the respective app as follows:

1. Start the SAP Fiori launchpad.

2. Open the SAP Fiori app in which you want to add the custom field. For our sample application, we open the Manage Sales Orders app (app ID F3893).
3. Open the user action menu that you see in Figure 18.8 via your profile in the SAP Fiori launchpad, and select the **Adapt UI** item.
4. You can now add a new UI element. To do this, right-click on the *UI element container*, that is, the specific UI area in which you want to add the field.
5. Click the plus button (**Add**) in the context menu (see Figure 18.9).
6. Select an available element; in our example, that's our custom **Voucher** field (see Figure 18.10).
7. Confirm your selection by clicking **OK**.

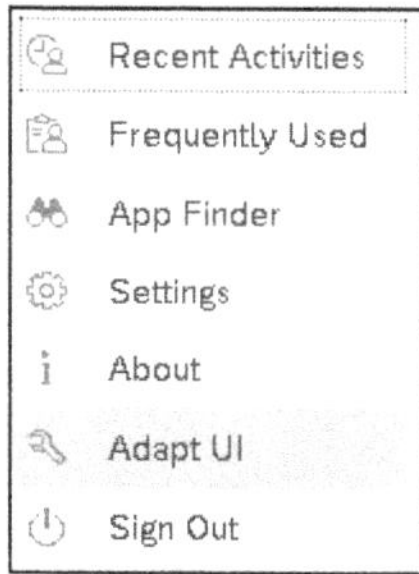

Figure 18.8 Selecting the Adapt UI Item in the User Action Menu

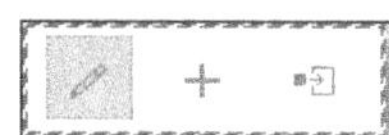

Figure 18.9 Context Menu in Edit Mode

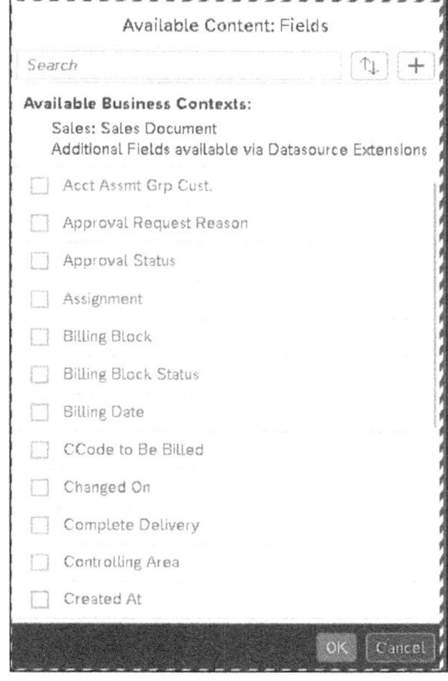

Figure 18.10 Selecting Available Content

Now you can see that the checkbox has been added to the app's UI (see Figure 18.11). Finally, you need to activate your customizations so that the checkbox is available in the processing view of a sales order.

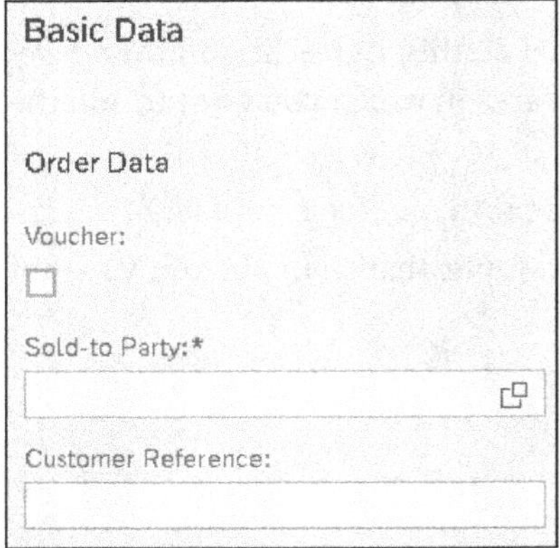

Figure 18.11 Custom Voucher Field

[»]

The Manage Sales Orders App

Information on the Manage Sales Orders (app ID F3893) can be found in the SAP Fiori apps reference library at *http://s-prs.de/v1064812*. There, you must first select your release. On the **IMPLEMENTATION INFORMATION** tab, you'll find all the information you need to configure the SAP Fiori app in your system, for example, which standard roles are available and which OData services need to be activated.

18.1.3 Custom Logic

The Custom Logic app

We can also add custom logic to the Manage Sales Orders app. Once the **Voucher** field has been added and is available in the app, it can be supplemented with custom logic. As an example, we want to implement that the **Voucher** field gets automatically checked. To define such logic, SAP provides the Custom Logic (app ID F6957). This app can be used to implement business add-ins (BAdIs) and behavioral extensions for business objects that have been released by SAP for this purpose.

[»]

The Custom Logic App

Information on the Custom Logic app (app ID F6957) can be found in the SAP Fiori apps reference library at *http://s-prs.de/v1064813*. There you must first select your release. On the **IMPLEMENTATION INFORMATION** tab, you'll then find all the information you need to configure the app in your system, such as which standard roles are available and which OData services need to be activated.

Adding custom logic

To add custom logic, follow these steps:

1. Start the SAP Fiori launchpad.
2. Open the Custom Logic app (app ID F6957). Make sure you're not using the discontinued version of the app (see Figure 18.12).

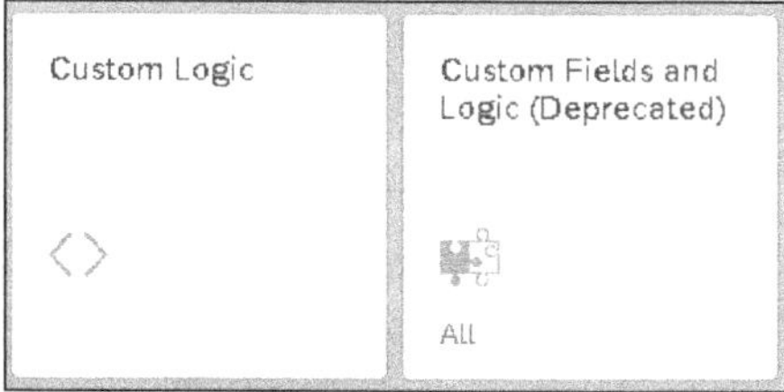

Figure 18.12 Tile for Opening the Custom Logic App

3. This takes you to the overview of the extension points. Select a business context; in our case, the context is **Sales: Sales Document**.
4. Select an extension point. You can open and view detailed documentation for each extension point by clicking on the **View Documentation** link (see Figure 18.13). For our example, we select extension point **SD_SLS_MODIFY_HEAD** because we want to modify the logic of the custom field in the sales order header data.

18

Figure 18.13 Selection of Extension Points

5. Add an implementation description and a unique ID for your extension. The implementation ID must be different from the fields you've added in the Custom Fields app. If that isn't the case, you'll be notified after you've clicked on the **Review** button (see Figure 18.14).

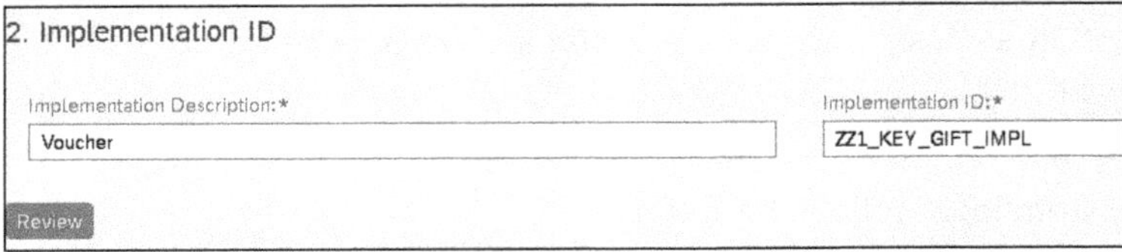

Figure 18.14 Assigning an Implementation ID

6. Click on **Create** (see Figure 18.15).

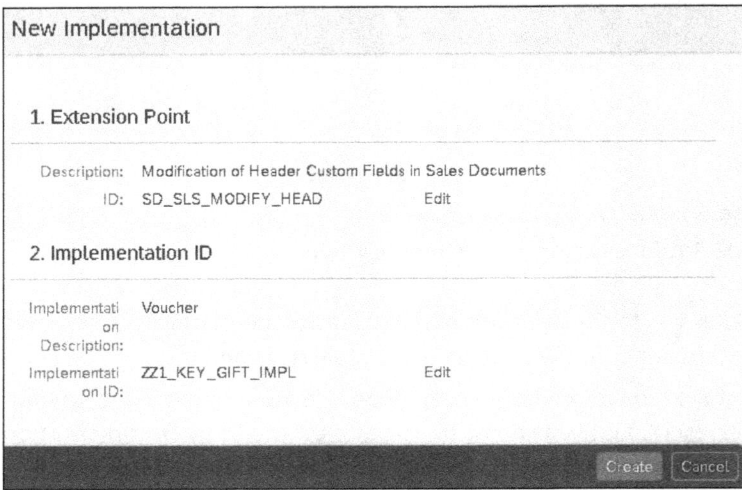

Figure 18.15 Creating an Implementation ID

7. The ID of the created implementation gets displayed again in the window that opens next (see Figure 18.16). To be able to open the code editor to implement your logic, you must make the implementation available using the **Publish** button.

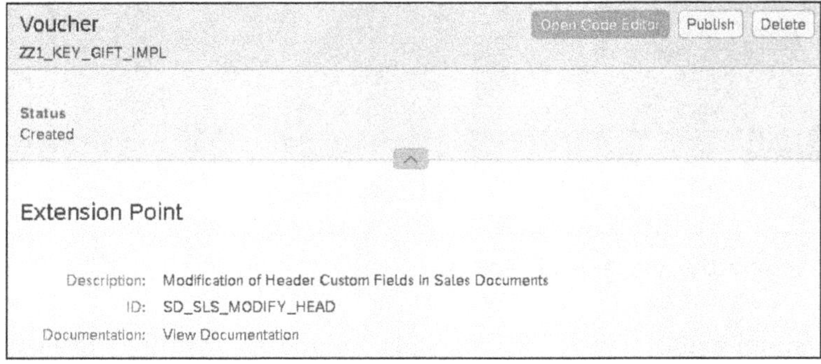

Figure 18.16 Display of the Extension Point

8. Click on the **Open Code Editor** button, and implement your custom logic.

Source code editor

You can now write your own code using the source code editor (see Figure 18.17). For this purpose, the ABAP language version option **ABAP for Key Users** is used.

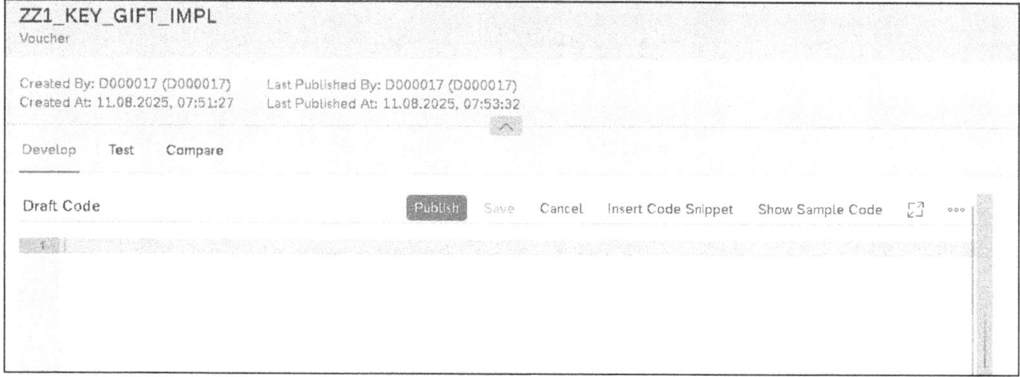

Figure 18.17 Source Code Editor for Defining the Custom Logic

[«]

ABAP for Key Users

ABAP for Key Users is a language version that contains a small subset of the statements in the **ABAP for Cloud Development** language version. It's used within the key user extensibility.

18

The editor is divided into three tabs that focus on different aspects of working with the code:

- **Develop**
  On this tab, you can write, save, and release ABAP code. Saving here means that a draft version of the code is written that can only be seen by you. As soon as the code is released, it's available in the system for all users. A sample code provided by SAP is displayed using the **Show Sample Code** button.
- **Test**
  On this tab, you can check the syntax of the code and test various parameter values. Both the draft and the released code can be checked in this way.
- **Compare**
  On this tab, you can view and compare different versions of the code.

For our example, we implement the following logic:

```
salesdocument_extension_out-zz1_key_gift_sdh =
  salesdocument_extension_in-zz1_key_gift_sdh.
salesdocument_extension_out-zz1_key_gift_sdh = 'X'.
```

When a new sales order is created, the **Voucher** field is automatically selected (see Figure 18.18).

Order Data
Voucher:
Requested Delivery Date: 12.09.2025
Customer Group:
Sold-to Party:*
Document Date: 12.09.2025
Shipping Conditions:
Customer Reference:
Order Reason:

Figure 18.18 Addition of Custom Logic

18.1.4 Transporting Key User Extensions

To transport the key user extensions from the development system to the production system, more steps are necessary, which we describe in this section.

Configuring software packages

Up to this point, the extensions created via the various key user apps have been assigned to a local package in accordance with the configuration of the Adaptation Transport Organizer. To be able to transport these extensions, the objects must later be assigned to a package, which in turn is assigned a transport layer (see Figure 18.19). This package can be created via the ABAP development tools. For our example, we create a package named ZTRANSPORT in **Transport Layer Z000**.

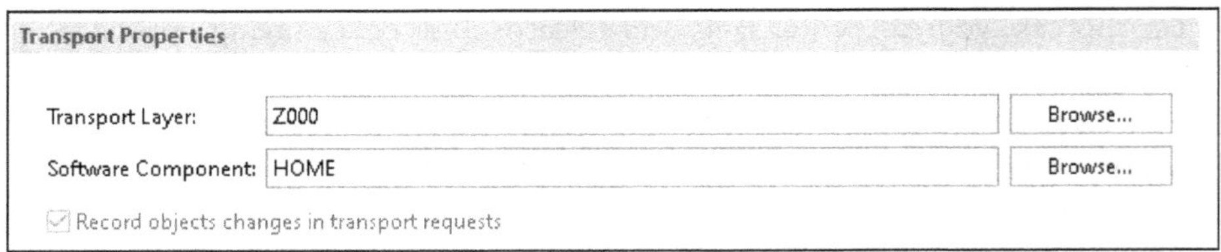

Figure 18.19 Package with Transport Layer

Registering a package for key user extensions

The package must then be registered as a package for key user extensions. To do so, follow these steps:

1. Start the SAP Fiori launchpad.
2. Open the Configure Software Packages app (app ID F1590; see Figure 18.20).

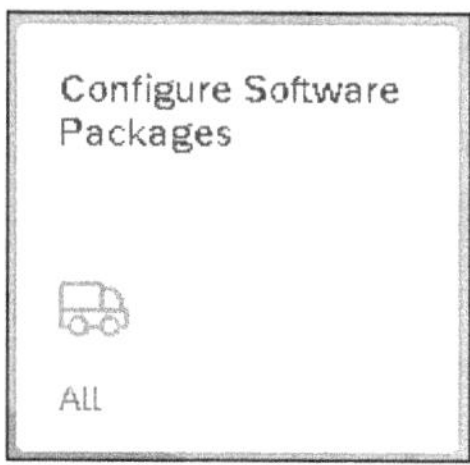

Figure 18.20 Tile for Opening the Configure Software Packages App

The Configure Software Packages App (Version 2)

Information on the Configure Software Packages (app ID F1590) can be found in the SAP Fiori apps reference library at *http://s-prs.de/v1064814*. There you must first select your release. On the **IMPLEMENTATION INFORMATION** tab, you'll then find all the information you need to configure the app in your system, such as which standard roles are available and which OData services need to be activated.

3. You'll see an overview of all registered software packages (none so far). To register a new package, click on the **Add Registration** button (see Figure 18.21).

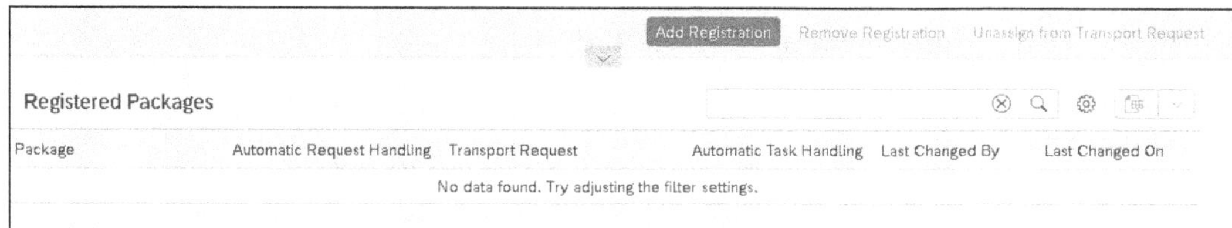

Figure 18.21 Overview of All Registered Packages

4. In the dialog box that follows, search for the package you want to register. Select it, and click on the **Select** button (see Figure 18.22).

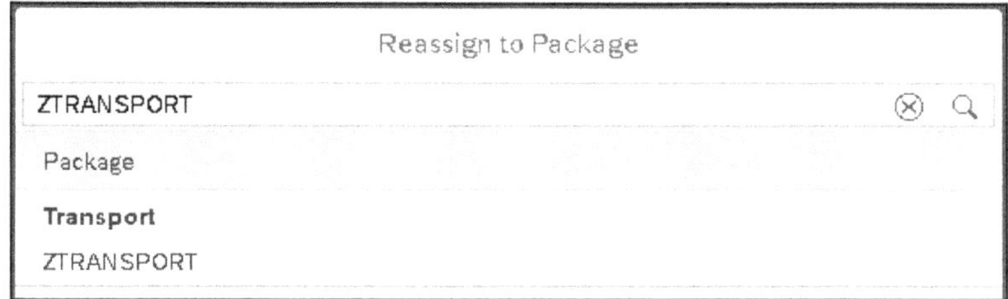

Figure 18.22 The Register Package Dialog Box

Now the package has been registered and can be used to transport extensions (see Figure 18.23).

Figure 18.23 Package After Successful Registration

Configuring the package

The following functions are now available for the package in the Configure Software Packages app:

- **Automatic order/task handling**
  If these options are activated, the transport requests and tasks will be managed by the system.
- **Transport request**
  A transport request can be selected manually. The transport request then contains all the extensions that have been assigned.
- **Removing the registration**
  Once a registration has been removed, the package can no longer be used to transport extensions.

Assigning an extension to a package

Everything is now ready to assign our extensions to the package:

1. Open the Register Extensions for Transport app (app ID F1589).

> **The Register Extensions for Transport App**
>
> Information on the Register Extensions for Transport app (app ID F1589) can be found in the SAP Fiori apps reference library at *http://s-prs.de/v1064815*. There you must first select your release. On the **IMPLEMENTATION INFORMATION** tab, you'll then find all the information you need to configure the app in your system, such as which standard roles are available and which OData services need to be activated.

   All extensions that have been created via the key user apps will get listed (see Figure 18.24).

2. Select our extension, and click on the **Assign to package** button.
3. In the popup window, search for your registered package, and select the respective item (see Figure 18.25).

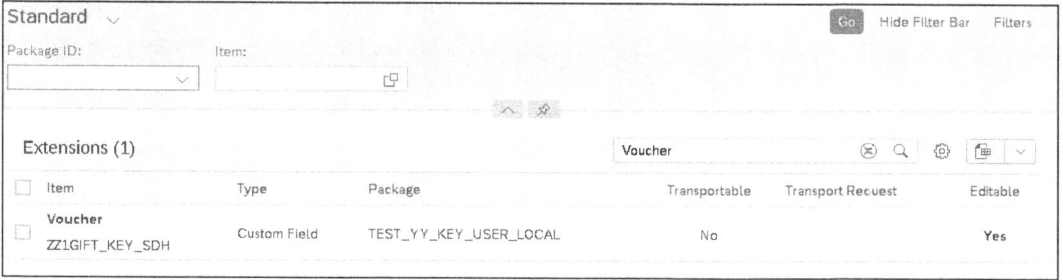

Figure 18.24 Overview of All Extensions

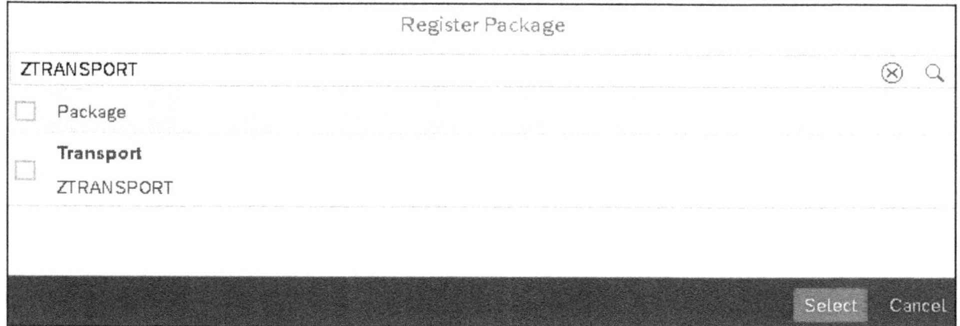

Figure 18.25 Assigning a Package

4. At this point, the extension has been assigned to the package. If you selected the **Automatic Request Handling/Automatic Task Handling** functions when registering the package (see Figure 18.23), a transport request will be generated right away. You can see the assignment in Figure 18.26.

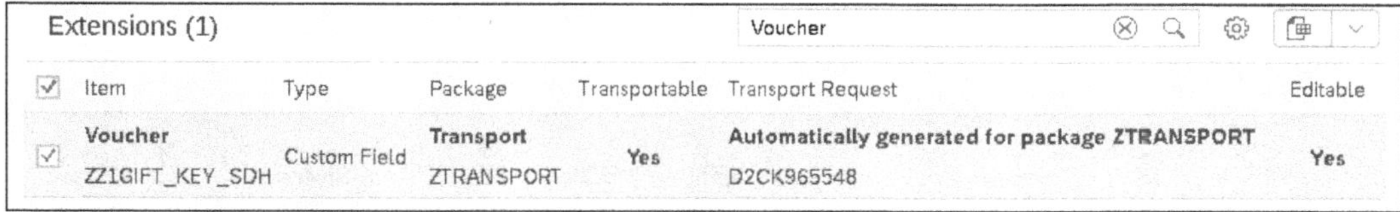

Figure 18.26 Assignment of the Extension to the Package

The extension can then be transported using the transport request and imported into the downstream systems.

18.1.5 Custom CDS Views

You can also create custom CDS views for default SAP objects using a key user app. In contrast to what has been shown so far in this book, you would then work on a form-based basis with the help of a wizard that supports you, for example, in selecting the fields for the CDS view. The source code of the CDS view and the app for displaying the data selected by the view are then created automatically.

The Custom CDS Views app

For this purpose, you want to use the Custom CDS Views app (app ID F1866A). Make sure you don't select the discontinued app version tile in your SAP Fiori launchpad (see Figure 18.27).

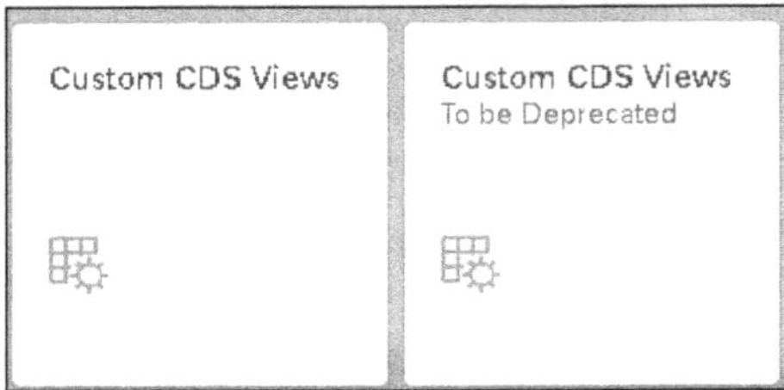

Figure 18.27 Tile for Opening the Custom CDS Views App

[»]

The Custom CDS Views App (Version 2)

Information on the Custom CDS Views app (Version 2, app ID F1866A) can be found in the SAP Fiori apps reference library at *http://s-prs.de/v1064816*. There you must first select your release. On the **IMPLEMENTATION INFORMATION** tab, you'll then find all the information you need to configure the app in your system, such as which standard roles are available and which OData services need to be activated.

After starting the app, you'll first be shown all the user-defined CDS views that have already been created (none so far; see Figure 18.28).

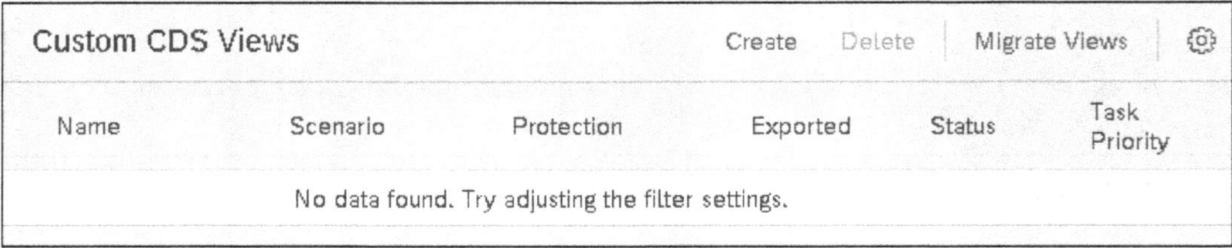

Figure 18.28 List of All Custom CDS Views

Creating a CDS view

We now want to create a new CDS view to create a list of all sales orders:

1. Click on the **Create** button.
2. In the dialog box that opens (see Figure 18.29), enter a unique name and a suitable descriptive identifier in the **Label** field. The label is displayed as the name of the CDS view when the results are displayed.

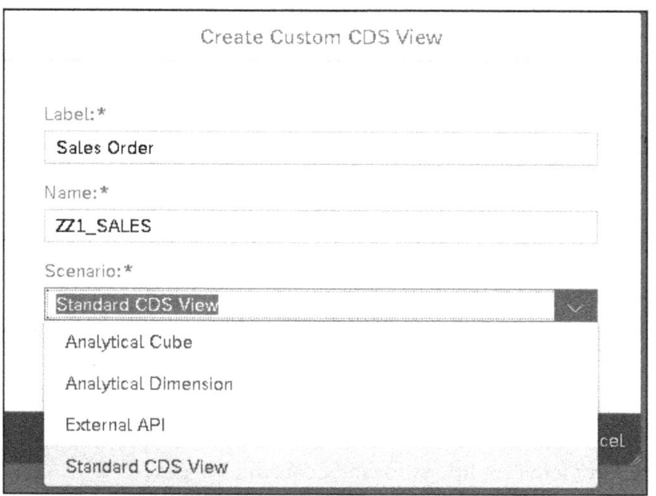

Figure 18.29 Creating a Custom CDS View

3. Select a view type in the **Scenario** field; here, it's **Standard CDS View.**
4. The custom CDS view is now created using a guided sequence of steps. First, you must specify the data sources. Specify a primary data source (here, it's `I_SalesOrder`) (see Figure 18.30). In addition to the primary data source, you can add additional data sources via **Add • Associated Data Source**. We haven't added an additional data source for our example.

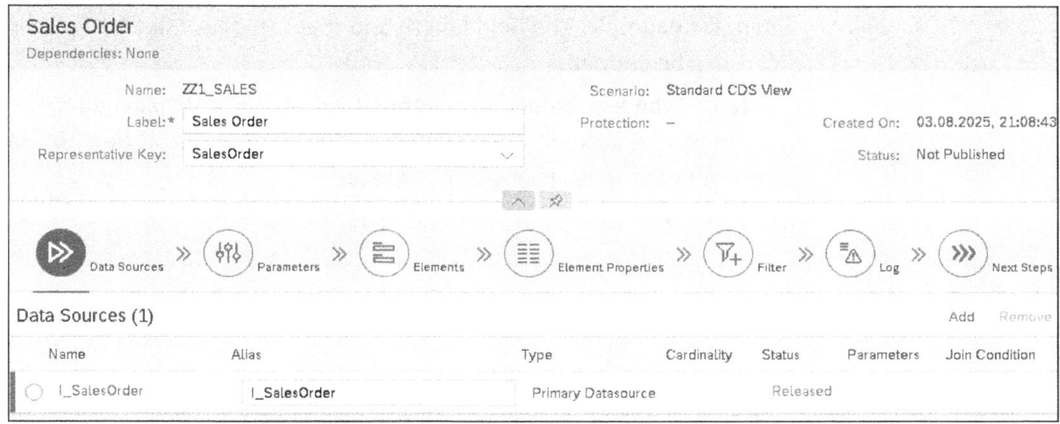

Figure 18.30 Specifying Data Sources for the Custom CDS View

5. These data sources can be linked to each other using join conditions (see Figure 18.31). As we haven't selected an additional data source, we don't need to define a join condition here.

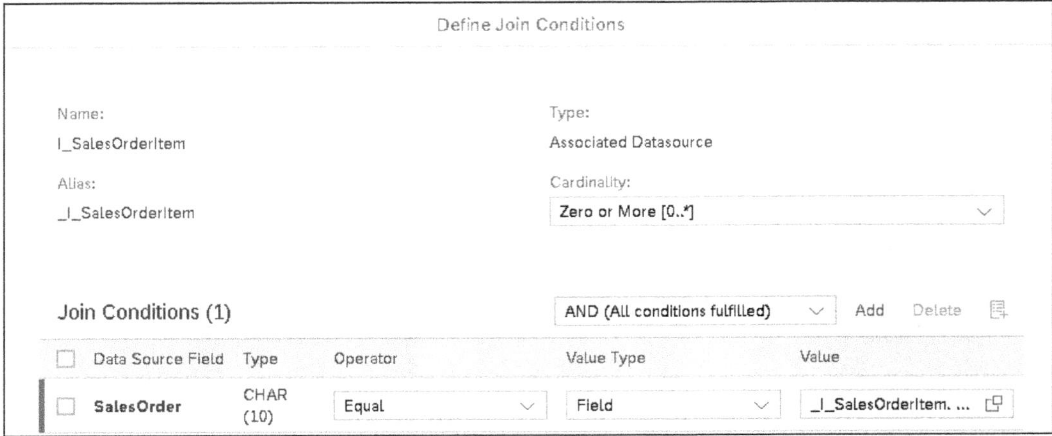

Figure 18.31 Maintaining Join Conditions for the Custom CDS View

6. The parameters can be defined. These selection parameters are options for users to filter the data from the various data sources. To add a parameter, click on the **Add** button. The following values must be specified for the parameter (see Figure 18.32):
   - **Name**: This is the technical name for the parameter.
   - **Label**: This will get displayed to the user.
   - **Data Type (Length and Decimals)**: The data type defines the behavior of the field. If the **DATS** data type is used, the input must be in date format, for example. The field length and the number of decimal places can also be defined.
   - **Default Type and Value**: A predefined value can be displayed for the parameter. A system value such as the system date can be entered here, for example, or simply a free value.
   - **Hidden**: The visibility of the field is specified.

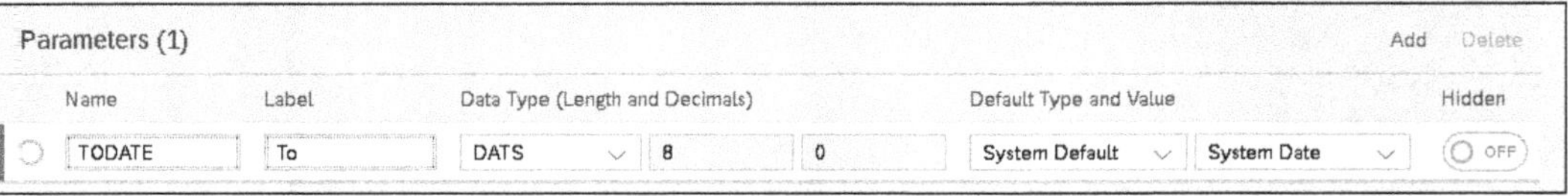

Figure 18.32 Maintaining Parameters for the Custom CDS View

7. Specify the fields that are to be visible as columns in the table-based data output of the CDS view, that is, the *elements*. You can freely define the labels for these fields (see Figure 18.33). For our example, these are the sales order number (`SalesOrder`) and the sales order type (`SalesOrder-Type`).

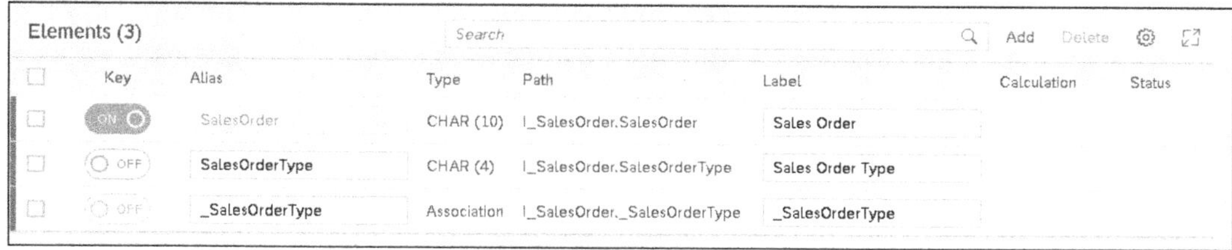

Figure 18.33 Defining Elements of the Custom CDS View

8. The condition is already preselected during creation. Data source fields can be entered there and preassigned with values. This means that the data sources are selected directly with the specified values the first time they are displayed (see Figure 18.34).

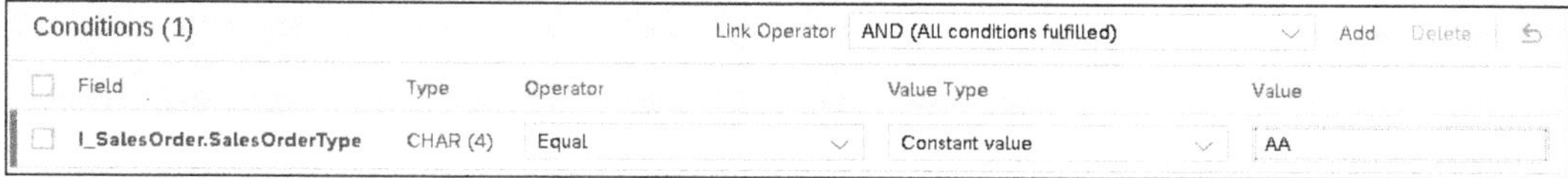

Figure 18.34 Defining Conditions for the Custom CDS View

9. Select which functions you want to be provided to users for the custom CDS view (see Figure 18.35):
   - **Publish**: The custom CDS view will be made available to anyone.
   - **Check**: This function checks the CDS view for syntax errors, which are then displayed on the **Logs** tab.
   - **Preview**: This function tests the CDS view.
   - **Cancel**: This function cancels editing, and all changes are lost.
   - **Data Browser**: This function displays the CDS view in the Customer Data Browser app (app ID F5746).
   - **Status**: This function allows you to view the currently released status of the CDS view so that you can adjust it if necessary in the event of errors.

Figure 18.35 Functions for the Custom CDS View

10. Keep in mind to make the CDS view available to all users using the **Publish** function.

[«]

The Customer Data Browser App

The Customer Data Browser app is discussed in Chapter 4.

Our custom CDS view looks like Figure 18.36. The columns that were selected in the **Elements** step are displayed. You can also see that the value for the **Until** parameter has been preset.

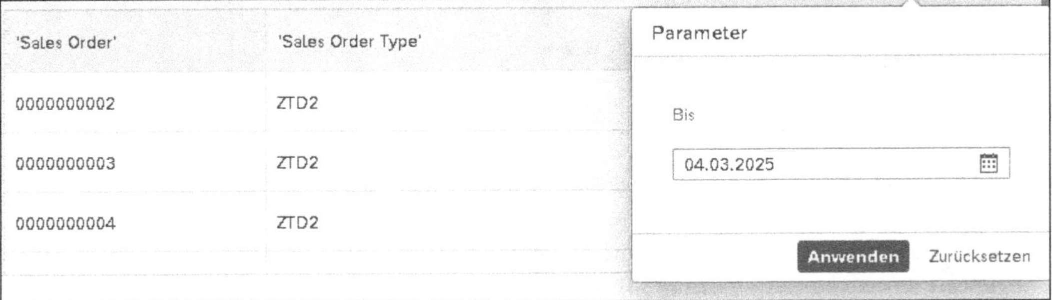

Figure 18.36 Preview of the CDS View

18.2 Developer Extensibility

The programming of standard SAP Fiori apps can be extended by means of developerextensibility. Developer extensibility supports the development of cloud-capable and upgrade-stable custom ABAP code. You only use APIs that have been approved and released by SAP to comply with the clean core specifications. Developer extensibility makes it possible to take advantage of custom ABAP code, taking into account the necessary restrictions for cloud capability and the new programming model.

For our application scenario of extending sales orders, we first search for the released APIs for sales orders. You can find out how to find released APIs in Chapter 17, Section 17.1.

The released `I_SALESORDERTP` API is suitable for our application scenario. You can find out everything you need to know about extending this API in the associated knowledge transfer document. We've described the handling of knowledge transfer documents in Chapter 15, Section 15.2.

Table 18.1 shows the API objects to be extended.

Extended Repository Object	Type
Table `VBAK`	Database table extension
`I_SalesDocumentBasic`	Entity extension
`I_SalesDocument`	Entity extension
`I_SalesOrder`	Entity extension

Table 18.1 Extension Objects of the I_SALESORDERTP API

Extended Repository Object	Type
R_SalesOrderTP	Entity extension
I_SalesOrderTP	Entity extension

Table 18.1 Extension Objects of the I_SALESORDERTP API (Cont.)

18.2.1 Extending Database Tables

Extensibility of a database table

First, we need to extend the database table. To do this, you must first check whether the SAP database table can be extended. In Listing 18.3, you can see that the database table contains the `@AbapCatalog.enhancement. category` annotation. This annotation determines whether and how a database table can be extended. It can have the following values:

- **#EXTENSIBLE_ANY**
  All kinds of extension types, such as an additional field or a structure, are permitted.
- **#EXTENSIBLE_CHARACTER**
  Only character-type extension types are permitted.
- **#EXTENSIBLE_CHARACTER_NUMERIC**
  In addition to the character-type extension types, numeric extension types are also permitted.
- **#NOT_EXTENSIBLE**
  No extensions are permitted.

18

```
@EndUserText.label : 'Sales Document: Header Data'
@AbapCatalog.enhancement.category : #EXTENSIBLE_CHARACTER_NUMERIC
@AbapCatalog.tableCategory : #TRANSPARENT
@AbapCatalog.deliveryClass : #A
@AbapCatalog.dataMaintenance : #RESTRICTED
...
  ext : include sdsalesdoc_incl_eew_ps;
  ...
}
```

Listing 18.3 Basic Database Table

Extension structure

SAP recommends using the structure with the `incl_eew_ps` ending in the released ABAP RESTful application programming model objects. Listing 18.4 shows this using the example of our database table extension. This extension structure can be used to add fields to a database table. In our example, we add fields to database table `VBAK`.

```
@EndUserText.label : 'Extension Include Persistent (VBAK)'
@AbapCatalog.enhancement.category : #EXTENSIBLE_CHARACTER_NUMERIC
@AbapCatalog.enhancement.fieldSuffix : 'SDH'
@AbapCatalog.enhancement.quotaMaximumFields : 204
@AbapCatalog.enhancement.quotaMaximumBytes : 4080
@AbapCatalog.enhancement.quotaShareCustomer : 50
@AbapCatalog.enhancement.quotaSharePartner : 50
define structure sdsalesdoc_incl_eew_ps {
  dummy_salesdoc_incl_eew_ps : dummy;extd
}
```

Listing 18.4 Extension Structure for Extending the Database Table

SAP used the following annotations in this extension structure:

- **@AbapCatalog.enhancement.fieldSuffix**
  This annotation can be used to avoid name conflicts for fields and associations. All fields contained in an extension must end with the specified suffix.
- **@AbapCatalog.enhancement.quotaMaximumFields**
  This annotation limits the number of fields that can be added.
- **@AbapCatalog.enhancement.quotaMaximumBytes**
  This annotation defines the maximum byte capacity reserved for extensions.
- **@AbapCatalog.enhancement.quotaShareCustomer**
  This annotation indicates for which group (SAP partner or customer) which quota of bytes and maximum number of fields is reserved.

In Listing 18.4, you can see that the extensions can contain a maximum of 204 fields, of which 50% are reserved for SAP partner fields and the other 50% for customer fields. In addition, the fields in the extensions must end with the _SDH suffix.

Creating a custom extension structure

Up to this point, we've shown how database tables or structures can be marked as extensible. Now, we'll add fields to the database table. To do this, we create a new structure that extends the sdsalesdoc_incl_eew_ps structure (see Listing 18.5). This adds the field ZZGIFT_DEV_SDH to the database table.

```
@EndUserText.label : 'ZEXT_STRUCT'
@AbapCatalog.enhancement.category : #NOT_EXTENSIBLE
extend type sdsalesdoc_incl_eew_ps with zext_gift {
```

```
  zzgift_dev_sdh : char1 not null;

}
```

Listing 18.5 Database Table Extension

18.2.2 Extending CDS Entities

Next, we extend the CDS entities based on the database table. Extensibility for CDS entities is based on an approach that gives SAP complete control over possible extensions and their quantity. Annotations are used to define whether the addition of new data sources or compositions is permitted. Every CDS entity that is later extended with extension fields and associations must be enriched with these annotations. The extensions must run through all CDS entities of the complete CDS stack.

Annotation for extensibility

Listing 18.6 shows the corresponding `@AbapCatalog.extensibility` annotation for CDS view `E_SalesDocumentBasic` based on database table `VBAK`.

```
@AbapCatalog.extensibility: {
  extensible: true,
  elementSuffix: 'SDH',
  allowNewDatasources: false,
  dataSources: ['Persistence'],
  quota: {
    maximumFields: 204,
    maximumBytes: 4080
  }
}
define view entity E_SalesDocumentBasic
  as select from vbak as Persistence
{
  key Persistence.vbeln as SalesDocument
}
```

Listing 18.6 Specification of the Extension Options for a CDS View

The data model must be supplemented by corresponding extension annotations that define the type and scope of possible extensions. The CDS entities must be marked as extensible to enable extensions of any kind. The value of the `@AbapCatalog.extensibility.extensible` annotation must be set to `true`. The following annotations define the extensibility of CDS views:

- `@AbapCatalog.extensibility.extensible`
  This annotation makes it possible to define a suffix that must be used by extension fields and associations. If an extension association doesn't use this suffix, a syntax error occurs.
- `@AbapCatalog.extensibility.quota`
  The CDS entity must contain a definition that specifies the capacities for extension content:
  - `@AbapCatalog.extensibility.quota.maximumFields:` Defines the maximum number of possible extension fields. The maximum number of fields should be set higher than the number of fields specified at database level.
  - `@AbapCatalog.extensibility.quota.maximumBytes:` Defines the maximum byte capacity reserved for extension fields.
- `@AbapCatalog.extensibility.dataSources`
  This defines the corresponding alias, which means the CDS view can be called directly from other CDS views and the technical name of the CDS view doesn't have to be entered.

Creating an extension

An extension for a CDS entity can be created using templates in the ABAP development tools in a similar way to how you've learned about other objects. You have two options:

- You create a data definition in the ABAP development tools. The template for the extension can be selected (see Figure 18.37).

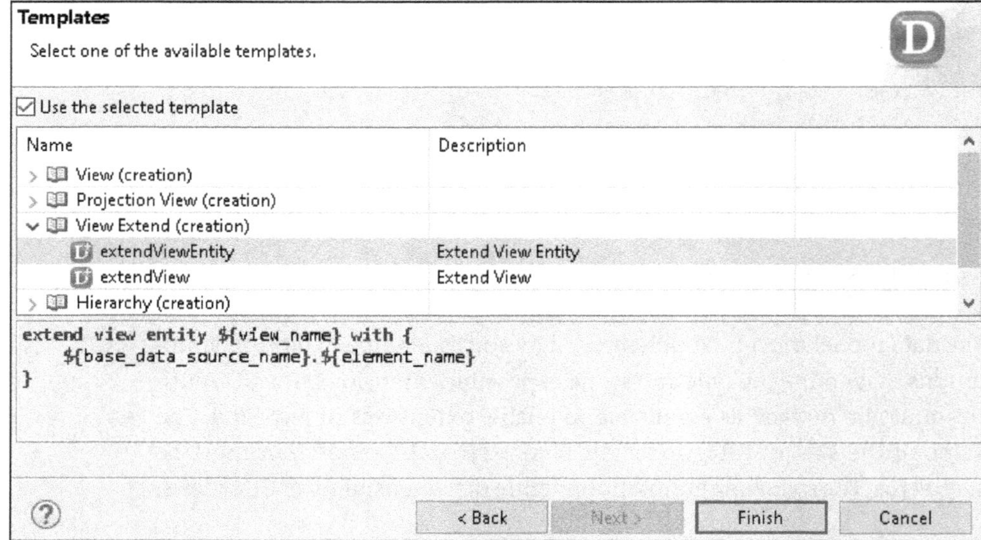

Figure 18.37 Selecting an Extension Template in the Creation Wizard for a Data Definition

- In the source code editor, you can use the **extendViewEntity** template via the quick fix function. To do this, simply enter "extendViewEntity" in the source code editor, and use the keyboard shortcut [Ctrl]+[space bar]. Select the appropriate template, and confirm your selection by pressing the [Enter] key (see Figure 18.38).

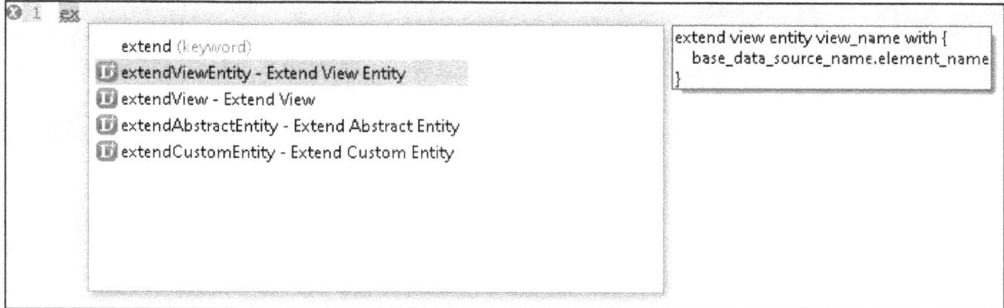

Figure 18.38 Selecting a Template in the Source Code Editor

Both variants generate the code shown in Listing 18.7.

Generated code

```
@AbapCatalog.sqlViewAppendName: ''
@EndUserText.label: 'Z_GIFT_DEV_EXT'
extend view entity view_name with Z_GIFT_DEV_EXT
{
    base_data_source_name.element_name
}
```

Listing 18.7 Generated "extend view" Code

With such a CDS entity extension, you can add custom view elements without making any changes to the original view. The extension is defined via CDS statement `extend view entity`.

[!]

Obsolete "extend view" Variant

The `extend view` variant was used to extend CDS views based on the ABAP Dictionary. However, these CDS views are obsolete. When you create new CDS views, you should use CDS view entities.

Field assignment

We now add the following extension fields to our template in the context of a field assignment. The field assignment determines which field is to be extended (`Persistence.zztest_sdh`) and can be supplemented with an alias (`as zzgift_dev_sdh`). It's therefore necessary to activate the field mapping for extension fields. Listing 18.8 shows this mapping for our example.

```
extend view entity E_SalesDocumentBasic with
{
  Persistence.zztest_sdh as zzgift_dev_sdh
}
```

Listing 18.8 Field Mapping for the Extension Fields of an Entity

Providing an extension at UI level

It's important that you now extend all entities from Table 18.1 accordingly. Once you've done this, the new field isn't yet available in the UI by default. It must still be made available in the app, as we've already shown for the custom field in Section 18.1.2. The field can now be used in the Manage Sales Orders app. Figure 18.39 shows the extended UI.

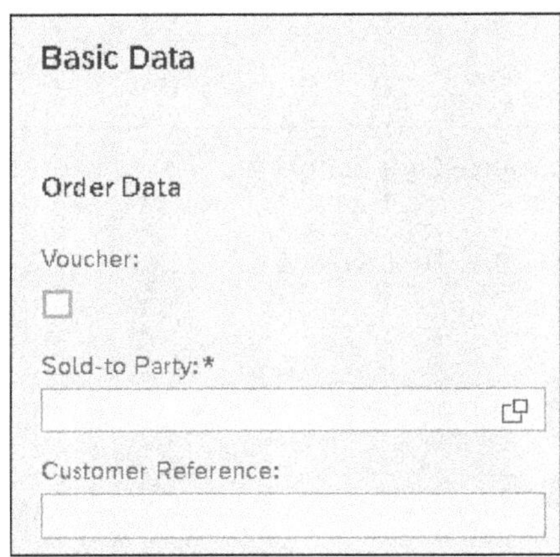

Figure 18.39 Extending the Sales Order to Include the Voucher Field

18.2.3 Implementing Custom Logic

Presetting a field value via events

You can use the standard events provided by the ABAP RESTful application programming model to set our checkbox as activated again by default.

Creating the implementation class

To do this, you must define a global class that waits for events of the Sales Order business object:

1. For this purpose, right-click on your package, and select **New • ABAP Class** from the context menu.
2. Enter a name and a description for the class. We call our class "ZCL_SALESORDER_EVENT" (see Figure 18.40). Confirm this by clicking the **Next** button.
3. In the window that opens, select a transport request. Click **Finish** to start generating the class.

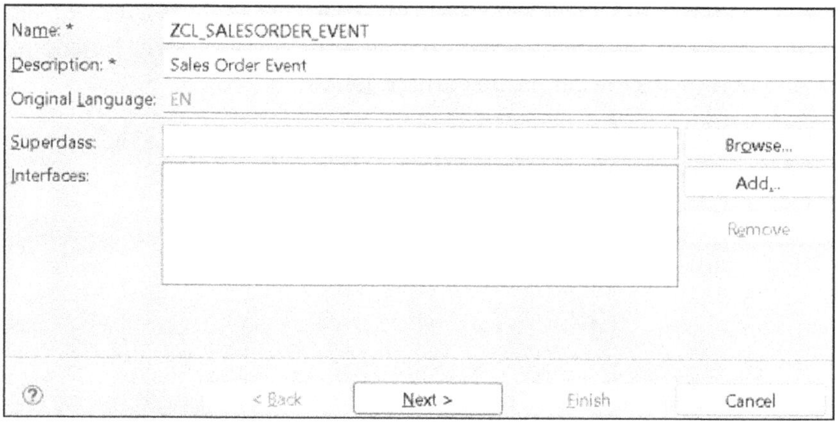

Figure 18.40 Creating a New ABAP Class

You can see the generated class in Listing 18.9.

```
CLASS zcl_salesorder_event DEFINITION
  PUBLIC
  FINAL
  CREATE PUBLIC.

  PUBLIC SECTION.
  PROTECTED SECTION.
  PRIVATE SECTION.
ENDCLASS.
CLASS zcl_salesorder_event IMPLEMENTATION.
ENDCLASS.
```

Listing 18.9 Generating a Class

Supplementing the implementation logic

Add the FOR EVENTS command to the implementation, in which you specify the behavior definition R_SalesOrderTP (see Listing 18.10). Now the ZCL_SALESORDER_EVENT class can respond to events for CDS entity R_SalesOrderTP.

```
CLASS zcl_salesorder_event  DEFINITION
  PUBLIC
  FINAL
  FOR EVENTS OF R_SalesOrderTP.

  PUBLIC SECTION.
  PROTECTED SECTION.
  PRIVATE SECTION.
ENDCLASS.
```

Listing 18.10 The FOR EVENTS Addition

Defining a local class

The next step is to add a local class. This local class inherits from the CL_ABAP_BEHAVIOR_EVENT_HANDLER class to access the event handler functions. Listing 18.11 shows the definition of this class.

```
*"* use this source file for the definition and implementation of
*"* local helper classes, interface definitions, and type
*"* declarations
CLASS lzcl_salesorder_event DEFINITION INHERITING FROM
  cl_abap_behavior_event_handler.
ENDCLASS.

CLASS lzcl_salesorder_event IMPLEMENTATION.
ENDCLASS.
```

Listing 18.11 Addition of a Local Class That Inherits from the Behavior Implementation

In this local class, you define a method to process specific events (see Listing 18.12). In our case, the SalesOrder.Created event is to be processed.

```
CLASS lzcl_salesorder_event DEFINITION INHERITING FROM
  cl_abap_behavior_event_handler.

    METHODS consume_created
      FOR ENTITY EVENT created FOR salesorder~created.

ENDCLASS.


CLASS lzcl_salesorder_event IMPLEMENTATION.

  METHOD consume_created.
  ENDMETHOD.

ENDCLASS.
```

Listing 18.12 Definition of a Method for Event Processing

In this method, you now implement the logic to automatically prefill the **Voucher** checkbox (see Listing 18.13). The ZZGIFT_DEV_SDH field is therefore assigned the value X by default.

```
CLASS lzcl_salesorder_event IMPLEMENTATION.

    MODIFY ENTITIES OF i_salesordertp
    ENTITY salesorder
```

```
    UPDATE FIELDS ( zzgift_dev_sdh )
    WITH VALUE #(  ( zzgift_dev_sdh = 'X' %key-salesorder = created[
                    1 ]-salesorder ) )
    FAILED DATA(failed)
    REPORTED DATA(reported).

ENDCLASS.
```

Listing 18.13 Implementation Logic

Finally, you need to save and activate the LZCL_SALESORDER_EVENT class.

Testing the SAP Fiori app

You can now view the result in the Manage Sales Orders app. The checkbox is automatically ticked (see Figure 18.41).

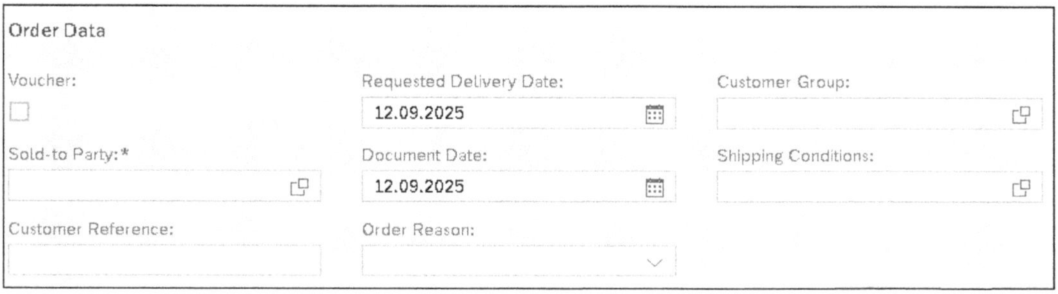

Figure 18.41 Custom Logic in the Manage Sales Orders App

18.3 Summary

In this chapter, you've learned about two extension options:

- The key user extensibility should be used for extensions by business experts (e.g., for layout changes, form and field extensions, and simple logic).
- Developer extensibility should be used for extensions that require professional ABAP development (interfaces with APIs, extension of complex BAdIs).

In the context of key user extensibility, you first learned how to set up the Adaptation Transport Organizer to transport the extensions. Then you've become familiar with the Custom Fields app, which you can use to add new fields. We've also introduced you to the Custom Logic app, which you can use to add your own logic for the fields. Finally, we've described the transport of key user extensions.

In the context of developer extensibility, you've learned about options for declaring database tables and CDS entities as extensible. We've also described how these objects can be extended and supplemented with custom logic.

Key user extensibility provides a way to implement extensions for a business object quickly and automatically. With developer extensibility, the extensions to many development objects, such as database tables or CDS views, have to be added manually.

Chapter 19
Outlook

ABAP, and ABAP Cloud in particular, are evolving. In this chapter, you'll learn which tools you can use to handle release-specific functions and which aspects you should consider with regard to compatibility between the different runtime environments. We also take a look at the road map.

Review

This book has provided an overview of modern ABAP development with ABAP Cloud as the development model. After introducing the basic principles in Chapter 1 and the introduction to the application scenario in Chapter 2, Chapter 3 to Chapter 16 dealt with specific problems from everyday developer life and demonstrated how you can solve them using current techniques that work in all runtime environments. In Chapter 17 and Chapter 18, we've broadened the perspective somewhat, from a custom development detached from the standard to the integration of custom code with the standard SAP system via APIs (Chapter 17) and extensions (Chapter 18).

Applicability of the techniques

In summary, it can be said that ABAP Cloud can now be used to solve the majority of everyday programming tasks. The degree of technological coverage is high, and the technologies provided as APIs or object types can be used very flexibly. The ABAP RESTful application programming model promotes software architectures that separate application logic, presentation, and persistence; make optimum use of the SAP HANA database; and are highly reusable. In user-oriented scenarios, SAP Fiori apps are largely generated using SAP Fiori elements or at least the Open Data (OData) protocol services are automatically provided for a freestyle SAPUI5 development.

Lack of functionality

In some places, however, there are still gaps or perhaps even steps backward compared to earlier techniques. We've already addressed some of them in this book. For example, the standard case of a maintenance view for a table can be mapped very well using a Business Configuration app with business configuration maintenance objects, including language-dependent texts and integration into the transport system. However, the classic table maintenance generator provides such a wide range of functions (with time-dependent data, integration into the central address management, and complex view clusters) that a lot of implementation work is still required for these advanced cases in the new model.

[»]

Switching to Classic ABAP

In SAP S/4HANA and SAP S/4HANA Cloud Private Edition, you can switch to classic ABAP. Formally speaking, these are applications that don't comply with the clean core concept because they violate at least the cloud capability criterion. You can use the three-tier model to manage such outliers. The use of classic ABAP isn't possible in SAP S/4HANA Cloud Public Edition and in the SAP BTP ABAP environment. In these two environments, however, you receive updates much faster, so that functions added by SAP reach you more quickly.

On-premise and cloud release statuses

In some constellations, a specific function may already be available, but not yet in the release you're using. SAP S/4HANA 2023 and SAP S/4HANA Cloud Private Edition 2023 have already created a good basis for ABAP cloud development. The ABAP RESTful application programming model in particular is now very mature. However, due to the two-year cycle of new on-premise releases, SAP S/4HANA Cloud Public Edition and SAP BTP ABAP environment were already a year and a half ahead of the on-premise release in terms of development status when this book was written.

The resulting deviations can mean, for example, that a class hasn't yet been released for use in ABAP Cloud. In such a case, it may even be possible to import a corresponding SAP Note. In other cases, however, the functional backlog can be a deal breaker. In Chapter 14, for example, it was planned to write the processing status of the change to a recipe in the file uploaded by the user and to provide that user with the updated file. However, changing existing Excel files isn't yet available in SAP S/4HANA 2023 and SAP S/4HANA Cloud Private Edition 2023 on the basis of the API that can be used in ABAP Cloud.

Release-dependent search

When you research the new technologies, it's not immediately clear which functions are available in which release. It can be very frustrating to find the solution to a problem on the internet, only to realize that it's not applicable to your system. You can use multiple sources for release-dependent research, which we present here.

ABAP Feature Matrix

Unfortunately, the relevant information is very scattered in SAP road maps, What's New viewers, blog posts, the SAP Help Portal, and the keyword documentation. Fortunately, however, the Software-Heroes initiative in the SAP community has done the work of aggregating and visualizing the data from these various sources. The resulting *ABAP Feature Matrix* is a tool from Software-Heroes that provides you with a clear overview of the availability of functions in ABAP technology. You can easily find the relevant

ABAP release for a function in the individual columns of the table overview (see Figure 19.1).

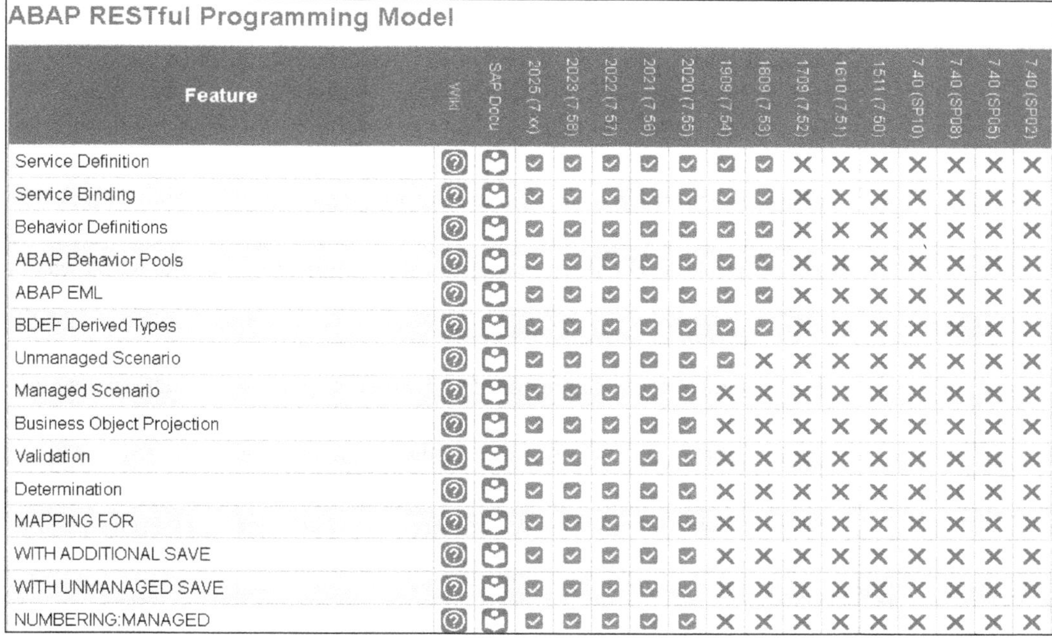

ABAP RESTful Programming Model

Feature	Wiki	SAP Docu	2025 (7.xx)	2023 (7.58)	2022 (7.57)	2021 (7.56)	2020 (7.55)	1909 (7.54)	1809 (7.53)	1709 (7.52)	1610 (7.51)	1511 (7.50)	7.40 (SP10)	7.40 (SP08)	7.40 (SP05)	7.40 (SP02)
Service Definition	?		✓	✓	✓	✓	✓	✓	✓	✗	✗	✗	✗	✗	✗	✗
Service Binding	?		✓	✓	✓	✓	✓	✓	✓	✗	✗	✗	✗	✗	✗	✗
Behavior Definitions	?		✓	✓	✓	✓	✓	✓	✓	✗	✗	✗	✗	✗	✗	✗
ABAP Behavior Pools	?		✓	✓	✓	✓	✓	✓	✓	✗	✗	✗	✗	✗	✗	✗
ABAP EML	?		✓	✓	✓	✓	✓	✓	✓	✗	✗	✗	✗	✗	✗	✗
BDEF Derived Types	?		✓	✓	✓	✓	✓	✓	✓	✗	✗	✗	✗	✗	✗	✗
Unmanaged Scenario	?		✓	✓	✓	✓	✓	✓	✗	✗	✗	✗	✗	✗	✗	✗
Managed Scenario	?		✓	✓	✓	✓	✓	✗	✗	✗	✗	✗	✗	✗	✗	✗
Business Object Projection	?		✓	✓	✓	✓	✓	✗	✗	✗	✗	✗	✗	✗	✗	✗
Validation	?		✓	✓	✓	✓	✓	✗	✗	✗	✗	✗	✗	✗	✗	✗
Determination	?		✓	✓	✓	✓	✓	✗	✗	✗	✗	✗	✗	✗	✗	✗
MAPPING FOR	?		✓	✓	✓	✓	✓	✗	✗	✗	✗	✗	✗	✗	✗	✗
WITH ADDITIONAL SAVE	?		✓	✓	✓	✓	✓	✗	✗	✗	✗	✗	✗	✗	✗	✗
WITH UNMANAGED SAVE	?		✓	✓	✓	✓	✓	✗	✗	✗	✗	✗	✗	✗	✗	✗
NUMBERING:MANAGED	?		✓	✓	✓	✓	✓	✗	✗	✗	✗	✗	✗	✗	✗	✗

Figure 19.1 Section of the ABAP Feature Matrix

The individual functions are listed in the rows. These are grouped according to different areas, such as ABAP RESTful application programming model, ABAP SQL, or ABAP Cloud. You can go directly from an entry to the relevant SAP documentation and additional information. You can access the ABAP Feature Matrix at *https://software-heroes.com/en/abap-feature-matrix*.

[+]

ABAP Learning Matrix

You can find a different format for the data in the *ABAP Learning Matrix*. This tool from Software-Heroes shows release deltas. For example, when you upgrade, you can find out directly which additional features you've received that you should look into. You can access the ABAP Learning Matrix at *https://software-heroes.com/en/abap-learning-matrix*.

Side effects of the on-premise solution

In addition to a lack of functionality, too much functionality can also lead to problems. In SAP S/4HANA and SAP S/4HANA Cloud Private Edition, you can use both development models: classic ABAP and ABAP Cloud. You don't have to use just one or the other either; they can be used in combination with each other where appropriate. An ABAP report with selection screen and SAP List Viewer (ALV) can use a core data services (CDS) view as a data source and change the data of a business object via Entity Manipulation Language (EML) just as well as a direct database access and a Business Application Programming Interface (BAPI) call. However, such combination options don't appear to be technically accounted for in all areas by SAP. The fact that the ABAP repository object generator for business configuration maintenance objects can be used by default to edit the underlying Customizing tables by bypassing the transport system and the validation checks in Transactions SE16 and SE16N, unless sufficiently strict authorization is also provided in the development system, indicates that the focus wasn't directly on use outside SAP S/4HANA Cloud Public Edition and SAP BTP ABAP environment.

The now more central concept of software components in custom developments can also be problematic in SAP S/4HANA and SAP S/4HANA Cloud Private Edition, as it's still technically optional there, while some APIs in ABAP Cloud already expect the assignment. One of the problematic concepts in this respect is the use of temporary packages. They aren't permitted in SAP S/4HANA Cloud Public Edition and in the SAP BTP ABAP environment.

Compatibility between runtime environments

These side effects can also cause problems with regard to compatibility between the runtime environments. ABAP Cloud guarantees as much as possible that developments can be used across the board, especially if you use ABAP Cloud in your on-premise system and want to switch to the cloud systems later. Conversely, the problem of release dependency described earlier arises. There are also components in ABAP Cloud that aren't or aren't yet standardized across the board. This particularly affects the area of connectivity, as network communication in cloud environments is largely managed via SAP BTP, while the on-premise systems can communicate more or less directly with each other and with the outside world. The authorization concept is also different and difficult to standardize.

Migrating from classic ABAP to ABAP Cloud

If you want to migrate your existing code to ABAP Cloud, the *ABAP Test Cockpit* and the Custom Code Migration app (app ID F3191) will help you. You can use this SAP Fiori app to monitor cloud compatibility and distribute organizational work packages. In terms of operations, you can use ABAP Test Cockpit findings to adapt the code piece by piece and renovate the codebase without directly obtaining syntax errors. You only change the ABAP language version after all findings for a component have been processed. In addition to the currently available quick fixes for the (partially) automated elimination of such ABAP Test Cockpit findings, the digital SAP assistant *Joule* should also be able to support and simplify the process with the help of generative AI in the future.

CDS table entities

The *CDS table entities* will also be of interest in the future. These objects aren't yet available in SAP S/4HANA 2023 and SAP S/4HANA Cloud Private Edition 2023, but they have been available in SAP BTP ABAP environment and SAP S/4HANA Cloud Public Edition since release 2502. With the CDS table entities, one of the last object types from the ABAP Dictionary that is relevant for ABAP Cloud, the database table, is transferred natively to CDS. In combination with the successors for data elements (i.e., the *CDS simple types*) and domains (i.e., the *CDS enumerations*), all relevant dictionary object types for database tables have thus been given a successor concept. This means that most use cases for the ABAP Dictionary in ABAP Cloud will no longer apply. Cloud Release 2502 is only the first version of CDS table entities available, and there's still a lack of integration into existing technologies. With SAP S/4HANA 2025 and SAP S/4HANA Cloud Private Edition 2025, the CDS table entities should also be available in the on-premise solution, even if the scope of the integration into existing technologies that will then be available isn't yet foreseeable.

Roadmap

This leaves us with a final look at the SAP road map. This book has given you an impression of how much has changed in ABAP development in recent years and how much will continue to change. It's worth taking a look at the road map to stay up-to-date and make architecture decisions that are compatible with future concepts. The road map information on the ABAP platform can be found in the SAP Help Portal at *http://s-prs.de/v1064817*.

The road map is divided into specifics for the various runtime environments and overarching information for ABAP Cloud that applies to all environments (see Figure 19.2).

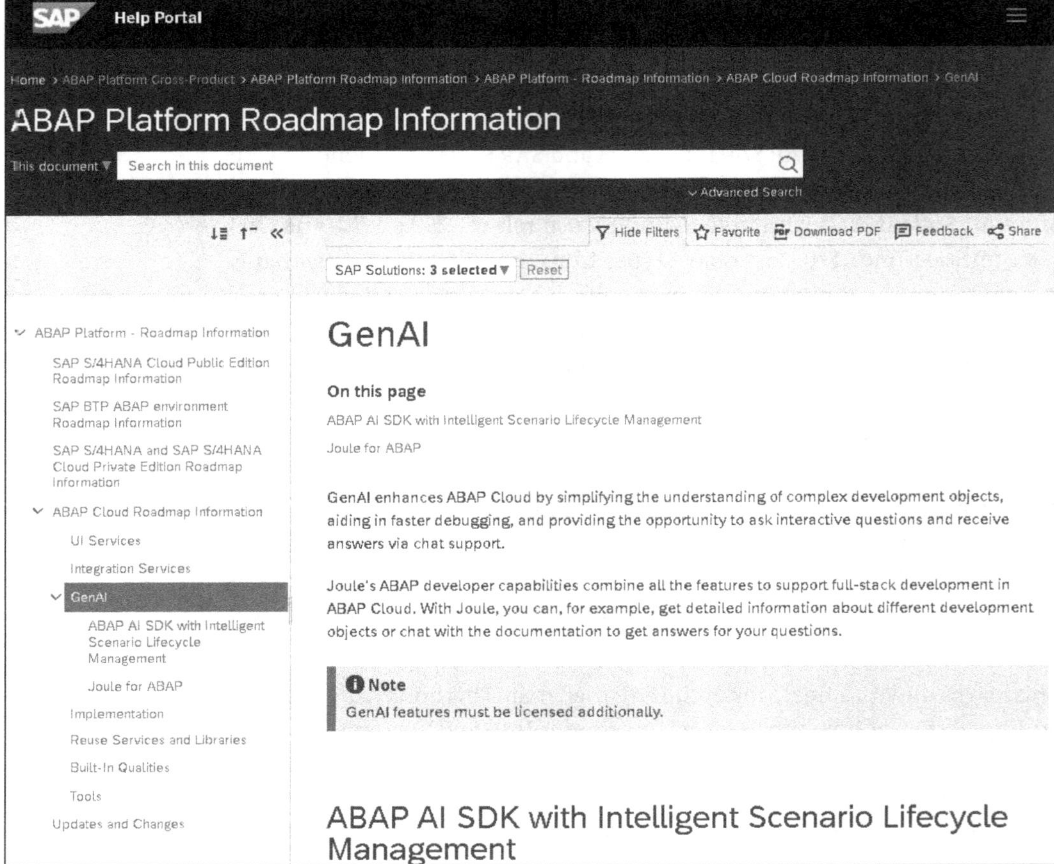

Figure 19.2 Excerpt from the Road Map for the ABAP Platform

You should pay particular attention to any changes in the road map. These are communicated in the **Updates and Changes** section. Road map information is always provided by SAP without any guarantee.

Appendices

Appendix A
Installing the Sample Application

On the publisher's website at *www.sap-press.com/6198* under the **Materials** tab, you will find the recipe portal, which serves as a sample application for this book and is created and expanded in the chapters. You can also access it at the following URL on GitHub: *http://s-prs.co/v619800*.

Download and installation

You can install the application in your system so that you can retrace the steps directly in your development environment or analyze details in the debugger. You should consult your system administrator before importing the objects into the system. The application is provided as an abapGit repository. Detailed installation instructions can be found in the *README.md* file at the top level of the archive or repository.

Alternatively, the application can also be opened for display in a text editor or a development environment of your choice. The use of a syntax highlighting plug-in for ABAP is recommended. You can find such a plug-in for Visual Studio Code at *http://s-prs.de/v1064818*.

Chapter references for the sample application

The overview in Table A.1 shows which subpackages are described in the sample application and in which chapters of this book. The DATAMODEL package is used across the board and is partially extended in other packages. The full package name depends on your chosen installation method. For example, it can be $COOKBOOK_DATAMODEL or ZCOOKBOOK_DATAMODEL.

No.	Chapter	Subpackage
2	The Application Scenario	DATAMODEL
3	Handling System Fields and Runtime Information	RUNTIME
4	Table Analysis	TABLEDISPLAY
5	Table Maintenance Using Business Configuration Maintenance Objects	CONFIG
6	Application Logs	LOG
7	Change Documents	CHANGEDOCUMENTS

Table A.1 Chapter References for the Sample Application

No.	Chapter	Subpackage
8	Lock Objects	LOCK
9	Number Range Objects	NUMBER
10	Background Processing	JOB
11	Email Dispatch	MAIL
12	Parallelizing Application Logic	PARALLEL
13	File Upload	DATAMODEL (extension)
14	Using Excel Files	EXCEL
15	Documenting Development Objects	DOCUMENTATION
16	Authorizations	AUTHORIZATION

Table A.1 Chapter References for the Sample Application (Cont.)

Appendix B
Naming Conventions for the Sample Application

This appendix contains the naming conventions used in our sample application that can serve as a guide. They are specified in the customer namespace, here with Z, and in a reserved namespace, here with /NSPC/ as an example.

ABAP Dictionary objects

Table B.1 lists the conventions for objects in the ABAP Dictionary.

Object Type	Naming Convention
Database table (normal)	ZACB_RECIPE /NSPC/ACB_RECIPE
Database table (ABAP RESTful application programming model, active data)	ZACB_RECIPE /NSPC/ACB_RECIPE Alternative: ZACB_A_RECIPE /NSPC/ACB_RECIPE
Database table (ABAP RESTful application programming model, draft data)	ZACB_RECIPE_D /NSPC/ACB_RECIPE_D Alternative: ZACB_D_RECIPE /NSPC/ACB_D_RECIPE
Data element	ZACB_RECIPE_ID /NSPC/ACB_RECIPE_ID
Domain	ZACB_RECIPE_ID /NSPC/ACB_RECIPE_ID
Structure type	ZACB_S_RECIPE_DATA /NSPC/ACB_S_RECIPE_DATA
Table type	ZACB_T_RECIPE_IDS /ACB/ACB_T_RECIPE_IDS

Table B.1 Naming Conventions for ABAP Dictionary Objects

Object Type	Naming Convention
Lock object	EZACB_RECIPE /NSPC/E_ACB_RECI

Table B.1 Naming Conventions for ABAP Dictionary Objects (Cont.)

ABAP RESTful application programming model objects and CDS entities

Table B.2 lists the conventions for objects of the ABAP RESTful application programming model and ABAP core data services (ABAP CDS) views.

Object Type	Naming Convention
Restricted view/root entity	ZACB_R_Recipe /NSPC/ACB_R_Recipe
Interface view entity	ZACB_I_Recipe /NSPC/ACB_I_Recipe
Projection view entity	ZACB_C_Recipe /NSPC/ACB_C_Recipe
Metadata extension	With the same name as the associated CDS entity
Behavior definition	With the same name as the associated CDS entity
Behavior implementation class	ZBP_ACB_R_RECIPE /NSPC/BP_ACB_R_RECIPE
Service definition	ZACB_UI_RECIPE /NSPC/ACB_UI_RECIPE (alternatively, API or no classification)
Service binding	ZACB_UI_RECIPE_04 /NSPC/ACB_UI_RECIPE_04 (alternatively, API)

Table B.2 Naming Conventions for ABAP RESTful Application Programming Model Objects and CDS Entities

ABAP objects

Table B.3 lists the conventions for ABAP objects.

Object Type	Naming Convention
Class (general)	ZCL_ACB_RECIPE /NSPC/CL_ACB_RECIPE LCL_RECIPE

Table B.3 Naming Conventions for ABAP Objects

Object Type	Naming Convention
Interface (general)	ZIF_ACB_USER_CONVERTER /NSPC/IF_ACB_USER_CONVERTER LIF_USER_CONVERTER
Exception class	ZCX_ACB_RECIPE_NOT_FOUND /NSPC/CX_ACB_RECIPE_NOT_FOUND LCX_RECIPE_NOT_FOUND

Table B.3 Naming Conventions for ABAP Objects (Cont.)

Appendix C
Installing the ABAP Development Tools for Eclipse

The ABAP development tools for Eclipse are provided by SAP as a plug-in for the Eclipse development environment. The installation instructions can be found at *https://tools.hana.ondemand.com/#abap*.

A detailed tutorial on installing and connecting to an SAP system can be found on the SAP Learning website at *https://developers.sap.com/tutorials/abap-install-adt.html*.

Appendix D
The Authors

Fabian Lupa has been a senior software engineer and trainer at adesso since 2022. He is primarily responsible for the training and further education of employees in the context of ABAP development as well as developer enablement in customer projects. Prior to this, he trained as an IT specialist with a part-time degree in business informatics and worked in-house for six years as an application developer in the areas of SAP Finance and SAP Retail. He is involved in the open-source ABAP community and actively shapes modern development guidelines. Fabian places particular emphasis on clean code and unit tests as well as the automation of development processes with modern tools.

Sven Treutler is a passionate ABAP developer. Since 2010, he has been working on ABAP development topics at rku.it GmbH in Herne, Germany. He is responsible for new technologies and quality assurance in the ABAP environment. Sven completed his master's degree at the Westphalian University of Applied Sciences in Gelsenkirchen, Germany. In his private life, he visits beautiful places with his wife and is a proud cat owner and fan of the Borussia Dortmund soccer team.

Index

A

B

C

M

N

O

P

Q

R

S

T

U

V

W

X

- An everyday reference for ABAP programmers of all levels
- Learn language elements, syntax, concepts, and more
- Explore modularization, modifications, and enhancements

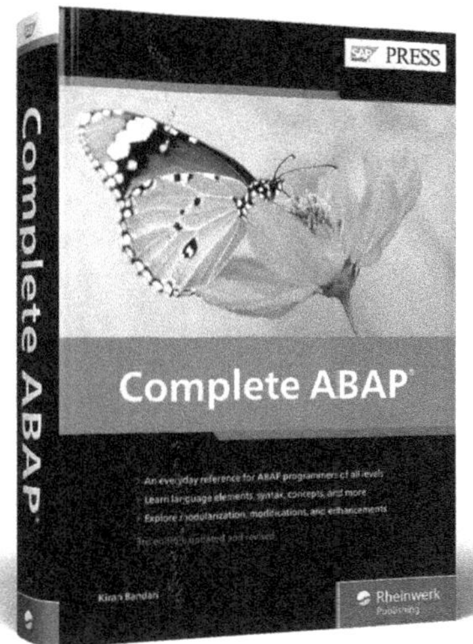

Kiran Bandari

Complete ABAP

Get everything you need to code with ABAP, all in one place! Are you a beginner looking for a refresher on the basics? You'll get an overview of SAP architecture and learn syntax. Already an experienced programmer and looking to improve your ABAP skills? Dive right into modifications and code enhancements. Understand the programming environment and build reports, interfaces, and applications with this complete reference to coding with ABAP!

912 pages, 3rd edition, pub. 09/2022
E-Book: $84.99 | **Print:** $89.95 | **Bundle:** $99.99

www.sap-press.com/5567

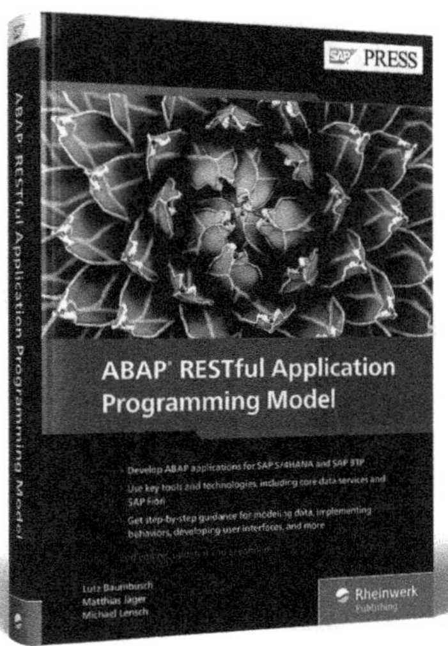

- Develop ABAP applications for SAP S/4HANA and SAP BTP
- Use key tools and technologies, including core data services and SAP Fiori
- Get step-by-step guidance for modeling data, implementing behaviors, developing user interfaces, and more

Baumbusch, Jäger, Lensch

ABAP RESTful Application Programming Model

The ABAP RESTful application programming model (RAP) is the cornerstone of modern development for SAP—get on the cutting edge with this guide! Develop applications from the ground up, from data modeling with CDS to interface generation with SAP Fiori elements. Walk through concrete use cases, including managed and unmanaged scenarios, and then adapt your applications to the SAP BTP, ABAP environment. You're on your way to working with RAP!

576 pages, 2nd edition, pub. 07/2025
E-Book: $84.99 | **Print:** $89.95 | **Bundle:** $99.99

www.sap-press.com/6161

www.sap-press.com

- Develop data models with ABAP core data services
- Create and extend models for analytical and transactional applications
- Define annotations and associations, implement access controls, work with hierarchies, and more

Colle, Dentzer, Hrastnik

Core Data Services for ABAP

If you're developing ABAP applications, you need CDS expertise. This book is your all-in-one guide, updated for SAP S/4HANA 2023! Start by learning to create and edit CDS views. Walk through CDS syntax and see how to define associations and annotations. Further refine your model by implementing access controls, service bindings, and table functions. Understand the CDS-based virtual data model, and then follow step-by-step instructions to model analytical and transactional applications. From modeling to testing to troubleshooting, this is the only book you need!

754 pages, 3rd edition, pub. 10/2023
E-Book: $84.99 | **Print:** $89.95 | **Bundle:** $99.99

www.sap-press.com/5642

- Build your ABAP skills and write your first program
- Use the ABAP Dictionary to read from and update a database
- Make programs modular, trouble-shoot errors, and more

Brian O'Neill, Jelena Perfiljeva

ABAP

An Introduction

Step into ABAP with this beginner's guide. First understand ABAP syntax and find out how to add data and logic to your applications. Then delve into backend programming: learn to work with the ABAP data dictionary, create database objects, and process and store data. Round out your skill set by practicing error handling, modularization, and string manipulation, and more. With guided examples, step-by-step instructions, and detailed code you'll become an ABAP developer in no time!

684 pages, 2nd edition, pub. 11/2019
E-Book: $64.99 | **Print:** $69.95 | **Bundle:** $79.99

www.sap-press.com/4955

www.sap-press.com

Interested in reading more?

Please visit our website for all new book and e-book releases from SAP PRESS.

www.sap-press.com